ELECTROMAGNETICS

B.B. LAUD

A HALSTED PRESS BOOK

JOHN WILEY & SONS
New York Chichester Brisbane Toronto Singapore

Published in the Western Hemisphere
by Halsted Press, a Division of
John Wiley & Sons, New York

Library of Congress Cataloging in Publication Data

Laud, B.B.

Electromagnetics.

"A Halsted Press Book."
Bibliography: P.
Includes index.

1. Electromagnetism. I. Title.
QC750. L 38 1983 537 83–3577
ISBN 0-470-27427-1

Printed in India at Prabhat Press, Meerut.

Preface

Electromagnetics has become a part of the foundations of Physics and as such is an essential ingredient of the Physics curriculum at different levels. The present book is intended as a text-book in electromagnetics for the B.Sc. and M.Sc. students in Physics.

The method of approach is dictated by the desire to meet the needs of the students. Hence, the primary concern of the book is to enable the students to obtain a satisfactory intuitive grasp of the subject and to make them realize that the subject is relevant and useful. The basic ideas are developed along familiar lines.

Electromagnetics is essentially mathematical in character. Without mathematics as an aid to thought the development of electromagnetism would have been almost impossible. However, no attempt has been made in this book, to achieve mathematical rigour because it would be difficult to be much more rigorous without causing the student to lose sight of the pragmatic physical content. No more mathematical background is required of the student than the customary undergraduate courses in Calculus and Vector analysis. More advanced techniques are outlined as they arise.

Problems at the end of the chapters provide details for which there is no room in the body of the text and call for additional applications. They also tend to test the students' understanding of the concepts discussed in the chapter.

The references at the end indicate the author's indebtedness to the ideas of others, but the list is, by no means, exhaustive.

B.B. Laud

Contents

A Guide to Symbols

Symbol	*Explanation*
α	Polarizability of the atom
A_m	Molecular refractivity
$\mathbf{A}$	Vector potential
$\mathcal{A}$	Four vector potential
$\mathbf{B}$	Magnetic flux density
χ	Electric susceptibility
χ_m	Magnetic susceptibility
c	Velocity of light
$\mathbb{C}$	Capacitance
δ	Skin depth
δ_{nm}	Kronecker delta
$\delta(r)$	Dirac delta function
∇	$\hat{\mathbf{e}}_x \frac{\partial}{\partial x} + \hat{\mathbf{e}}_y \frac{\partial}{\partial y} + \hat{\mathbf{e}}_z \frac{\partial}{\partial z}$
$\mathbf{D}$	Electric displacement
ϵ	Permittivity
ϵ_r	Relative permittivity, dielectric constant
$\mathcal{E}$	E.M.F.
e	Electronic charge
$\hat{\mathbf{e}}_r$	Unit vector in the direction r
$\mathbf{E}$	Electric field strength
ϕ	Electrostatic potential
Φ	Magnetic flux
ϕ_m	Magnetic scalar potential
$\mathbf{F}$	Force
$F_{\mu\nu}$	Electromagnetic field tensor
γ	Damping coefficient
$\mathcal{G}(x, x')$	Green's function
$\mathbf{H}$	Magnetic field strength
I	Current
$\mathbf{J}$	Current density
J_ν	Bessel's function
$\mathcal{J}$	Four-vector current-density
$\mathbf{k}$	Propagation vector

$*\lambda$	Charge per unit length, wavelength
λ_g	Gravitational constant
$\mathcal{L}$	Differential operator
$*L$	Lagrangian; Self-inductance
$*\mu$	Magnetic moment; permeability
m	Magnetic dipole moment
M	Mutual inductance
n	Refractive index
$\mathbf{N}$	Poynting vector
$*\Omega$	Solid angle; ohm
p	Dipole moment;
$\mathbf{P}$	Macroscopic polarization density
$\mathcal{P}$	Magnetic pressure
P_l	Legendre polynomial
P_l^m	Associated Legendre polynomial
q, Q	Charge
ρ	Charge density
$*R$	Reflection coefficient; resistance
$*\sigma$	Conductivity; scattering cross-section
S_l	Surface harmonic
$*\tau$	Relaxation time; torque
T	Transmission coefficient
U	Energy
$\mathcal{U}$	Four-vector velocity
V	Potential
$\mathbf{V}$	Volume
ω	Angular frequency
W	Energy
Z_0	Impedance

*These symbols are used to represent different quantities. The context generally makes clear which quantity is under discussion.

Chapter 1

Force, Field and Energy in Electorstatics

The world around us is made up of atoms which consist of positive and negative charges. The dominant force between atomic particles, therefore, is electrostatic. An understanding of a few basic laws of forces of nature can go a long way towards guiding us through the mysteries of science. One such law—the electrostatic law of force—was provided by Coulomb in 1785. Atomic reactions can be explained with great precision by Coulomb's Law.

1.1 Coulomb's Law

Coulomb showed experimentally that in free space oppositely charged bodies attract each other, while similarly charged bodies repel with a force that varies directly as the magnitude of each charge and inversely as the square of the distance between them, the force being directed along the line joining the charges.

Let us describe these observations of Coulomb in a mathematical form. We will use vector notation throughout this book. This has some advantages:

(i) the arbitrariness associated with the choice of the coordinate systems disappears and the physical content becomes more clear, and

(ii) the equations of electrodynamics become more concise and vivid if written in vector notation.

If '1' and '2' are two particles carrying charges q_1 and q_2 respectively and separated by a distance r_{12} in vacuum (Fig. 1.1), then the electric force exerted by the particle '1' on the particle '2' as given by Coulomb's law is

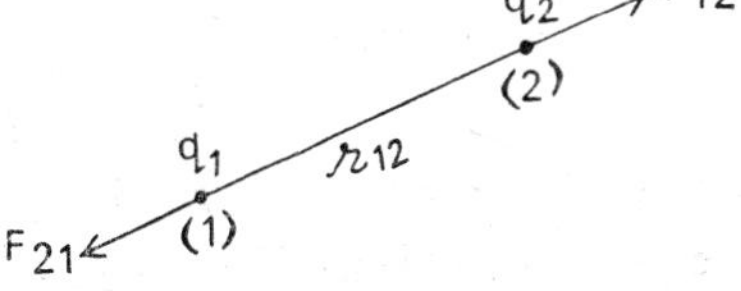

Fig. 1.1

$$F_{12} \propto \frac{q_1 q_2}{r_{12}^2}$$

i.e.

$$F_{12} = K\frac{q_1 q_2}{r_{12}^2} = -F_{21} \tag{1.1}$$

where K is the constant of proportionality and F_{21} is the force exerted by the particle '2' on the particle '1'.

In vector notation this can be written as

$$\mathbf{F}_{12} = K\frac{q_1q_2}{|\mathbf{r}_{12}|^2}\hat{\mathbf{e}}_r = K\frac{q_1q_2}{|\mathbf{r}_{12}|^3}\mathbf{r}_{12} \tag{1.2}$$

where $\hat{\mathbf{e}}_r = \frac{\mathbf{r}_{12}}{|\mathbf{r}_{12}|}$ is the unit vector along r_{12}.

The sign of the charges decides whether the force is attractive or repulsive. If the charges are similar (i.e. both positive or both negative), $\mathbf{F}_{12}$ is positive and represents the force of repulsion; while if one is positive and the other negative, $\mathbf{F}_{12}$ is negative and is the force of attraction.

We have two alternatives for choosing the value of K and the units of charge: either

(i) K is arbitrarily given some fixed numerical value and Eq. (1.2) used to determine the units of charge; or

(ii) the unit of charge is taken as some arbitrary value and the constant K determined experimentally.

In the Gaussian system (CGS) distances are measured in centimetres, mass in grams, time in seconds, force in dynes and the electrical units are defined by assuming the constant of proportionality K to be unity (first alternative). That is,

$$\mathbf{F}_{12} = \frac{q_1q_2}{|\mathbf{r}_{12}|^3}\mathbf{r}_{12} \text{ (dynes)} \tag{1.3}$$

Thus, the force between two unit charges separated by a distance of one centimetre is one dyne. The unit of charge, therefore, is that charge which experiences a force of 1 dyne when placed 1 cm from an identical charge. The unit of charge, thus, defined, is called the *statcoulomb* or electrostatic unit (esu).

In SI system (*Système Internationale*) of units, the distances are measured in metres, mass in kilograms, time in seconds, force in newtons and the charge in coulombs (second alternative). The constant K is then equal to $\frac{1}{4\pi\epsilon_0}$ and the Coulomb's law is written as

$$\mathbf{F}_{12} = \frac{1}{4\pi\epsilon_0}\frac{q_1q_2}{|\mathbf{r}_{12}|^3}\mathbf{r}_{12} \quad \text{(newtons)} \tag{1.4}$$

Thus, the force between two particles each carrying a charge of one coulomb and separated by a distance of one metre, is one *newton*.

The factor 4π has been introduced in order to simplify the form of some important relations occurring in the electromagnetic theory. In developing the various formulae, we will have often to deal with spherical shapes and hence a factor 4π is bound to appear. It will be advantageous, therefore, to use a constant containing 4π explicitly. The constant ϵ_0 has the value

$$\epsilon_0 = 8.85\times10^{-12} \text{ coulomb}^2 \text{ newton}^{-1} \text{ metre}^{-2} \tag{1.5}$$

You will see later that this is the so-called *permittivity of free space*.

Which of these two systems shall we use? One would think it best to use CGS system as in this system $K = 1$ and the relation connecting the force with the charges is relatively simple. However, if we adopt this system, the electric current will have to be measured in clumsy units of $\frac{1}{3 \times 10^9}$ amperes. Thus, although we thought we have chosen the units judiciously so that the constant is banished from Eq. (1.1), it appears in a different guise in another place. Any of these systems, therefore, is as good as the other. However, as the ordinary measuring instruments are calibrated in SI units, we shall use SI system in this book.

Since Coulomb's law is based primarily on experiments, one may ask whether the law is exactly that of inverse square? That is, if the force is proportional to $1/r^n$, is n exactly equal to 2? Cavendish showed that $n = 2 \pm 0.02$. Plimpton and Lawton (1936) found that n differs from 2 by not more than one part in 10^9. More recently Lamb and Rutherford (1947) also confirmed from their measurements of the energy levels of the hydrogen atom that the exponent in Coulomb's law is correct to one part in 10^9 at distances of the order of 10^{-10} metres. Evidence from nuclear physics has shown that the electrostatic forces vary approximately according to the inverse square law even at distances of the order of 10^{-15} metres. Considering the areas of our interest, we may use the inverse square law with complete confidence. It must be mentioned here that the logical construction of a consistent theory of electromagnetism that is developed in this book is based on the laws such as the one due to Coulomb, which are the results of experiments and as such are probably approximate. However, these are very good approximations and lead to the correct results.

Notice the similarity between Coulomb's law and the Newton's law of gravitation

$$F = \lambda_g \frac{m_1 m_2}{r^2}$$

where λ_g is the gravitational constant.

It must be mentioned here that although the charge on the electron is extremely small, electrostatic forces are immensely strong. A comparison between the force of attraction between two electrons due to gravitation and the force of repulsion due to electronic charges, will give an idea about the relative magnitudes of these forces.

$$\frac{\text{Electrostatic repulsion}}{\text{Gravitational attraction}} = \frac{1}{4\pi\epsilon_0} \cdot \frac{e^2}{r^2} \cdot \frac{r^2}{\lambda_g m^2}$$

$$= \frac{e^2}{4\pi\epsilon_0 \lambda_g m^2} = 9 \times 10^9 \times \frac{(1.60 \times 10^{-19})^2}{(6.67 \times 10^{-11}) \times (9.1 \times 10^{-31})^2}$$

$$= 4.17 \times 10^{42} \qquad \left(\because \frac{1}{4\pi\epsilon_0} = 9 \times 10^9\right)$$

(How is it, then, that we normally do not notice electrostatic forces, which are so powerful?)

EXAMPLE 1.1. In Rutherford's scattering experiment gold nuclei were bombarded with α-particles energetic enough to approach within 2×10^{-14} metres of the nucleus. Find the electrostatic force of repulsion experienced by the α-particles.

The charge on the gold nucleus is '79 e' and that on an α-particle is '2 e' where 'e' is the charge on an electron.

$$\therefore F = \frac{1}{4\pi\epsilon_0} \times \frac{2 \times 79 \times e^2}{(2 \times 10^{-14})^2}$$
$$= 91.2 \text{ newtons.}$$

1.2 Principle of Superposition

If there are more than two particles present with charges, say, q_1, $q_2, \ldots$, the total force on any one particle is the vector sum of forces it experiences due to all other particles separately. This is known as the *Principle of superposition*. For example, in Fig. 1.2, in which there are three charges, force on q_3 is given by

$$\mathbf{F} = \mathbf{F}_{13} + \mathbf{F}_{23} = \frac{1}{4\pi\epsilon_0}\frac{q_1 q_3}{|\mathbf{r}_{13}|^3}\mathbf{r}_{13} + \frac{1}{4\pi\epsilon_0}\frac{q_2 q_3}{|\mathbf{r}_{23}|^3}\mathbf{r}_{23}$$

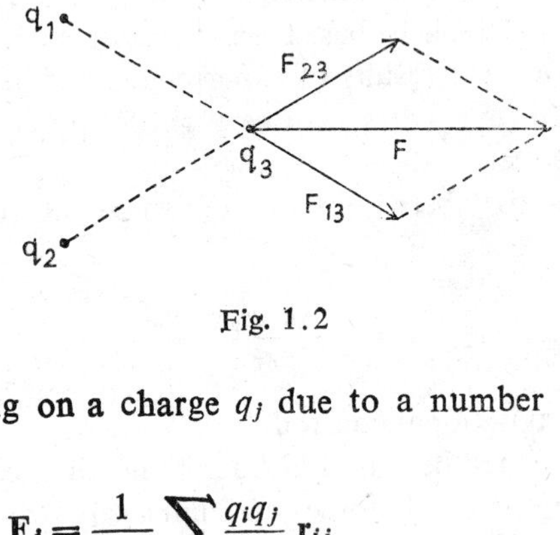

Fig. 1.2

The force acting on a charge q_j due to a number of other charges present is

$$\mathbf{F}_j = \frac{1}{4\pi\epsilon_0}\sum_{i \neq j}\frac{q_i q_j}{|\mathbf{r}_{ij}|^3}\mathbf{r}_{ij} \tag{1.6}$$

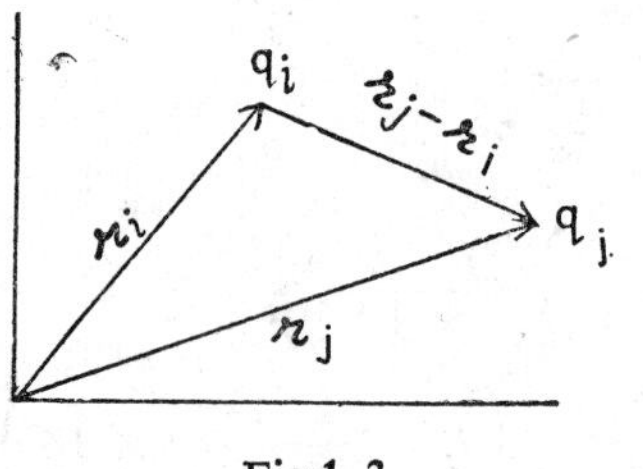

Fig 1.3

Equation (1.6) is often stated in terms of the position vectors. If $\mathbf{r}_i$ and $\mathbf{r}_j$ (Fig. 1.3) are the vectors giving the location of the charges q_i, q_j respectively, Eq. (1.6) can be written as

$$\mathbf{F}_j = \frac{1}{4\pi\epsilon_0}\sum_{i \neq j}\frac{q_i q_j}{|\mathbf{r}_j - \mathbf{r}_i|^3}(\mathbf{r}_j - \mathbf{r}_i) \tag{1.7}$$

The principle of superposition has facilitated considerably the mathematical handling of the theory. The principle fails in nuclear interactions and this is one of the reasons why the nuclear theory is somewhat more troublesome than the theory of atomic interactions.

1.3 Electric Field

In the treatment of physical problems, the concept of a 'field' is found to be of great utility. There are as many kinds of fields as there are types of mathematical or physical quantities that can be represented at a point. One finds, for instance, scalar, vector and tensor fields in the ordinary three dimensional space of classical physics and four dimensional vector fields in the four dimensional space-time domain of relativity theory. In the study of classical electromagnetics we shall be concerned with scalar and vector fields in three dimensions.

The space in which electrostatic forces act is called electrostatic field. This, however, is merely a qualitative description of the field. How do we define it quantitatively? Consider a system of charges distributed in space (Fig. 1.4).

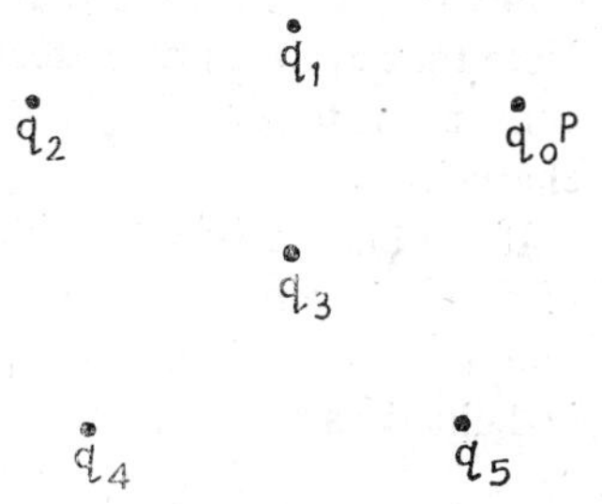

Fig. 1.4

What is the field due to these charges at a point P? To answer this question we put a charge q_o at P. We assume the charge q_o to be so small that it does not disturb the properties of the field, i.e. it exerts negligible force on the other charges. We call it a test charge. The force acting on this charge is

$$\mathbf{F}_0 = \frac{1}{4\pi\epsilon_0} \sum_i \frac{q_0 q_i}{|\mathbf{r}_0 - \mathbf{r}_i|^3} (\mathbf{r}_0 - \mathbf{r}_i) \tag{1.8}$$

where $\mathbf{r}_0$ is the location vector of the point P, and $\mathbf{r}_i$ the vectors giving the location of the other charges.

The force per unit charge experienced by the test particle at the point of interest is

$$\frac{\mathbf{F}_0}{q_0} = \frac{1}{4\pi\epsilon_0} \sum_i \frac{q_i}{|\mathbf{r}_0 - \mathbf{r}_i|^3} (\mathbf{r}_0 - \mathbf{r}_i)$$

We have assumed that q_0 does not disturb the properties of the field. This, however, is not possible in practice. We, therefore, assume q_0 to be vanishingly small and define *electric field* as

$$\mathbf{E}(\mathbf{r}_0) = \lim_{q_0 \to 0} \frac{\mathbf{F}_0}{q_0} = \frac{1}{4\pi\epsilon_0} \sum \frac{q_i}{|\mathbf{r}_0 - \mathbf{r}_i|^3} (\mathbf{r}_0 - \mathbf{r}_i) \tag{1.9}$$

The strength of this field $|\mathbf{E}(\mathbf{r})|$ is called the *electric field intensity*.

From the definition of the electric field we see that if a charge q is placed at a point at which the field is $\mathbf{E}$, the force acting on the charge is

$$\mathbf{F} = q\,\mathbf{E} \tag{1.10}$$

The electric field $\mathbf{E}(r)$ is a function of position and is itself a vector. An electric field, therefore, is a vector field. Examine carefully the Eq. (1.9) for the field $\mathbf{E}(r)$. It shows that the field at a point, due to the distribution of several charges, is the vector sum of the fields due to all the charges except the charge, if any, at the point under consideration. For the calculation of the electric field, we do not consider any charge to be at the point except the test charge. Otherwise, the contribution to the field at the point due to the charge present there would be infinite because of the singularity ($r = 0$) in the inverse square law and the theory would be useless. The concept such as that of a 'point' charge is meaningful if one accepts that measurements, in practice, are never made closer to such a charge than the distance of the order of atomic radii. Note that even a single electron has a finite size. However, it is often convenient to regard a small region of charged particles as a 'point charge'.

If the charge is not confined to a point, but is distributed over a region of space, it is possible to consider it as a continuous quantity and talk about a charge density or charge per unit volume. The charge density ρ is defined as

$$\rho = \lim_{V \to 0} \left(\frac{Q}{V} \right) \tag{1.11}$$

Consider, for example, the charge distribution within a hydrogen atom. We know that the electrons are not stationary point charges and their positions cannot be sharply defined. It is convenient and quite appropriate to consider the charge on the electron to be smeared out in a cloud around the nucleus (Fig. 1.5). If the charge density at a point specified by the position vector $\mathbf{r}$ is $\rho(\mathbf{r})$, the charge contained in a small volume element $d\tau$ at $\mathbf{r}$ is $\rho(\mathbf{r})\,d\tau$ and the total charge in the atom is given by

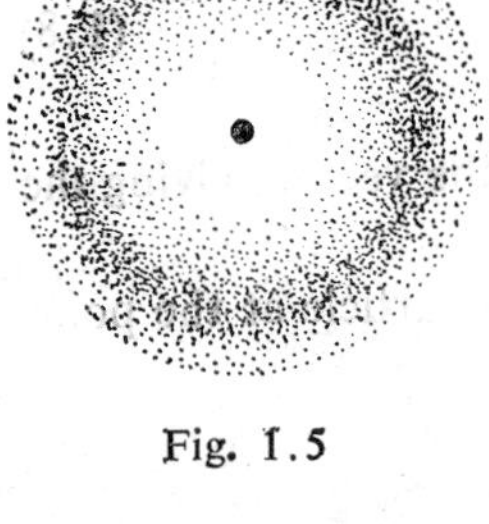

Fig. 1.5

$$\int \rho(\mathbf{r})\,d\tau = -e \tag{1.12}$$

The charge density also is a function of position, but it is a scalar quantity and its field is a scalar field.

In the case of continuous charge distribution the electric field is given by

$$\mathbf{E}(\mathbf{r}) = \frac{1}{4\pi\epsilon_0} \int \frac{\rho(\mathbf{r})\,(\mathbf{r} - \mathbf{r}_i)}{|\mathbf{r} - \mathbf{r}_i|^3}\,d\tau \tag{1.13}$$

When the charge is distributed over surface, we talk of surface charge density or charge per unit area $\sigma(\mathbf{r})$. In this case

$$\mathbf{E}(\mathbf{r}) = \frac{1}{4\pi\epsilon_0}\int \frac{\sigma(\mathbf{r})\,(\mathbf{r} - \mathbf{r}_i)}{|\mathbf{r} - \mathbf{r}_i|^3}\, dS \tag{1.14}$$

where dS is an element of area.

Similarly, when the charge is distributed over a curve, we talk of linear charge density or charge per unit length $\lambda(\mathbf{r})$ and

$$\mathbf{E}(\mathbf{r}) = \frac{1}{4\pi\epsilon_0}\int \frac{\lambda(\mathbf{r})\,(\mathbf{r} - \mathbf{r}_i)}{|\mathbf{r} - \mathbf{r}_i|^3}\, dl \tag{1.15}$$

where dl is an element of length.

It is sometimes convenient to regard a point charge as a 'fictitious' continuous charge distribution. This may be done with the help of Dirac delta function $\delta(\mathbf{r} - \mathbf{r}_i)$ which has among other the following properties.

(i) $\delta(\mathbf{r} - \mathbf{r}_i) = 0$ for $\mathbf{r} \neq \mathbf{r}_i$ (1.16)

(ii) $\int\delta(\mathbf{r} - \mathbf{r}_i)\, d\tau = 1$ if the region includes $\mathbf{r} = \mathbf{r}_i$ (1.17)

$= 0$ otherwise

(iii) $\int_V f(\mathbf{r}_i)\, \delta(\mathbf{r} - \mathbf{r}_i) d\tau = f(\mathbf{r})$ (1.18)

Suppose q' is a point charge at the point P represented by the position vector $\mathbf{r}'$. How can this charge be replaced by a 'fictitious' continuous charge distribution?

Consider the charge distribution

$$\rho(\mathbf{r}) = q'\delta(\mathbf{r} - \mathbf{r}') \tag{1.19}$$

From the definition of $\delta(\mathbf{r} - \mathbf{r}')$ we know that

$$\rho(\mathbf{r}) = 0 \qquad \text{for} \quad \mathbf{r} \neq \mathbf{r}'$$

and

$$\int\rho(\mathbf{r})d\tau = \int q'\delta(\mathbf{r} - \mathbf{r}')d\tau = q'$$

when integrated over the whole space. We see that $\rho(\mathbf{r})$ represents a charge distribution which is zero everywhere except at the point whose position vector is $\mathbf{r}'$ and is such that the total charge present is q'. Thus it represents a point charge q' situated at the point represented by the position vector $\mathbf{r}'$.

1.4 Lines and Tubes of Force

The continuous lines drawn in such a way that everywhere they are parallel to the direction of the field are called *lines of force* or *field lines*. Thus, a line of force is a curve in the electric field, such that the tangent at every point is in the direction of the electric field at that point. Since there is a single direction for the electric field at every point of the field there is usually just one field line through any given point, i.e. two lines of force can never intersect. Figure 1.6 shows the lines of force around point charges and near a pair of point charges of equal magnitude one positive and one negative.

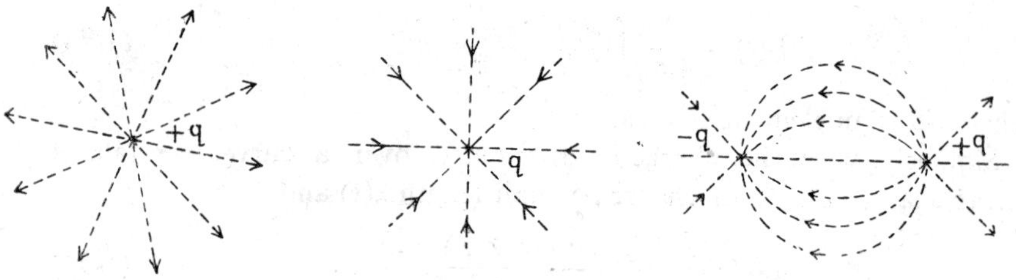

Fig. 1.6

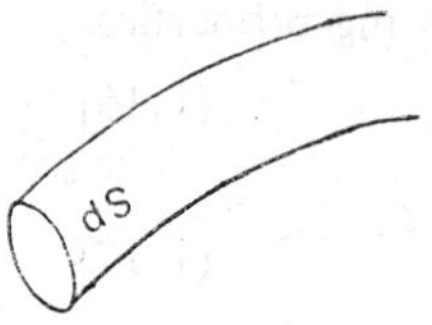

Fig. 1.7

Consider an element of area dS in the field, small enough to suppose the electric intensity to be the same in magnitude and direction at every point of dS, so that the lines of force at all points on the boundary will be approximately all parallel. The "tubular" surface (Fig 1.7) thus obtained is called a *tube of force.* The normal cross-section of the tube of force forms a part of an equipotential surface.

1.5 Electric Flux

Imagine that a fluid is flowing with a speed v through a small flat surface dS in a direction normal to the surface (Fig. 1.8a). The rate of flow of the fluid i.e. the volume of the fluid crossing the area per unit time is called the flux of the fluid and is equal to vdS. If the normal to the surface is not parallel to the direction of the flow but makes an angle θ with it (Fig. 1.8b), the projected area in a plane perpendicular to v is $dS \cos \theta$ and the flux F is

$$F = vdS \cos \theta \tag{1.20}$$

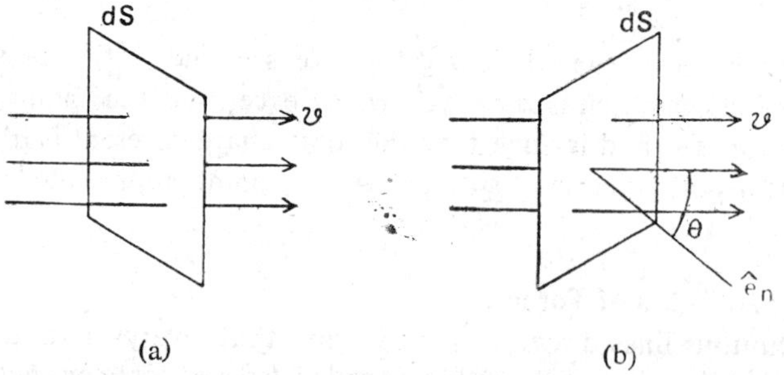

(a) (b)

Fig. 1.8

If we represent the surface by vector $\hat{\mathbf{e}}_n\, dS$ having magnitude dS and direction along the normal to the surface, $\hat{\mathbf{e}}_n$ being the unit vector in this direction, the flux can be written in vector notation as

$$\mathbf{F} = \mathbf{v} \cdot \hat{\mathbf{e}}_n dS \tag{1.21}$$

Although there is nothing actually flowing in an electrostatic field, we can mathematically define a quantity analogous to the flux of a fluid. We call it *electric flux* and define it as

$$\text{Electric flux} = \mathbf{E} \cdot \hat{\mathbf{e}}_n dS \tag{1.22}$$

1.6 Gauss' Law (Integral form)

The relation (1.22) leads to an important law in electrostatics which relates the flux through any closed surface to the net amount of charge enclosed within the surface.

Imagine a closed surface S surrounding a charge q_i (Fig. 1.9). Let dS be an element of area around the point A on the surface and $\hat{\mathbf{e}}_n$ an outward unit vector normal to it. Let θ be the angle between the electric field at A and the unit vector $\hat{\mathbf{e}}_n$. The flux through the element dS is

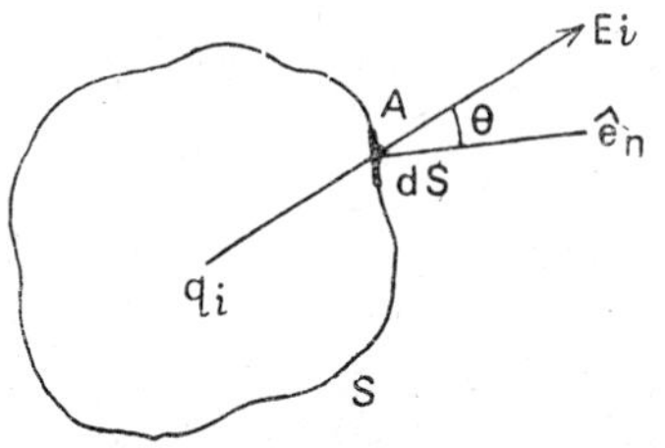

Fig. 1.9

$$dF = \mathbf{E}_i \cdot \hat{\mathbf{e}}_n dS = E_i \cos\theta \, dS = \frac{1}{4\pi\epsilon_0} \frac{q_i}{r^2} \cos\theta \, dS$$

where r is the distance of the element dS from q_i.

Now $d\Omega = \dfrac{dS \cos\theta}{r^2}$ is the solid angle subtended by dS at q_i.

$$\therefore \qquad dF = \mathbf{E}_i \cdot \hat{\mathbf{e}}_n dS = \frac{1}{4\pi\epsilon_0} q_i d\Omega \tag{1.23}$$

Hence, the flux through the entire surface S is

$$F = \int_S \mathbf{E}_i \cdot \hat{\mathbf{e}}_n dS = \frac{1}{4\pi\epsilon_0} q_i \int d\Omega = \frac{q_i}{\epsilon_0} \tag{1.24}$$

If there is an arbitrary distribution of charges within S, by the principle of superposition

$$\mathbf{E} = \Sigma \, \mathbf{E}_i$$

This leads to the relation

$$\int_S \mathbf{E}_i \cdot \hat{\mathbf{e}}_n dS = \Sigma q_i/\epsilon_0 = \frac{Q}{\epsilon_0} \tag{1.25}$$

where Q is the total charge within the closed surface.

In the case of a continuous distribution of charge within S, the relation (1.25) changes to

$$\int_S \mathbf{E}_i \cdot \hat{\mathbf{e}}_n dS = \frac{1}{\epsilon_0} \int \rho d\tau \tag{1.26}$$

Suppose now that the charge q_i is outside the closed surface S as shown in Fig. 1.10. The electric field vector passes through two elements of area dS_1 and dS_2 cut out by a cone with its apex at q_i. Let $\hat{e}_{n1}$, $\hat{e}_{n2}$ be the outward drawn unit vectors normal to the two elements of area making angles θ_1, θ_2 respectively with the field. The net outward flux through the two elements of area is

$$dF = \frac{1}{4\pi\epsilon_0}\frac{q_i}{r_1^2}\cos(180+\theta_1)\,dS_1 + \frac{1}{4\pi\epsilon_0}\frac{q_i}{r_2^2}\cos\theta_2\,dS_2$$

where r_1, r_2 are the distances of the elements dS_1, dS_2 from q_i.

$$\therefore \quad dF = \frac{q_i}{4\pi\epsilon_0}\left(-\frac{\cos\theta_1 dS_1}{r_1^2} + \frac{\cos\theta_2 dS_2}{r_2^2}\right)$$

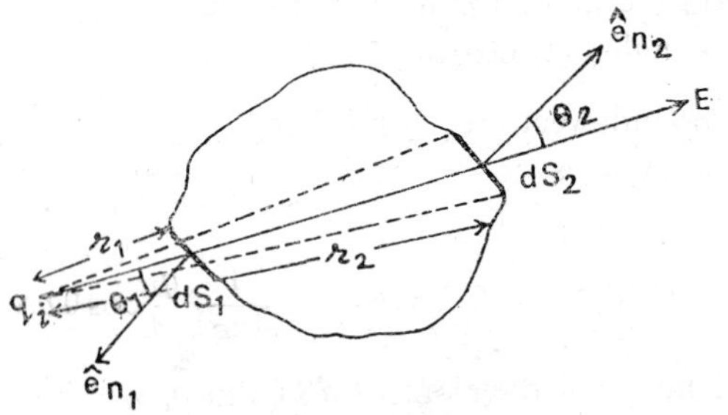

Fig. 1.10

Since the solid angles subtended by dS_1 and dS_2 at q_i are the same, i.e.

$$\frac{\cos\theta_1 dS_1}{r_1^2} = \frac{\cos\theta_2 dS_2}{r_2^2} = d\Omega$$

$$dF = 0$$

This is true for all the cones drawn from q_i through the surface and, hence, the total flux F is equal to zero.

These results constitute what is known as Gauss' law which can be stated as:

"The total outward flux through a closed suface S in an electric field is given by

$$\left.\begin{aligned}\int \mathbf{E}\cdot\hat{\mathbf{e}}_n\, dS &= \frac{Q}{\epsilon_0} && \text{for charge } Q \text{ inside } S \\ &= 0 && \text{for charges outside } S\end{aligned}\right\} \quad (1.27)$$

This is the integral form of the Gauss' law.

1.7 Gauss' Law (Differential form)

The Gauss' law can as well be expressed in a differential form.

In vector calculus we define the divergence of a vector **A** as

$$\text{div } \mathbf{A} = \lim_{d\tau \to 0} \frac{\int_S \mathbf{A} \cdot \hat{\mathbf{e}}_n dS}{d\tau} \tag{1.28}$$

where S is the surface enclosing the volume element. Integrating over a finite volume, we have,

$$\int_V \text{div } \mathbf{A} d\tau = \int_S \mathbf{A} \cdot \hat{\mathbf{e}}_n \, dS \tag{1.29}$$

This is the well-known *Divergence theorem* usually used for transforming a volume integral into a surface integral and vice-versa.

Using this, the Gauss' law can be written as

$$\int_S \mathbf{E} \cdot \hat{\mathbf{e}}_n \, dS = \int_V \text{div } \mathbf{E} d\tau = \frac{1}{\epsilon_0} \int_V \rho d\tau$$

i.e.

$$\int_V (\text{div } \mathbf{E} - \rho/\epsilon_0) \, d\tau = 0$$

This is true for any arbitrary volume V.

$$\therefore \quad \text{div } \mathbf{E} - \rho/\epsilon_0 = 0$$

i.e.

$$\text{div } \mathbf{E} = \nabla \cdot \mathbf{E} = \rho/\epsilon_0 \tag{1.30}$$

which is the differential form of the Gauss' law.

In electrostatics Coulomb's law has been adopted as the fundamental law. The Gauss' law is a consequence of the fact that the electric force between charged particles is Coulomb's inverse square law. Any other law, say, $\frac{1}{r^n}$ with $n \neq 2$ would not give Gauss' law (why?). Hence, Gauss' law can as well be taken as the fundamental law of electrostatics.

1.8 Some Applications of Gauss' Law

We have seen that the charge inside a Gaussian surface determines the electric flux through the surface. If the charge distribution has a simple symmetry, electric field can be evaluated easily using Gauss' law. We shall discuss in this section some such useful problems.

(i) The field due to an infinite layer of positive charge with uniform surface density σ:

Let $ABCD$ be the plane (Fig. 1.11). By symmetry, the field lines are perpendicular to the plane. Consider a right circular cylinder XYZ with the cross-section dS. By Gauss' law

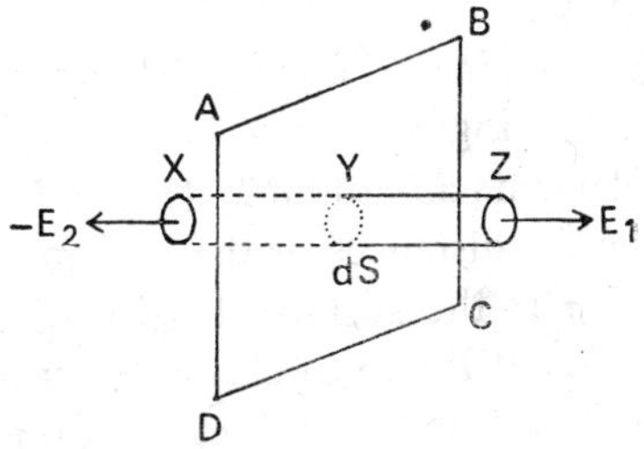

Fig. 1.11

$$(E_1 - E_2) dS = \frac{\sigma dS}{\epsilon_0}$$

$$\therefore \quad E_1 - E_2 = \frac{\sigma}{\epsilon_0} \tag{1.31}$$

That is the field E changes across a charge layer and the change is σ/ϵ_0

(ii) The field outside an isolated charged sphere:

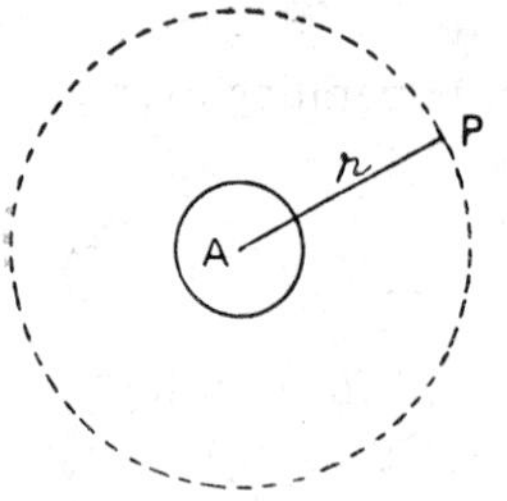

Fig. 1.12

Suppose A (Fig. 1.12) is a sphere filled with a uniform charge distribution. What will be the field at a point P outside the sphere?

Imagine a spherical surface concentric with the charged sphere and passing through the point P. If r is the radius of the imaginary sphere, its area is $4\pi r^2$. By symmetry, the electric intensity E is the same at every point of the surface. Hence, the flux outwards through the surface is

$$\int \mathbf{E}\cdot\hat{\mathbf{e}}_n\, dS = 4\pi r^2 E = \frac{Q}{\epsilon_0} \qquad \text{(by Gauss' law)}$$

where Q is the total charge within the sphere.

$$\therefore \quad E = \frac{1}{4\pi\epsilon_0}\frac{Q}{r^2} \tag{1.32}$$

This is the same as the field produced by the point charge Q at the centre of the sphere. If there is a continuous charge distribution within the sphere,

$$Q = \frac{4}{3}\pi a^3 \rho$$

where 'a' is the radius of the sphere and 'ρ' the charge density.

Hence,
$$E = \frac{1}{4\pi\epsilon_0}\frac{4\pi a^3\rho}{3r^2} = \frac{a^3\rho}{3\epsilon_0 r^2} \tag{1.33}$$

(iii) Field due to a spherical shell of charge:

The arguments as applied in case (ii) above show that the field outside a thin spherical shell of charge is the same as if the total charge on the shell is concentrated at the centre. What will be the field at a point P inside the shell?

Imagine a cone with apex at P and extending on either side to cut out surface elements dS_1 and dS_2 (Fig. 1.13). Let r_1, r_2 be the distances of these elements from P. If σ is the surface density of charge, the fields at P due to elements are $\sigma dS_1/r_1^2$ and $\sigma dS_2/r_2^2$ and will act in opposite directions. But

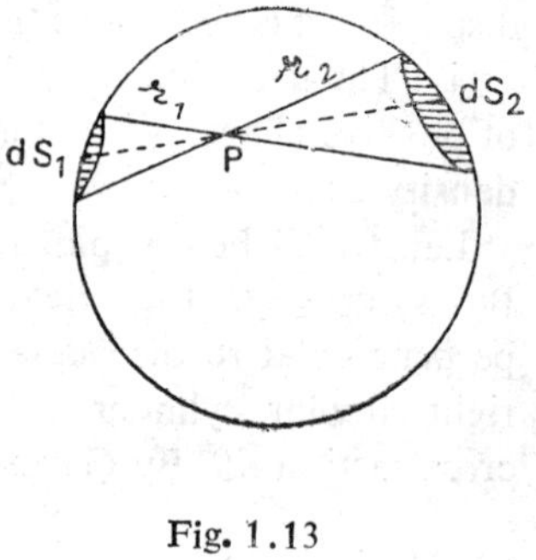

Fig. 1.13

$$\frac{dS_1}{r_1^2} = \frac{dS_2}{r_2^2} = d\Omega$$

(solid angle subtended at P by the two elements of area)

The contributions to the field due to the two elements being thus equal and opposite, cancel exactly. One can proceed to cover the entire sphere in this fashion by balancing off field contributions of opposite differential areas. Since each pair of differential areas gives a zero contribution we conclude that the field at P is zero.

(iv) The field at a point P inside a uniformly charged sphere of radius R:

Let the point P be at a distance r from the centre (Fig. 1.14). As before imagine a sphere passing through P. The charge in the shell of thickness $R-r$ does not contribute to the field at P since the point lies within the shell (case iii). If E is the field at P, the flux through the imaginary surface is

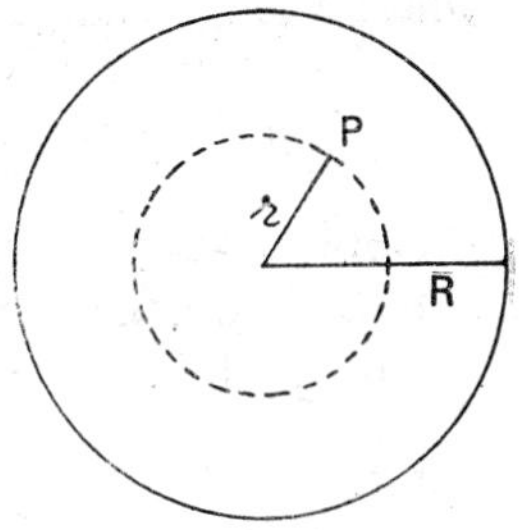

Fig. 1.14

$$4\pi r^2 E = \frac{1}{\epsilon_0}\frac{4\pi r^3\rho}{3} \quad \text{(by Gauss' law)}$$

$$\therefore \quad E = \frac{r\rho}{3\epsilon_0} \tag{1.34}$$

Equations (1.33) and (1.34) give the field outside and inside the sphere respectively. You can see that the two forms match, as they should, when $r = a$. Figure 1.15 gives the variation of the electric field inside and outside the sphere of radius 'a'.

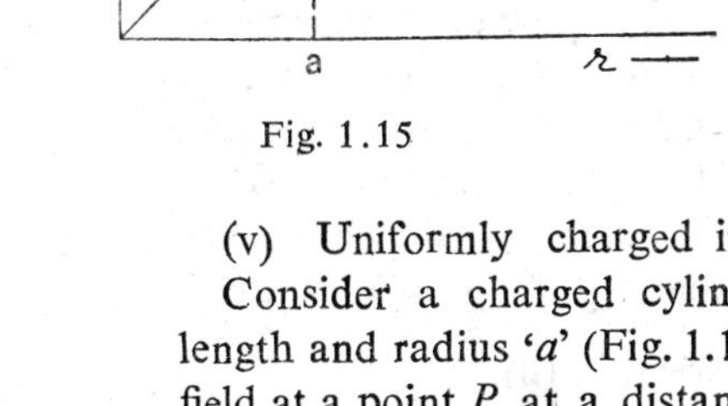

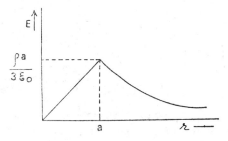

Fig. 1.15

(v) Uniformly charged infinite cylinder:

Consider a charged cylinder of infinite length and radius 'a' (Fig. 1.16). To find the field at a point P at a distance 'd' from the axis, imagine a closed cylindrical surface coaxial with the cylinder and passing through P, its two end faces being perpendicular to the axis of the cylinder. We can easily surmise by symmetry, that the field will be normal to the axis directed away from the axis and is the same at equal distances from

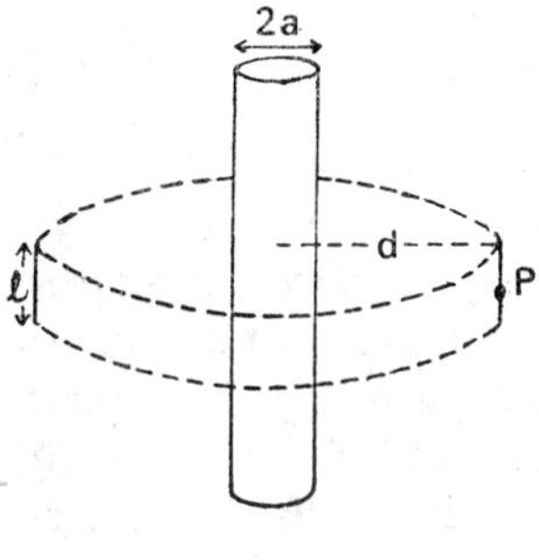

Fig. 1.16

the axis. If 'l' is the height of the cylindrical surface, the flux through this surface is $2\pi dlE$. Note that the end faces do not contribute to the field since E is tangential to these faces. By Gauss' law

$$2\pi dlE = \frac{\lambda l}{\epsilon_0}$$

where λ is the charge per unit length.

$$\therefore \quad E = \frac{\lambda}{2\pi d\epsilon_0} \tag{1.35}$$

Notice that the result is independent of the radius of the charged cylinder and hence holds also for the rectilinear distribution of charges.

This result could have been obtained by direct integration. Consider a filament of infinite length of constant line density of charge λ coulomb/metre (Fig. 1.17). The field at a point P due to a line element dz is

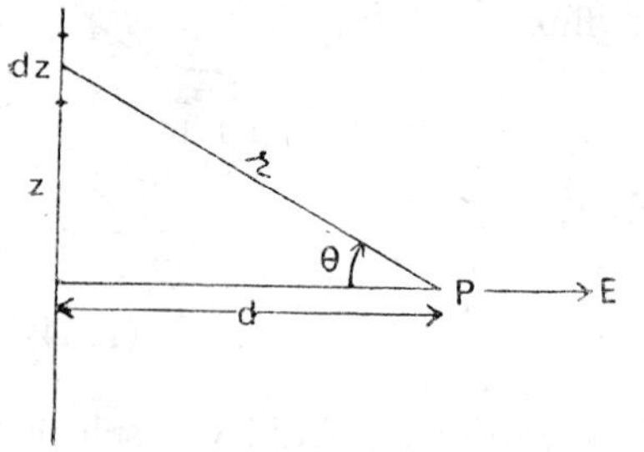

Fig. 1.17

$$dE_d = \frac{\lambda dz \cos\theta}{4\pi\epsilon_0 r^2}$$

Note that we have assumed the filament to coincide with the z-axis. The perpendicular distance of P from the filament is 'd' and the distance of 'P' from the element dz is 'r'. The total field is given by

$$E_d = \frac{\lambda}{4\pi\epsilon_0}\int_{-\infty}^{\infty} \frac{\cos\theta}{r^2}\, dz$$

Since $z = d\tan\theta$ and $r = d\sec\theta$

$$E_d = \frac{\lambda}{4\pi\epsilon_0}\int_{-\pi/2}^{\pi/2} \frac{\cos\theta}{d}\, d\theta = \frac{\lambda}{2\pi\epsilon_0 d}$$

(vi) Field between two concentric spheres which have equal and opposite charges:

Imagine a Gaussian surface drawn through the point P (Fig. 1.18). The field at P due to the outer sphere is zero as shown in (iii) above and that due to the inner sphere is $E = \frac{1}{4\pi\epsilon_0}\frac{q}{r^2}$ as shown in (ii).

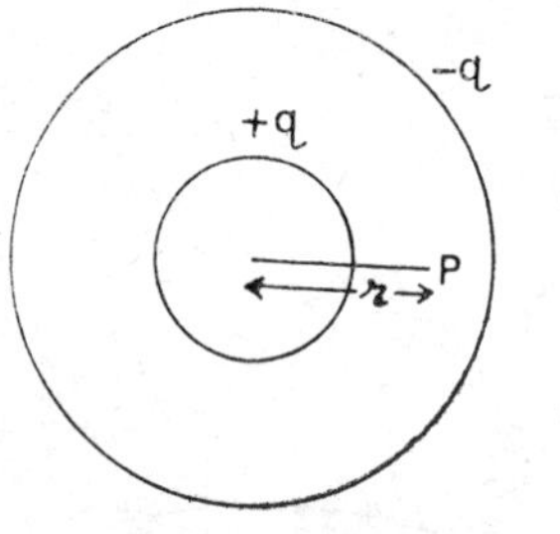

Fig. 1.18

It is indeed true that the charge inside a Gaussian surface determines the flux of the electric field. It is also true that the field at a point is determined by Gauss' law when there is a symmetry such that the normal component of the field is constant in each surface of the Gaussian volume. In each of this case the field at the Gaussian surface is determined by the charge within the Gaussian surface. However, in the case of asymmetrical charge distribution, the field is determined by charges both inside and outside the Gaussian surface. Consider now two non-concentric charged spheres (Fig. 1.19). The redistribution of charge and field lines would be as shown in Fig. 1.19. We draw a cone as before with P as the apex. Since the two charge densities on dS_1 and dS_2 are not equal $dE_2 > dE_1$ and, hence, the two different areas yield a net dE in the negative x direction. If the entire sphere is covered with this process, we will find a net $\mathbf{E}_-$, due to the outer sphere in the negative x direction. The field due to the inner sphere at P is $\mathbf{E}_+$ and hence, the effective field at P is $\mathbf{E}=\mathbf{E}_+ + \mathbf{E}_-$.

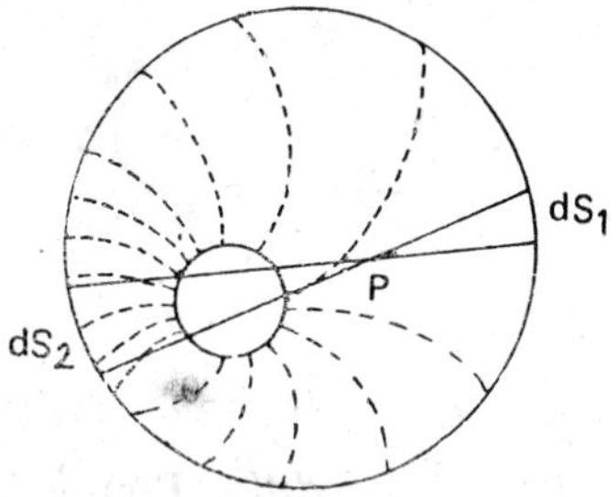

Fig. 1.19

1.9 A Useful Theorem in Electrostatics

The following theorem has been found to be very useful in electrostatics.

Theorem : The average value of the electrostatic field over the volume V of a sphere, due to a point charge 'q' somewhere within the sphere is

$$\langle E \rangle_{av} = -\frac{q r_0}{3\epsilon_0 V} \tag{1.36}$$

Proof : For any field point $\mathbf{r}$ the field is given by

$$\mathbf{E}(\mathbf{r}) = \frac{q}{4\pi\epsilon_0}\left[\frac{(\mathbf{r}-\mathbf{r}_0)}{|\mathbf{r}-\mathbf{r}_0|^3}\right]$$

where $\mathbf{r}_0$ gives the position of the charge.

Thus,

$$\langle E(r) \rangle_{av} = \frac{q}{4\pi\epsilon_0 V}\int_V \frac{(\mathbf{r}-\mathbf{r}_0)}{|\mathbf{r}-\mathbf{r}_0|^3}\,d\tau$$

$$= -\left\{\frac{q}{4\pi\epsilon_0 V}\int_V \frac{(\mathbf{r}_0-\mathbf{r})}{|\mathbf{r}_0-\mathbf{r}|^3}\,d\tau\right\} = -\frac{q r_0}{3\epsilon_0 V}$$

because the bracketed expression is simply the field at r_0 due to a uniform charge density $\frac{q}{V}$ in the sphere.

1.10 Electrostatic Potential

A test charge will not move in opposition to the force it experiences in an electrostatic field, unless external work is done on it. The charge moved in this way acquires potential energy.

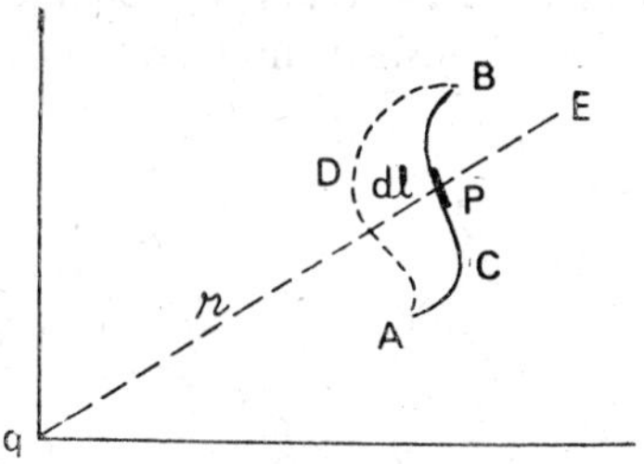

Fig. 1.20

Suppose a unit test charge is carried from A to B along the line joining A and B as shown in Fig. 1.20 in the field of point charge q. If both the charges are positive the work is done by the repulsive force as the unit charge moves from A to B. the work done when the charge is moved through an infinitesimal section dl of the path is

$$dW = \mathbf{E}(\mathbf{r})\cdot\hat{\mathbf{e}}_l\, dl = \frac{q\,dl\cos\theta}{4\pi\epsilon_0 r^2} = \frac{q\,dr}{4\pi\epsilon_0 r^2}$$

where $\mathbf{r}$ is the position vector of the point P. The total work done in carrying the charge from the point A distant r_1 to B distant r_2 is

$$W = \int_A^B \mathbf{E}\cdot\hat{\mathbf{e}}_l\, dl = \frac{q}{4\pi\epsilon_0}\int_{r_1}^{r_2}\frac{dr}{r^2}$$

$$= \frac{q}{4\pi\epsilon_0}\left[\frac{1}{r_1} - \frac{1}{r_2}\right] \qquad (1.37)$$

We see that the line integral $\int_A^B \mathbf{E}\cdot\hat{\mathbf{e}}_l\, dl$ depends on the end positions only and is independent of the path joining the points A and B. This shows that the work done is the same whether the charge is taken along the path ACB or ADB or some other.

The work done by the electrostatic force when a test charge is carried from a point A to a point B is equal to the loss of potential energy of the test charge. That is

$$\text{Potential energy at } B - \text{Potential energy at } A = -\int_A^B \mathbf{E}\cdot\hat{\mathbf{e}}_l dl$$

What would happen if the charge which is transported from A to B is brought back to A by any path of your choice? Since, the energy lost in going from A to B is regained in the reverse movement, the net charge in the potential energy is zero, i.e.

$$W = \int_c \mathbf{E}\cdot\hat{\mathbf{e}}_l\, dl = 0 \qquad (1.38)$$

A force which has the property of doing no work and thus dissipating no energy around a closed path is called a *conservative force*. Electrostatic force is a conservative force.

In vector calculus the line integral of a vector round a closed curve is called the *circulation* or the *curl* of the vector field. Equation (1.38) indicates that the circulation of the electric field is zero. Its sources are given by (1.30). Other examples of such fields are the gravitational field, the magnetic field in region free from electric current and the velocity field of the flow of incompressible fluids free from viscosity. Such fields are known as *irrotational* fields.

If the work done is measured from infinity, $r_1 = \infty$ and

$$W = -\int_{\infty}^{r_2} \frac{q}{4\pi\epsilon_0 r^2}\, dr = \frac{q}{4\pi\epsilon_0 r_2}$$

i.e. the work done in bringing a unit charge from infinity to the point B is $\frac{q}{4\pi\epsilon_0 r_2}$. Assuming that the potential energy at infinity is zero, we see that the potential energy at any point is a function of position only. We call it electrostatic potential of the charge q at a point distant r from it and represent it by the symbol $\Phi(r)$

$$\Phi(r) = \frac{q}{4\pi\epsilon_0 r} \tag{1.39}$$

We can thus associate with every point of the field an electrostatic potential $\Phi(r)$ defined by (1.39). The potential has only one value at any point in the field. It is a single-valued function of the space coordinate for any stationary distribution of charges.

If we have a set of point charges q_j, the potential at a point r_i is given by the algebraic sum of the individual potential, i.e.

$$\Phi(r_i) = \frac{1}{4\pi\epsilon_0} \sum_j \frac{q_j}{r_{ij}} \tag{1.40}$$

In the case of a continuous distribution of charges

$$\Phi(r_i) = \frac{1}{4\pi\epsilon_0} \int \frac{p\, d\tau_j}{r_{ij}} \tag{1.41}$$

1.11 Relation Between the Field and the Potential

We have defined electrostatic field and electrostatic potential in the preceding section. We shall now show how these are related to each other.

Suppose, for convenience, that the charge q is at the origin of the frame of reference. The field at a point distant r is

$$\mathbf{E} = \frac{1}{4\pi\epsilon_0} \frac{q}{|\mathbf{r}|^2} \hat{\mathbf{e}}_r$$

where $\hat{\mathbf{e}}_r$ is the unit vector in the direction of the field. We also know that

$$\text{grad}\left(\frac{1}{r}\right) = \nabla\left(\frac{1}{r}\right) = -\frac{1}{|\mathbf{r}|^2}\hat{\mathbf{e}}_r$$

$$\therefore\ \mathbf{E} = -\frac{q}{4\pi\epsilon_0}\nabla\left(\frac{1}{r}\right) = -\nabla\left(\frac{q}{4\pi\epsilon_0 r}\right)$$

$$= -\nabla\Phi \quad \text{from (1.38)} \tag{1.42}$$

Thus, once the potential distribution is known, the field at any point can be found by taking its gradient. It is usually simpler to work in terms of potential rather than field because potential is a scalar quantity.

EXAMPLE 1.2. Determine the potential and the field produced by a ring of charges at a point on the axes of the ring.

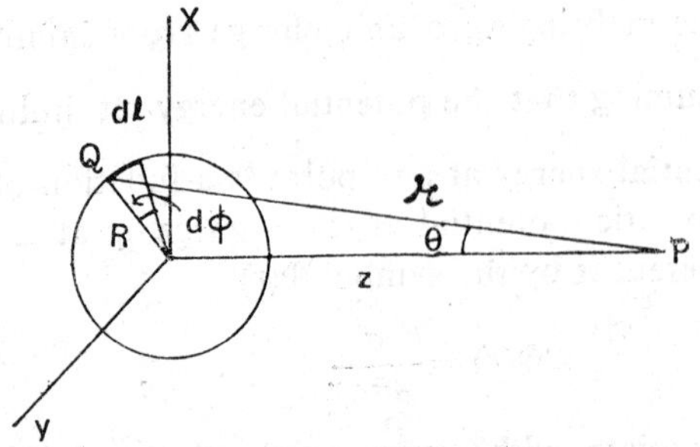

Fig. 1.21

Let λ be the linear charge density and R the radius of the ring, For convenience we assume the ring to be situated in the $z=0$ plane as shown in Fig. 1.21. The charge in a small element of length dl at Q is $dq=\lambda dl=\lambda R d\phi$ and the distance of this element from the point P at which the potential is to be found is

$$r=\sqrt{R^2+z^2}$$

The potential at P is

$$\Phi=\frac{1}{4\pi\epsilon_0}\oint\frac{\lambda dl}{\sqrt{R^2+z^2}}=\frac{1}{4\pi\epsilon_0}\oint\frac{\lambda R d\phi}{\sqrt{R^2+z^2}}$$

$$=\frac{\lambda R}{4\pi\epsilon_0}\frac{2\pi}{\sqrt{R^2+z^2}}=\frac{Q}{4\pi\epsilon_0\sqrt{R^2+z^2}} \qquad (1.43)$$

where $Q=2\pi R\lambda$, the total charge on the ring.

The resulting field is

$$\mathbf{E}=-\nabla\Phi=-\hat{\mathbf{e}}_z\frac{\partial\Phi}{\partial z}=\hat{\mathbf{e}}_z\frac{Q}{4\pi\epsilon_0}\frac{z}{(R^2+z^2)^{3/2}} \qquad (1.44)$$

Plots of the magnitude of the potential and the field are given in Fig. 1.22 (a) and (b).

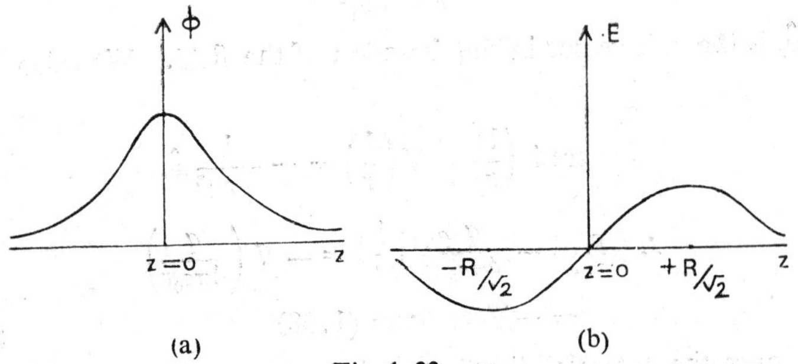

(a) (b)

Fig. 1.22

If $z \gg R$, Φ and $\mathbf{E}$ reduce to

$$\Phi \approx \frac{Q}{4\pi\epsilon_0 \mid z \mid} \tag{1.45}$$

and

$$\mathbf{E} \approx \begin{cases} \hat{\mathbf{e}}_z \dfrac{Q}{4\pi\epsilon_0 z^2} & \text{for} \quad z \gg R \\ -\hat{\mathbf{e}}_z \dfrac{Q}{4\pi\epsilon_0 z^2} & \text{for} - z \ll -R \end{cases} \tag{1.46}$$

The result (1.44) could have been obtained by direct evaluation using (1.15) which, taking the origin at the centre of the ring, reduces to

$$\mathbf{E} = \frac{1}{4\pi\epsilon_0} \int \frac{\lambda R d\phi}{(R^2+z^2)} \hat{\mathbf{e}}_r = \frac{\lambda R}{4\pi\epsilon_0 (R^2+z^2)} \int_0^{2\pi} \hat{\mathbf{e}}_r \, d\phi$$

where $\hat{\mathbf{e}}_r$ is the unit vector in the direction QP. Now

$$\hat{\mathbf{e}}_r = -\hat{\mathbf{e}}_x \sin\theta \cos\phi - \hat{\mathbf{e}}_y \sin\theta \sin\phi + \hat{\mathbf{e}}_z \cos\theta$$

By the symmetry of the system along the z-axis, the field would be directed along z-axis only

$$\therefore \quad \mathbf{E} = \frac{\lambda R}{4\pi\epsilon_0 (R^2+z^2)} \int_0^{2\pi} \hat{\mathbf{e}}_z \cos\theta \, d\phi$$

$$= \hat{\mathbf{e}}_z \frac{\lambda R 2\pi \cos\theta}{4\pi\epsilon_0 (R^2+z^2)} = \hat{\mathbf{e}}_z \frac{Q}{4\pi\epsilon_0} \frac{z}{(R^2+z^2)^{3/2}}$$

This procedure, in general, is quite laborious and the evaluation of the potential Φ first, followed by $\mathbf{E} = -\nabla\Phi$ is almost always the preferable procedure. However, one must be careful while using this method not to overlook components of $\mathbf{E}$ that may be missing from $-\nabla\Phi$ along particular axis of symmetry as will be shown in the next example.

EXAMPLE 1.3. Find the field of a semicircular uniform distribution of line charge of linear charge density λ (Fig. 1.13).

We first follow the same procedure as in the preceding example.

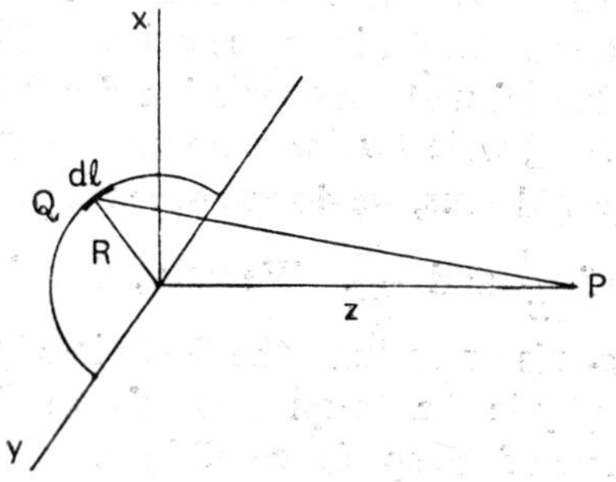

Fig. 1.23

$$\Phi = \frac{1}{4\pi\epsilon_0}\int \frac{\lambda dl}{r_{QP}} = \frac{\lambda R}{4\pi\epsilon_0 (R^2+z^2)^{1/2}}\int_{-\pi/2}^{\pi/2} d\phi$$

$$= \frac{\lambda R\pi}{4\pi\epsilon_0(R^2+z^2)^{1/2}} = \frac{Q}{8\pi\epsilon_0(R^2+z^2)^{1/2}} \tag{1.47}$$

where $\frac{Q}{2} = \pi R\lambda$, net charge in the semi-circle. Now

$$\mathbf{E} = -\nabla\Phi = -\hat{\mathbf{e}}_z \frac{\partial\Phi}{\partial z} = \hat{\mathbf{e}}_z \frac{Q}{8\pi\epsilon_0}\frac{z}{(R^2+z^2)^{3/2}} \tag{1.48}$$

The field, therefore, is z-directed and exactly one-half of the value found for the complete ring.

Let us now evaluate $\mathbf{E}$ directly using (1.15)

$$\mathbf{E} = \frac{1}{4\pi\epsilon_0}\int \hat{\mathbf{e}}_r \frac{\lambda dl}{r^2} = \frac{\lambda R}{4\pi\epsilon_0(R^2+z^2)}\int_{-\pi/2}^{\pi/2} \hat{\mathbf{e}}_r d\phi$$

$$= \frac{\lambda R}{4\pi\epsilon_0(R^2+z^2)}\left\{-\hat{\mathbf{e}}_x \int_{-\pi/2}^{\pi/2} \sin\theta\cos\phi\, d\phi - \hat{\mathbf{e}}_y\int_{-\pi/2}^{\pi/2} \sin\theta\sin\phi\, d\phi\right.$$

$$\left. + \hat{\mathbf{e}}_z \int_{-\pi/2}^{\pi/2} \cos\theta\, d\phi\right\}$$

$$= \frac{\lambda R}{4\pi\epsilon_0(R^2+z^2)}\left\{-\hat{\mathbf{e}}_x 2\sin\theta + \hat{\mathbf{e}}_z \pi\cos\theta\right\}$$

$$= \frac{\lambda R}{4\pi\epsilon_0(R^2+z^2)}\left\{-\hat{\mathbf{e}}_x \frac{2R}{(R^2+z^2)^{1/2}} + \hat{\mathbf{e}}_z\pi\frac{z}{(R^2+z^2)^{1/2}}\right\}$$

$$= \frac{Q}{8\pi\epsilon_0(R^2+z^2)^{3/2}}\left\{-\hat{\mathbf{e}}_x\frac{2R}{\pi} + \hat{\mathbf{e}}_z z\right\} \tag{1.49}$$

This does not agree with (1.48). Obviously we have done something wrong in applying one of the procedures. What is it? Considering the symmetry of the system we see that although the y-component of $\mathbf{E}$ is zero, there must be an E_x component since all the charges are positive in value and located above the $x=0$ plane. The expression (1.48) lacks this component and, hence, is not complete. The reason for this omission is that we have found the expression for Φ for points on the z-axis i.e. the path along which we have no knowledge of the potential dependence on x or y and thus, we do not have the necessary information for the evaluation of $\frac{\partial\Phi}{\partial x}$ and $\frac{\partial\Phi}{\partial y}$. We must, therefore, either evaluate Φ at a point off the z-axis or evaluate the field directly using (1.15). One has to consider the physical consequences of the field $\mathbf{E}$ when Φ does not contain all the three of the special conditions.

1.12 Two Important Relations

Taking the curl of the vectors in (1.42), we find

$$\nabla \times \mathbf{E} = -\nabla \times \nabla \Phi = 0 \tag{1.50}$$

This result can also be deduced from (1.38).

We can use Stokes' theorem (see Appendix A) to transform the line integral of (1.38) into a surface integral

$$\oint_c \mathbf{E} \cdot d\mathbf{l} = \int_S \nabla \times \mathbf{E} \cdot \hat{\mathbf{e}}_n \, dS$$

$$\therefore \quad \int_S \nabla \times \mathbf{E} \cdot \hat{\mathbf{e}}_n \, dS = 0$$

and since S is arbitrary $\nabla \times \mathbf{E} = 0$.

We, thus, have two important laws of electrostatics:

$$\text{(i)} \quad \nabla \cdot \mathbf{E} = \rho/\epsilon_0 \tag{1.51}$$

$$\text{(ii)} \quad \nabla \times \mathbf{E} = 0 \tag{1.52}$$

Of these, the first follows from Coulomb's inverse square law. The other does not depend upon this law. Any other dependence would have given this result. However, it does depend upon the central nature of the force. Since all predictions of electrostatics follow from these laws, they can be considered fundamental laws of electrostatics.

1.13 Equipotential Surfaces

A surface at every point of which the potential has the same value is called an equipotential surface. The equation $\Phi(r) = C$, where C is a constant, represents an equipotential surface. Different values of C generate a family of such surfaces. For example, for a point charge, the equipotential surfaces are spheres with their centre at the charge (Fig. 1.24).

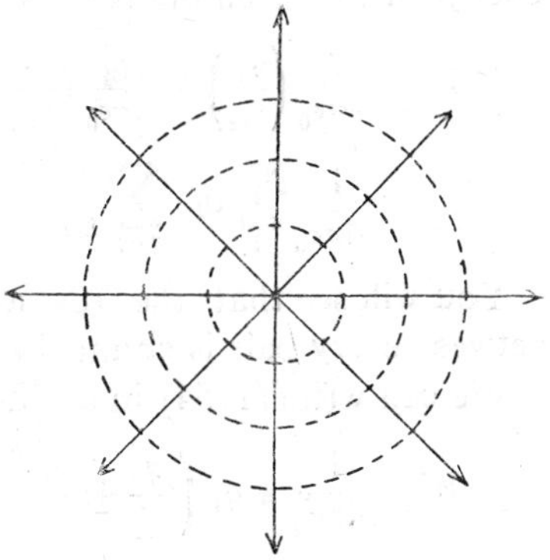

Fig. 1.24

Consider a displacement $d\mathbf{r}$ on an equipotential surface $\Phi(\mathbf{r}) = C$. Since it is an equipotential surface

$$\Phi(\mathbf{r}) - \Phi(\mathbf{r} + d\mathbf{r}) = 0$$

i.e.

$$\Phi(\mathbf{r}) - \left\{ \Phi(\mathbf{r}) + \frac{\partial \Phi}{\partial r} \cdot d\mathbf{r} \right\} = 0$$

$$\therefore \quad \frac{\partial \Phi}{\partial r} \cdot d\mathbf{r} = 0$$

i.e.

$$\nabla \Phi \cdot d\mathbf{r} = 0$$

or

$$\mathbf{E} \cdot d\mathbf{r} = 0 \tag{1.53}$$

This shows that if $d\mathbf{r}$ is on the surface, $\mathbf{E}$ must be normal to the surface i.e. the field lines are normal to the equipotential surface.

1.14 Electrostatic Energy

We shall discuss in this section the energy of electrostatic systems. Obviously, this energy is solely the potential energy arising from the interaction between the charges.

Let a charge q_1 be situated at a certain point in space and a charge q_2 which was initially at infinity, be brought up to a distance r_{12} from q_1. The potential of the charge q_1 at r_{12} is $\frac{q_1}{4\pi\epsilon_0 r_{12}}$. This being, by definition, the work done in bringing a unit charge from infinity to the point, the work done in bringing the charge q_2 from infinity to the point specified by r_{12} is $\frac{q_1 q_2}{4\pi\epsilon_0 r_{12}}$. If a third charge q_3 is added to the system, work will have to be done against the field of q_1 and q_2. If r_{13}, r_{23} are the distances of q_3 from q_1 and q_2 respectively, the additional contribution to the potential energy is

$$\frac{q_1 q_3}{4\pi\epsilon_0 r_{13}} + \frac{q_2 q_3}{4\pi\epsilon_0 r_{23}}$$

Continuing to build up the assembly in this way, we find that the total energy of the assembly is

$$W = \frac{q_2}{4\pi\epsilon_0}\left(\frac{q_1}{r_{12}}\right) + \frac{q_3}{4\pi\epsilon_0}\left(\frac{q_1}{r_{13}} + \frac{q_2}{r_{23}}\right) + \frac{q_4}{4\pi\epsilon_0}\left(\frac{q_1}{r_{14}} + \frac{q_2}{r_{24}} + \frac{q_3}{r_{34}}\right) + \cdots$$

$$= \frac{1}{4\pi\epsilon_0}\sum_i q_i \sum_{j<i} \frac{q_j}{r_{ij}} \tag{1.54}$$

You will see that the restriction $j < i$ ensures that the interaction between every pair is counted only once.

We can write (1.54) in a different form

$$W = \frac{1}{4\pi\epsilon_0}\,\tfrac{1}{2}\,q_1\left(\frac{q_2}{r_{12}} + \frac{q_3}{r_{13}} + \cdots\right) + \frac{1}{4\pi\epsilon_0}\,\tfrac{1}{2}\,q_2\left(\frac{q_1}{r_{21}} + \frac{q_3}{r_{23}} + \cdots\right)$$

$$+ \frac{1}{4\pi\epsilon_0}\,\tfrac{1}{2}\,q_3\left(\frac{q_1}{r_{31}} + \frac{q_2}{r_{32}} + \frac{q_4}{r_{34}} + \cdots\right) + \cdots$$

$$= \frac{1}{4\pi\epsilon_0}\,\tfrac{1}{2}\sum_{i=1} q_i \sum_{j=1} \frac{q_j}{r_{ij}}; \quad i \neq j \tag{1.55}$$

The factor $\frac{1}{2}$ appears in the expression because in this case each pair is counted twice.

Since $\frac{1}{4\pi\epsilon_0}\sum_{j\neq i} \frac{q_j}{r_{ij}}$ is the potential Φ_i, produced by all the charges except ith one at the point at which q_i is situated

$$W = \tfrac{1}{2}\sum_i q_i \Phi_i \tag{1.56}$$

If the charges are not localized and there is a continuous distribution of charges.

$$W = \tfrac{1}{2} \int \Phi \rho d\tau \tag{1.57}$$

EXAMPLE 1.4. Calculate the energy of a sphere of charge of radius R in which the charge is uniformly distributed.

The energy we are going to calculate is equal to the work done in bringing the charges from infinity to the sphere. We may imagine that the sphere is formed by assembling various thin shells of charge. Consider a small sphere of charge of radius r (Fig. 1.25). If ρ is the density of charge, the total charge on the sphere is $\frac{4}{3}\pi r^3 \rho$. Suppose a small layer of charge dq in the form of thin shell of thickness dr is deposited on the sphere.

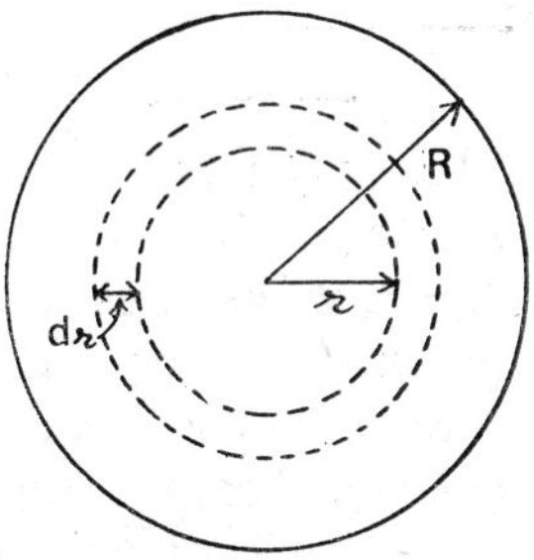

Fig. 1.25

$$\therefore \quad dq = \rho \, 4\pi r^2 dr$$

The work done in bringing this charge from infinity is

$$dW = \text{potential at } r \times dq$$

$$= \frac{1}{4\pi\epsilon_0} \frac{4/3\pi r^3 \rho}{r} \rho \, 4\pi r^2 dr$$

$$= \frac{4\pi\rho^2}{3\epsilon_0} r^4 dr$$

The total energy required to assemble charges so as to build up a sphere of radius R is

$$W = \frac{4\pi\rho^2}{3\epsilon_0} \int_0^R r^4 dr = \frac{4\pi\rho^2}{15\epsilon_0} R^5 \tag{1.58}$$

1.15 Electric Dipole

We have seen that the electric flux vector of a point charge is directed radially outwards and that its magnitude can be found by a simple application of Gauss' law. If, however, more than one charge is present, the field pattern becomes more complicated. Consider, for example, two

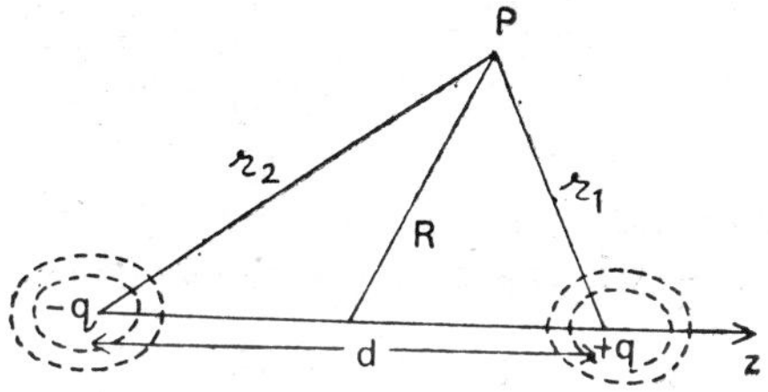

Fig. 1.26

equal and opposite charges $\pm q$ separated by a small distance 'd' (Fig. 1.26). Such a system is known as a **dipole**. We often come across such system in physics. For example —

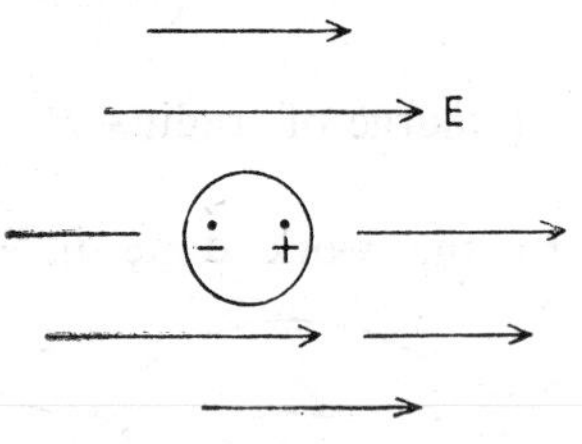

Fig. 1.27

(i) When an atom or a molecule is placed in an electric field, the positive and negative charges feel opposite forces and are displaced slightly forming a dipole (Fig. 1.27).

(ii) The structure of some molecules is such that even though there is no external field, the positive charge is slightly separated from the negative charge (e.g., H_2O) (Fig. 1.33).

Although this arrangement is not equivalent to the one we defined as dipole, we can treat them as such when we are considering the field they produce at a very long distance.

Suppose the charges are placed on the z-axis, one at $d/2$ and the other at $-d/2$ from the origin. What is the magnitude of the potential due to this system at a point P ? If the point is close to either of the charges, the potential is almost the same as if the charges were independent of one another and equipotential surfaces are nearly spherical as shown in the figure. (Notice that the charges are not shown at the centre of the sphere. Can you explain why?). If the point is at a distance r from the origin and $r \gg d$, the potential is given by

$$\Phi(r) = \frac{1}{4\pi\epsilon_0}\left(\frac{q}{r_1} - \frac{q}{r_2}\right) \tag{1.59}$$

where r_1, r_2 are the distances of P from $+q$ and $-q$ respectively.

$$\text{Now } r_1^2 = R^2 + \left(\frac{d}{2}\right)^2 - 2\left(\frac{d}{2}\right) R \cos\theta$$

if θ is the angle between the z-axis and the position vector of P.

$$\therefore \quad r_1^2 = R^2\left\{1 + \frac{d^2}{4R^2} - \frac{d\cos\theta}{R}\right\}$$

$$\text{Hence, } \frac{1}{r_1} = \frac{1}{R}\left\{1 + \frac{d^2}{4R^2} - \frac{d\cos\theta}{R}\right\}^{-1/2} = \frac{1}{R}\left\{1 + \frac{d\cos\theta}{2R} \cdots\right\}$$

$$\approx \frac{1}{R} + \frac{d\cos\theta}{2R^2}$$

neglecting terms in d^2 and higher order terms. This approximation is justified if we let $d/R \to 0$ keeping the product qd finite.

$$\text{Similarly } \frac{1}{r_2} \approx \frac{1}{R} - \frac{d\cos\theta}{2R^2}$$

Substituting in (1.59)

$$\Phi(r) = \frac{q}{4\pi\epsilon_0}\frac{d\cos\theta}{R^2} \tag{1.60}$$

The vector along the axis of the dipole pointing in the direction $-q$ to $+q$ and having magnitude qd is called the **dipole moment** and is represented by the symbol $\mathbf{p}$. Since $\mathbf{p}\cdot\mathbf{r} = pr\cos\theta = qdr\cos\theta$

$$\Phi(\mathbf{r}) = \frac{p\cos\theta}{4\pi\epsilon_0 r^2} = \frac{\mathbf{p}\cdot\mathbf{r}}{4\pi\epsilon_0|\mathbf{r}|^3} \tag{1.61}$$

Notice the following two points regarding the potential of a dipole:

(i) The potential defined as above is independent of the coordinate system used i.e. the formula (1.61) is correct even if $\mathbf{p}$ is not pointing in the z-direction.

(ii) The potential varies as $1/r^2$ and, hence, falls off more rapidly than that around an isolated point charge where it varies as $1/r$.

We can express dipole potential as a gradient. Since

$$\nabla\left(\frac{1}{r}\right) = -\frac{\mathbf{r}}{|\mathbf{r}|^3},$$

the relation (1.61) can be written as

$$\Phi(\mathbf{r}) = -\frac{\mathbf{p}}{4\pi\epsilon_0}\cdot\nabla\left(\frac{1}{r}\right) = -\mathbf{p}\cdot\nabla\Phi_0. \tag{1.62}$$

where $\Phi_0 = \dfrac{1}{4\pi\epsilon_0 r}$ is the potential of a unit charge.

Once the potential is found, electrostatic field can be obtained by taking the gradient of Φ. Thus, for a dipole oriented along the z-axis (Fig. 1.26), the three components of the field are

$$\left.\begin{aligned} E_z &= -\frac{\partial\Phi}{\partial z} = -\frac{p}{4\pi\epsilon_0}\frac{\partial}{\partial z}\left(\frac{r\cos\theta}{r^3}\right) \\ &= -\frac{p}{4\pi\epsilon_0}\frac{\partial}{\partial z}\left(\frac{z}{r^3}\right) = -\frac{p}{4\pi\epsilon_0}\left(\frac{1}{r^3} - \frac{3z^2}{r^5}\right) \\ &= \frac{p}{4\pi\epsilon_0 r^3}\left(\frac{3z^2}{r^2} - 1\right) = \frac{p}{4\pi\epsilon_0 r^3}\left(3\cos^2\theta - 1\right) \\ E_x &= \frac{p}{4\pi\epsilon_0}\frac{3zx}{r^5} \\ \text{and}\quad E_y &= \frac{p}{4\pi\epsilon_0}\frac{3zy}{r^5} \end{aligned}\right\} \tag{1.63}$$

It is sometimes convenient to express the field due to a dipole in spherical polar coordinates

$$\Phi(r) = \frac{p}{4\pi\epsilon_0}\frac{\cos\theta}{r^2}$$

$$\therefore \quad E_r = -\frac{\partial \Phi}{\partial r} = \frac{2p\cos\theta}{4\pi\epsilon_0 r^3},$$

$$E_\theta = -\frac{1}{r}\frac{\partial \Phi}{\partial \theta} = \frac{p}{4\pi\epsilon_0}\frac{\sin\theta}{r^3},$$

and

$$E_\phi = -\frac{1}{r\sin\theta}\frac{\partial \Phi}{\partial \phi} = 0 \tag{1.64}$$

The field of an electric dipole may also be expressed in the following way.

$$\mathbf{E} = -\nabla\Phi = -\frac{1}{4\pi\epsilon_0}\nabla\left(\frac{\mathbf{p}\cdot\mathbf{r}}{|\mathbf{r}|^3}\right)$$

$$= -\frac{1}{4\pi\epsilon_0}\left[\frac{1}{|\mathbf{r}|^3}\nabla(\mathbf{p}\cdot\mathbf{r}) + (\mathbf{p}.\mathbf{r})\nabla\left(\frac{1}{|\mathbf{r}|^3}\right)\right]$$

Since $\quad \nabla(\mathbf{p}\cdot\mathbf{r}) = \mathbf{p}$ and $\nabla\left(\frac{1}{|\mathbf{r}|^3}\right) = -\frac{3\mathbf{r}}{|\mathbf{r}|^5}$

$$\mathbf{E} = \frac{1}{4\pi\epsilon_0}\left[\frac{3(\mathbf{p}\cdot\mathbf{r})\mathbf{r}}{|\mathbf{r}|^5} - \frac{\mathbf{p}}{|\mathbf{r}|^3}\right]$$

$$= \frac{1}{4\pi\epsilon_0 r^3}\left[\frac{3(\mathbf{p}\cdot\mathbf{r})\mathbf{r}}{r^2} - \mathbf{p}\right] \tag{1.65}$$

1.16 Dipole in Uniform Electric Field

Suppose a dipole consisting of two charges $-q$ and $+q$ distance d apart is placed in a uniform field (Fig. 1.28). The potential energy of the dipole is

$$W = q(\Phi_B - \Phi_A) = -qd\cos\theta\, E$$

$$= -q\mathbf{d}\cdot\mathbf{E} = -\mathbf{p}\cdot\mathbf{E} \tag{1.66}$$

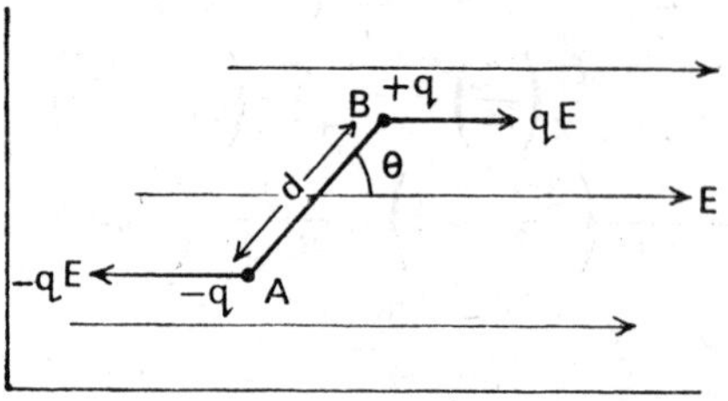

Fig. 1.28

This shows that the energy does not depend upon the position of the dipole, it depends only on the angle between **p** and **E**. We conclude that there is no translational force acting on the dipole. Since, however, the field exerts a force $+q\mathbf{E}$ on the charge $+q$ and $-q\mathbf{E}$ on $-q$, there is a couple **T** acting on it.

$$\mathbf{T} = -\frac{\partial W}{\partial \theta} = pE \sin \theta = \mathbf{p} \times \mathbf{E} \tag{1.67}$$

which tends to turn the dipole into a position parallel to the field.

1.17 Electric Dipole in a Non-uniform Electric Field

If, however, the field in which the dipole is placed is non-uniform, a translational force is exerted on it. Let us find the expression for this force.

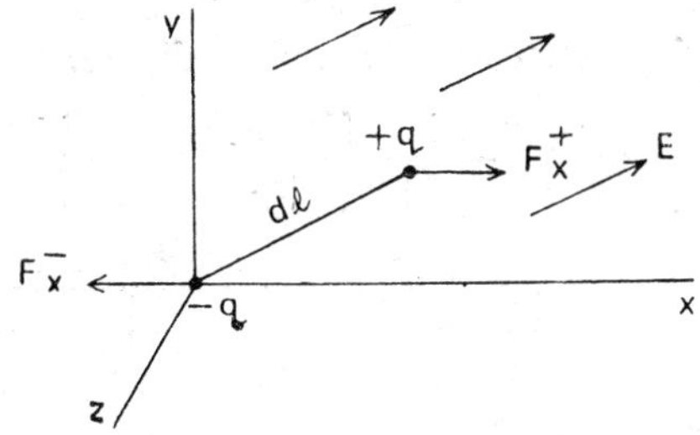

Fig. 1.29

Suppose the origin of the coordinate system coincides with the charge $-q$ as shown in Fig. 1.29. The force acting on the dipole in the positive x-direction is

$$F_x^+ = q(E_x + dE_x)$$

while in the negative direction the force is

$$F_x^- = qE_x$$

The net component of the force in the positive x-direction is

$$F_x = qdE_x$$

Since the field is generally non-uniform

$$dE_x = \left(\frac{\partial E_x}{\partial x}\right) dx + \left(\frac{\partial E_x}{\partial y}\right) dy + \left(\frac{\partial E_x}{\partial z}\right) dz$$

$$\therefore \quad F_x = q\left(dx\frac{\partial E_x}{\partial x} + dy\frac{\partial E_x}{\partial y} + dz\frac{\partial E_x}{\partial z}\right)$$

$$= q\left(dx\frac{\partial}{\partial x} + dy\frac{\partial}{\partial y} + dz\frac{\partial}{\partial z}\right)E_x$$

$$= q(\mathbf{l}\cdot\nabla)\, E_x = (\mathbf{p}\cdot\nabla)\, E_x \tag{1.68}$$

because $\quad d\mathbf{l} = \hat{\mathbf{e}}_x dx + \hat{\mathbf{e}}_y dy + \hat{\mathbf{e}}_z dz$ and

$$\nabla = \hat{\mathbf{e}}_x \frac{\partial}{\partial x} + \hat{\mathbf{e}}_y \frac{\partial}{\partial y} + \hat{\mathbf{e}}_z \frac{\partial}{\partial z}$$

In similar manner the other components can be determined. The force on the electric dipole, in general, is

$$\mathbf{F} = (\mathbf{p} \cdot \nabla)\mathbf{E} \tag{1.69}$$

1.18 Mutual Potential Energy of Two Dipoles

Let $\mathbf{p}_1$ and $\mathbf{p}_2$ be the dipole moments of the two dipoles. We may consider the first dipole to be placed in the field produced by the second, i.e.

$$W = -(\mathbf{p}_1 \cdot \mathbf{E}) \tag{1.70}$$

where $\mathbf{E}$ is the field produced by the second dipole.

$$\therefore \quad W = -\left[\mathbf{p}_1 \cdot \frac{1}{4\pi\epsilon_0 |\mathbf{r}|^3}\left\{\frac{3(\mathbf{p}_2 \cdot \mathbf{r})}{|\mathbf{r}|^2}\mathbf{r} - \mathbf{p}_2\right\}\right] \text{ (see 1.65)}$$

$$= \frac{1}{4\pi\epsilon_0 |\mathbf{r}|^3}\left[\mathbf{p}_1 \cdot \mathbf{p}_2 - \frac{3(\mathbf{p}_1 \cdot \mathbf{r})(\mathbf{p}_2 \cdot \mathbf{r})}{|\mathbf{r}|^2}\right] \tag{1.71}$$

1.19 Electric Double Layers

In some biological and colloid problems it has been observed that on a given surface there is a double layer of surface charges consisting of two neighbouring layers of charges of equal magnitude but opposite sign separated by an infinitesimal distance dx (Fig. 1.30 (a)). The product of the surface charge density σ and the local separation dx is called the strength of the layer:

$$D = \sigma dx$$

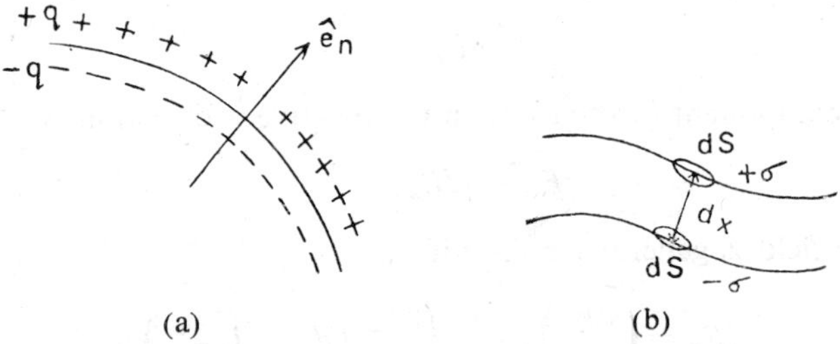

Fig. 1.30

In Fig. 1.30 (b) we have shown two surfaces with two small elemental areas on it. The two charges $\pm\,\sigma dS$ separated by a distance dx are equivalent to a dipole moment $p = \sigma dS\, dx = DdS$. We may, therefore, as well picture a layer of dipoles rather than think in terms of two layers with positive and negative charges. The direction of the dipole moment of the layer is normal to the surface. The potential of the dipole layer at a point distant r is

$$\Phi(r) = \frac{1}{4\pi\epsilon_0}\int \frac{\hat{\mathbf{e}}_n DdS \cdot \mathbf{r}}{|\mathbf{r}|^3} = \frac{D}{4\pi\epsilon_0}\int \frac{\hat{\mathbf{e}}_n dS \cdot \mathbf{r}}{|\mathbf{r}|^3} = \frac{Dd\Omega}{4\pi\epsilon_0} \tag{1.72}$$

when $d\Omega$ is the element of the solid angle subtended at the point by the elemental area dS.

1.20 Electric Quadrupole

At very large distances any collection of charge $q_1, q_2, \ldots, q_i, \ldots, q_r \ldots$ occupying a finite volume acts as if it were a point charge of magnitude Σq_i (where the sum is algebraic), and the electric field falls off as the inverse square of the distance. If it happens that $\Sigma q_i = 0$, we must not hasten to conclude that the field is zero, because the system could be a dipole where field falls off as the inverse cube of the distance. We now go a stage further and consider systems for which not only $\Sigma q_i = 0$ (a scalar sum) but also $\Sigma \mathbf{p}_i = 0$ (a vector sum). Consider, for example, the distribution of charges shown in Fig. 1.31 which is known as a **linear quadrupole**

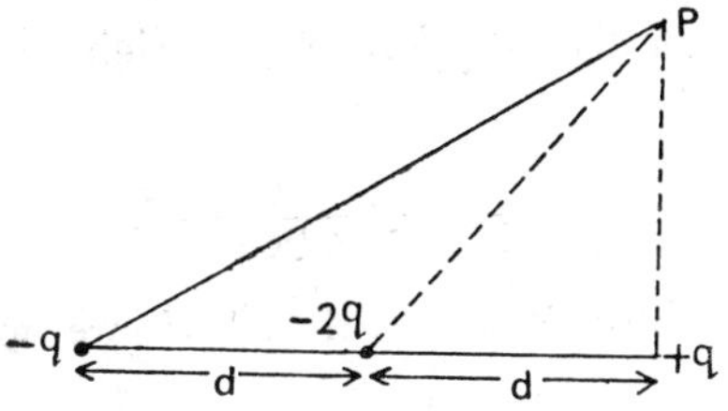

Fig. 1.31

We can regard this as derived from two equal and opposite dipole moments **p** placed in a line and separated by a distance d. The potential at a point P due to the quadrupole is given by

$$\Phi_P = -d\frac{\partial}{\partial x}\left(\frac{p\cos\theta}{4\pi\epsilon_0 r^2}\right) = -\frac{pd}{4\pi\epsilon_0}\frac{\partial}{\partial x}\left(\frac{\cos\theta}{r^2}\right)$$

$$= -\frac{pd}{4\pi\epsilon_0}\left\{-\frac{1}{r^2}\sin\theta\,\frac{\partial\theta}{\partial x} - \frac{2\cos\theta}{r^3}\frac{\partial r}{\partial x}\right\}$$

$$= -\frac{pd}{4\pi\epsilon_0}\left\{\frac{\sin^2\theta}{r^3} - \frac{2\cos^2\theta}{r^3}\right\} \quad \left\{\because \frac{\partial\theta}{\partial x} = -\frac{\sin\theta}{r}, \frac{\partial r}{\partial x} = \cos\theta\right\}$$

$$= \frac{pd}{4\pi\epsilon_0 r^3}(3\cos^2\theta - 1) \tag{1.73}$$

The quadrupole moment, defined analogously to the dipole moment, is given by

$$Q = pd = qd^2$$

$$\therefore \quad \Phi_P = \frac{Q}{4\pi\epsilon_0 r^3}(3\cos^2\theta - 1) \tag{1.74}$$

Clearly the field decreases as $1/r^4$.

1.21 Potential due to an Arbitrary Distribution of Charge

Suppose we are given an assembly of point charges distributed within a region in a complicated way (Fig. 1.32). What will be the potential due to this distribution at a point P distant R from the origin?

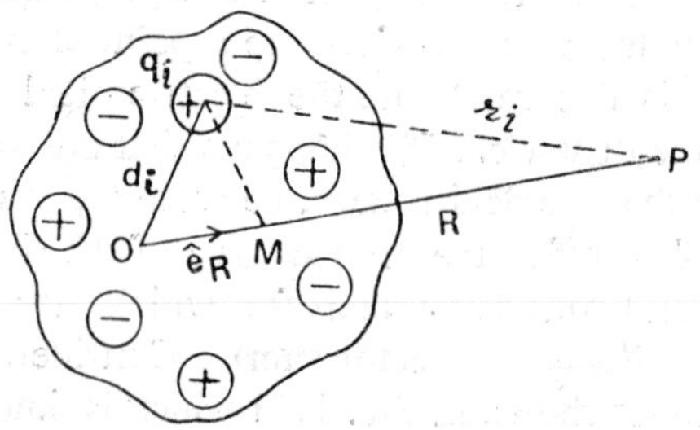

Fig. 1.32

Let q_i be a charge at a distance d_i from the origin, the potential at P due to this charge is

$$\Phi_i(r_i) = \frac{q_i}{4\pi\epsilon_0 r_i} = \frac{q_i}{4\pi\epsilon_0 |\mathbf{R} - \mathbf{d}_i|} \tag{1.75}$$

where r_i is the distance of q_i from P. If $R \gg d_i$ we may write to first approximation

$$\Phi_i(r_i) = \frac{q_i}{4\pi\epsilon_0 R} \quad \text{for each charge } q_i.$$

Hence, the total potential due to the entire distribution of charges is

$$\Phi(R) = \frac{\Sigma q_i}{4\pi\epsilon_0 R} = \frac{Q}{4\pi\epsilon_0 R} \tag{1.76}$$

The potential, therefore, is the same as that produced by a point charge Q at the origin where

$$Q = \sum_i q_i.$$

An interesting case arises when there are equal number of positive and negative charges in the region. In this case $Q = 0$ i.e. the object is neutral. One will be tempted to conclude that there will not be anything like potential around the object. However, there may be objects which are neutral and yet the charges are not all at one point and, hence, we should expect some effect of these separate charges at points not far away from the object. How do we find the potential due to such a distribution? Well, we will use the same formula (1.75) but will discard the approximation that d_i being much smaller than R can be discarded. Instead, we introduce a better approximation. We assume that r_i differs from R by the projection of d_i on R i.e. by OM as shown in the figure.

$$\therefore \quad r_i = R - \mathbf{d}_i \cdot \hat{\mathbf{e}}_R \tag{1.77}$$

where $\hat{\mathbf{e}}_R$ is the unit vector in the direction of R.

$$\therefore \quad \frac{1}{r_i} = \frac{1}{R\left(1 - \dfrac{\mathbf{d}_i \cdot \hat{\mathbf{e}}_R}{R}\right)} = \frac{1}{R}\left(1 + \frac{\mathbf{d}_i \cdot \hat{\mathbf{e}}_R}{R}\right) \quad \text{neglecting higher terms}$$

$$\therefore \quad \Phi(R) = \sum_i \frac{q_i}{4\pi\epsilon_0 R}\left(1 + \frac{\mathbf{d}_i \cdot \hat{\mathbf{e}}_R}{R}\right)$$

$$= \frac{Q}{4\pi\epsilon_0 R} + \sum_i q_i \frac{\mathbf{d}_i \cdot \hat{\mathbf{e}}_R}{4\pi\epsilon_0 R^2} \tag{1.78}$$

The first term is the same as in (1.76) and is equal to zero, since $Q = 0$

$$\therefore \quad \Phi(R) = \sum_i q_i \frac{\mathbf{d}_i \cdot \hat{\mathbf{e}}_R}{4\pi\epsilon R^2} \tag{1.79}$$

If we define the dipole moment of the distribution as $\mathbf{p} = \Sigma q_i \mathbf{d}_i$

$$\Phi(R) = \frac{\mathbf{p} \cdot \hat{\mathbf{e}}_R}{4\pi\epsilon_0 R^2} \tag{1.80}$$

Therefore, for any distribution of charges that is as a whole neutral, the potential is a dipole potential.

As an example of such a distribution, we may cite a water molecule which has a dipole moment because there is a net negative charge on the oxygen and a net positive charge on each of the hydrogen atoms (Fig. 1.33).

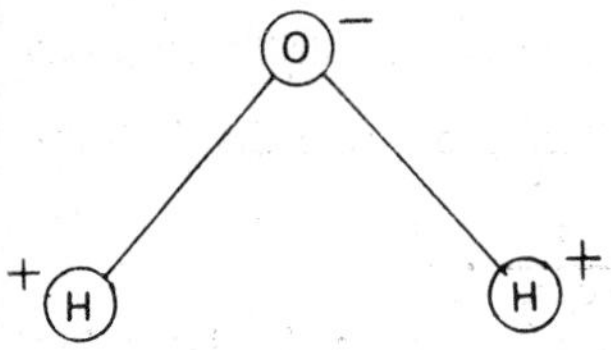

Fig. 1.33

Fig. 1.34

On the other hand, CO_2 is a symmetric linear molecule (Fig. 1.34) which beside being a neutral molecule does not have a dipole moment. (Does it mean that such a distribution will not produce any potential at all?).

PROBLEMS

1.1 Two particles each of mass m and charge q are separated by strings each of length l from the same point. Prove that the inclination θ of each string to the vertical is given by

$$q^2 \cos\theta = 16\pi\varepsilon_0 \, mgl^2 \sin^3\theta$$

1.2 Four equal small spheres, each of mass m and carrying a charge q, are suspended by light strings each of length l, from the same point. Find the side of the square that the spheres will form due to the force of repulsion. If the charge on each sphere and length of each string is increased by a factor k, what should be the mass of each sphere so as to make the side of the square ka?

1.3 α-particles of energy 5×10^6 eV are fired at heavy nuclei each carrying a charge sixty times the charge on the electron. Find the closest distance of approach of the α-particles and the nuclei.

1.4 The mass of the earth is 6×10^{24} kg, that of the moon is 7×10^{22} kg and their separation is 4×10^8 metres. Find the electric charge that would be required on the earth and the moon to balance their gravitational attraction if the charges were in the same ratio as the masses (The gravitational constant = 6.7×10^{-11} Nm^{2}kg^{-2}).

1.5 If the electric field everywhere within a uniformly charged spherical shell is zero, show without using the differential form of Gauss' law that the electrostatic law of force must vary as the inverse square of the distance.

1.6 A spherical region of space of radius 'a' is filled with positive charge of volume density ρ which varies with the distance r from the centre of the sphere but is constant for a given r. Find the total charge in the sphere if

(a) $\rho = \dfrac{\rho_0 a}{r}$

(b) $\rho = \rho_0 \left(1 - \dfrac{r^2}{a^2}\right)$

1.7 The electric field in the atmosphere 1400 metres above the surface of the earth acts in the downward direction and is equal to 20 V/m. The field at the surface of the earth is also directed downward and is equal to 200 V/m. Find the average charge density in the atmosphere below 1400 m and state whether it is due to positive or negative ions.

1.8 A square sheet is uniformly charged with charge density σ. Show that the potential at the centre of the square is $\Phi_0 = \dfrac{\sigma a}{\pi \epsilon_0} \ln (1 + \sqrt{2})$ where 'a' is the length of the side of the square.

1.9 The charge density for the electron in a hydrogen atom has a distribution given by $\rho(r) = \dfrac{e}{\pi a_0^3} e^{-2r/a_0}$ where a_0 is the Bohr radius. Show that the electron-proton interaction energy is given by $U = -\dfrac{e^2}{4\pi \epsilon_0 a_0}$.

1.10 A metal sphere of 10 litres capacity is filled with nitrogen, its density being 3×10^{22} molecules per litre. Find the potential of the sphere if one electron from each molecule is removed to a large distance from the sphere.

1.11 The uranium nucleus has a radius 10^{-14} m and contains 92 electrons. Assuming that the charge on the nucleus is uniformly distributed throughout the nucleus, calculate its electrostatic energy in MeV.

If the uranium nucleus splits into two equal fragments each of radius 8×10^{-15} m, find the electrostatic energy released.

1.12 Two charges of opposite sign but different magnitudes are placed in given positions. Show that the equipotential surface for which $V = 0$ is spherical. What happens if the two charges are of equal magnitude?

1.13 Two electric dipoles A and B are placed in such a way that the direction of A passes through B while the direction of B is perpendicular to that of A. Show that the actual force exerted by A on B is not in the same direction as the actual force exerted by B on A. Does this not contradict Newton's third law of motion?

1.14 Two charges, each $-Q$, are a distance $2a$ apart. A charge $+Q$ on a particle of mass m is placed midway between them. If the particle is given a small displacement perpendicular to the line of charges and then released, find the period of the resultant oscillatory motion. What happens if, instead, the particle is given a small displacement along the line of charges?

Chapter 2

Electrostatics in Dielectrics

We have dealt, so far, with the electric field produced by the charges in vacuum. It has long been known, however, that the medium plays an important part in determining the absolute value of the field. The electromagnetic fields that result in matter are, in general, very complex displaying considerable variation from particle to particle over distances of the order of separation of atoms. In dealing with the electromagnetic phenomena that range over large distances, we shall not be concerned with the detailed fluctuations in the field on an atomic scale. We will present in this chapter a consistent macroscopic model of matter that does predict in every instance the correct fields. One must note, however, that the macroscopic fields within matter are not necessarily the fields actually exerted on any of the particles in the matter.

2.1 Conductors and Insulators

On the basis of their behaviour under the influence of an external electric field, we can divide the materials as far as their electric properties are concerned, into two distinct groups:

(i) Those in which there are some electrons which are free to move in the presence of a field. Electrons are light particles and are, therefore, mobile when freed inside bodies. Such materials in which an electric field produces a steady drift of charge (i.e. electric current) through them, are called **conductors.** Examples of these are metals, carbon etc. (There are among these some substances, both elements and compounds, the resistance of which vanishes at a certain critical temperature and they behave as **superconductors** at that temperature, e.g. mercury at 4.2°K.)

(ii) Those in which the electrons are strongly bound to the atoms or molecules composing the material and cannot be detached by the application of an electric field to these materials. Hence, it does not cause any steady motion of the electrons; at the most they may be slightly displaced from their positions in response to the field applied. Such materials are called **insulators** or **dielectrics**, e.g. sulphur, porcelain, mica, etc. It must be mentioned, however, that if the field is too large, catastrophic breakdown may occur in such substances. A good insulator breaks down in fields of about 10^9 volts/metre.

We, thus, have perfect conductors and perfect insulators. Between these there is yet a third class of materials, the so called **semiconductors** which have properties intermediate between conductors and insulators. The present day technology depends upon such materials.

2.2 Conductor in an Electrostatic Field

Let us consider what happens when a conductor is placed in an electric field. Figure 2.1 shows a slab of conducting material placed in an electric field. Under the influence of the field, the electrons can move around freely in the material but cannot leave the surface because strong forces act to prevent them from leaving. They, thus, leave a net charge on the surface. The charges thus induced on the surfaces generate a field inside the conductor which is in a direction opposite to the external field. The electrons continue to move so long as there is any field left in the conductor and until the induced charge ensures that the resultant field inside the conductor is zero. Hence, no electrostatic field can exist within the body of a conductor. Since $\mathbf{E} = -\nabla\Phi = 0$, Φ does not vary from point to point. The conductor, therefore, is an equipotential region. Further, because $\mathbf{E} = 0$, $\nabla \cdot \mathbf{E} = \rho/\epsilon = 0$, i.e. $\rho = 0$. Hence, there cannot be a volume distribution of charges within the conductor and the charges must reside entirely on the surface. The surface of a conductor upon which charges are at rest, is an equipotential surface. If this was not the case, there would be an electrostatic field along the surface resulting into a current. The surface being an equipotential surface, the field is normal to the surface.

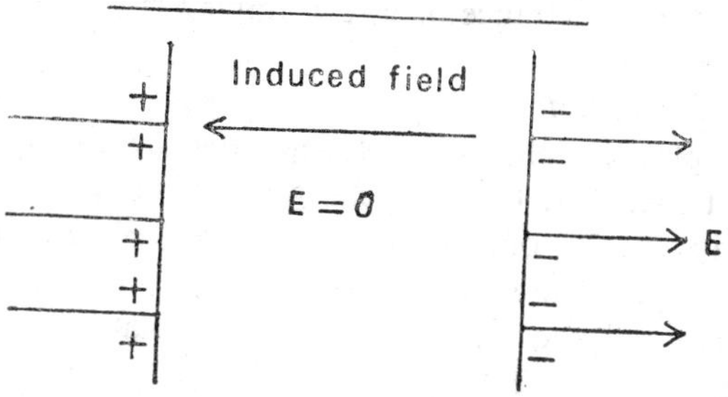

Fig. 2.1

2.3 Electric Field at the Surface of a Charged Conductor

Imagine a small disc enclosing an area dS on the surface of a conductor (Fig. 2.2). The volume enclosed by the outer surface of the disc is half in the conductor and half in the air. The flat surfaces of the disc are parallel to the surface of the conductor. If E is the field, the flux through the surface outside is EdS.

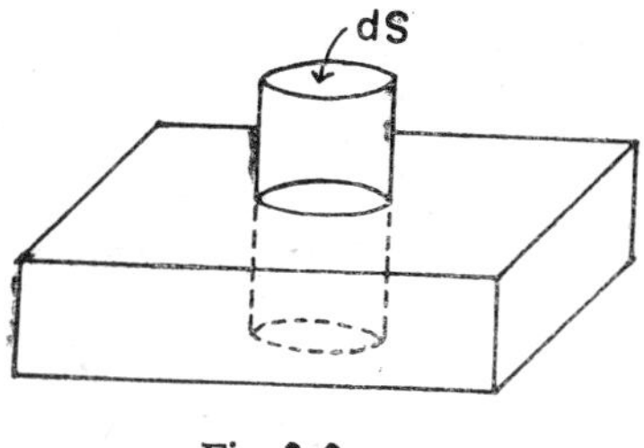

Fig. 2.2

The flux through the other surface is zero since the field inside a conductor is zero. Therefore, by Gauss' law

$$EdS = \frac{\sigma dS}{\epsilon_0}$$

where σ is the surface charge density. Hence

$$E = \sigma/\epsilon_0 \tag{2.1}$$

2.4 Capacitors

A capacitor is a device used for storing energy. It consists of a pair of conductors carrying equal and opposite charges. Suppose the two conductors in a capacitor carry charges $\pm Q$. The charges are distributed over the surfaces with the surface densities σ_+ and σ_- respectively. These may not be constant over the surfaces of the conductors; but the potential Φ_+ and Φ_- are, since the surfaces are equipotential surfaces. Let the potential difference between the conductors be V

$$V = \Phi_+ - \Phi_- \tag{2.2}$$

In a capacitor potential difference is always proportional to the charge Q, i.e.

$$Q = CV \tag{2.3}$$

where C is a constant and is called the capacitance (or capacity) of the capacitor. Capacitance is related to the ability of a capacitor to hold charges. The capacitance depends on the geometry of the conductors. Let us calculate the capacitance of a few simple type of capacitors.

(i) *Parallel Plate Capacitor*

Consider two plates, each of area A, separated by a small distance 'd' (Fig. 2.3). Let equal and opposite charges $\pm Q$ be put on the plates. The charges will spread out uniformly on the plates. If the lateral dimensions of the plates are much larger than this separation, the edge effects may be ignored and the field between the plates be assumed to be uniform. The potential difference between the plates is the work required to take a unit charge from one plate to the other.

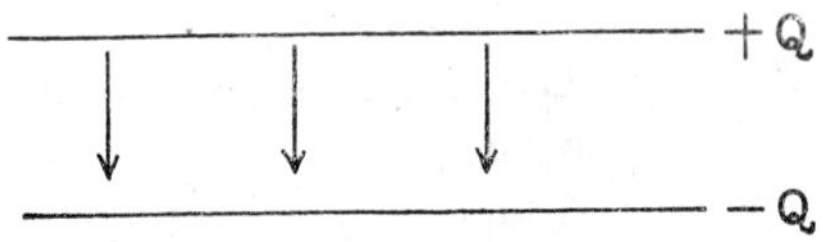

Fig. 2.3

$$V = Ed = \frac{\sigma}{\epsilon_0} d = \frac{Qd}{\epsilon_0 A} \left(\because \quad \sigma = \frac{Q}{A}\right)$$

$$\therefore \quad C = \frac{Q}{V} = \frac{\epsilon_0 A}{d} \tag{2.4}$$

(ii) *Capacitance of an Isolated Sphere*

Let the sphere have a radius 'a' and carry a charge Q. If we imagine another sphere of infinite radius carrying a charge $-Q$, the two together may be considered as forming a capacitor. The potential of the sphere V is given by

$$V = -\int_{\infty}^{a} E dr = -\int_{\infty}^{a} \frac{Q}{4\pi\epsilon_0 r^2} dr = \frac{Q}{4\pi\epsilon_0 a} \tag{2.5}$$

$$\therefore \quad C = \frac{Q}{V} = 4\pi\epsilon_0 a \tag{2.6}$$

(iii) "*Co-axial cable*" *Capacitor*

Consider two co-axial cylinders of infinite length. Let their radius be 'a' and 'b' as shown in Fig. 2.4.

If the charge per unit length on the inner cylinder is $+\lambda$, the field E is given by

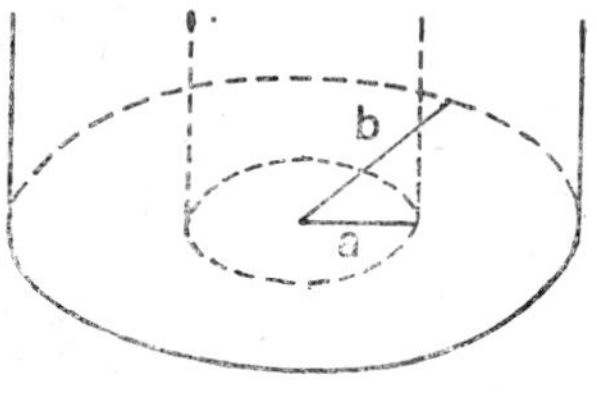

Fig. 2.4

$$E = \frac{\lambda}{2\pi\epsilon_0 r} \quad \text{(see 1.35)}$$

The potential difference between the cylinders is given by

$$V = \Phi_a - \Phi_b = -\int_b^a E dr = -\frac{\lambda}{2\pi\epsilon_0}\int_b^a \frac{dr}{r}$$

$$= -\frac{\lambda}{2\pi\epsilon_0}(\ln a - \ln b)$$

$$= \frac{\lambda}{2\pi\epsilon_0}\ln\frac{b}{a} \tag{2.7}$$

$\therefore$ C (the capacitance of length 'l') is

$$C = \frac{\lambda l}{\frac{\lambda}{2\pi\epsilon_0}\ln\frac{b}{a}} = \frac{2\pi\epsilon_0 l}{\ln\frac{b}{a}} \tag{2.8}$$

2.5 The Energy of a Capacitor

Since in the case of conductors charge resides only on the surfaces, the potential energy of a conductor can be found from (1.56) by replacing summation by integration.

$$W = \tfrac{1}{2}\int \sigma\Phi dS \tag{2.9}$$

In the case of capacitors, therefore, the potential energy will be

$$W = \tfrac{1}{2}\int \sigma_+\Phi_+ \, dS + \tfrac{1}{2}\int \sigma - \Phi_- \, dS$$

where Φ_+, Φ_- are the potentials and σ_+, σ_- the surface densities of charges on the two conductors.

$$W = \tfrac{1}{2}Q(\Phi_+ - \Phi_-) = \tfrac{1}{2}QV \tag{2.10}$$

This can also be expressed in terms of capacity as

$$W = \tfrac{1}{2} CV^2 = \tfrac{1}{2}\frac{Q^2}{C} \tag{2.11}$$

This energy can as well be expressed in a different way by regarding it to be distributed over the space between the charges occupied by their electric field. Consider two equipotential surfaces close enough together between two conductors as shown in Fig. 2.5.

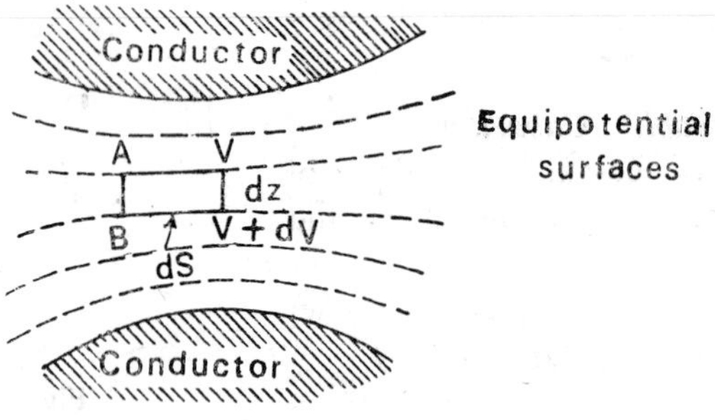

Fig. 2.5

Suppose the two surfaces differ in their potential by a small amount dV. If now two conducting surfaces with potentials exactly as on the equipotential surfaces are inserted so as to coincide with the surfaces, the condition of the problem will remain unaltered. The two conducting surfaces will form a capacitor with capacity $= \frac{\epsilon_0 dS}{dz}$, where dS is the area of each plate and dz their separation. The energy of the capacitor is

$$dW = \tfrac{1}{2} CV^2 = \tfrac{1}{2}\frac{\epsilon_0 dS}{dz}(dV)^2 = \tfrac{1}{2}\frac{\epsilon_0 dS}{dz}(Edz)^2$$

$$= \tfrac{1}{2}\epsilon_0 E^2 dS dz$$

$$= \tfrac{1}{2}\epsilon_0 E^2 d\tau$$

where $d\tau = dSdz$ is the volume occupied by the capacitor.

$$\therefore \quad W = \tfrac{1}{2}\epsilon_0 \int E^2 d\tau$$

We may, therefore, regard the energy as distributed throughout the field, the energy density at a point being $\tfrac{1}{2}\epsilon_0 E^2$.

We shall now derive this result more rigorously. Figure 2.6 shows an assembly of conductors enclosed by a surface so sufficiently far away that the field on it is negligible. Suppose there exists within the surface a volume distribution of charge of density ρ and surface distribution of density σ over the conductors. The electrostatic energy, is given by

$$W = \tfrac{1}{2}\int \rho\Phi d\tau + \tfrac{1}{2}\int \sigma\Phi dS_1 \tag{2.12}$$

where the surface integral is taken over the surfaces of all the conductors and volume integral over the entire region occupied by the field i.e. the

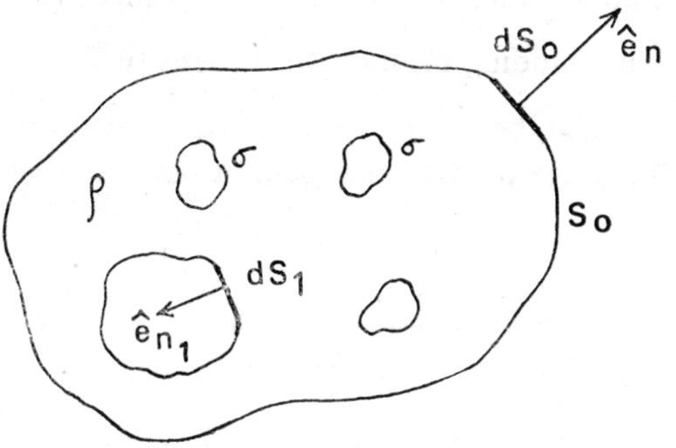

Fig. 2.6

volume bounded by the surface S_0. Using Gauss' differential form $\nabla \cdot \mathbf{E} = \rho/\epsilon_0$, we can write

$$\tfrac{1}{2}\int \rho\Phi d\tau = \tfrac{1}{2}\int \epsilon_0(\nabla \cdot \mathbf{E})\Phi d\tau$$

$$= \frac{\epsilon_0}{2}\int\left\{\text{div}\,(\Phi\mathbf{E}) - \mathbf{E}\cdot\nabla\Phi\right\}d\tau$$

$$= \frac{\epsilon_0}{2}\int \text{div}\,(\Phi E)\, d\tau + \frac{\epsilon_0}{2}\int E^2 d\tau$$

$$= \frac{\epsilon_0}{2}\int \Phi\mathbf{E}\cdot\hat{\mathbf{e}}_n dS + \frac{\epsilon_0}{2}\int E^2 d\tau$$

using Divergence theorem. The first integral is to be taken over the closed surface S_0 bounding the volume as well as over the surfaces of the conductors.

$$\therefore \quad \frac{1}{2}\int \rho\Phi d\tau = \frac{\epsilon_0}{2}\int \Phi\mathbf{E}\cdot\hat{\mathbf{e}}_n dS_0 + \frac{\epsilon_0}{2}\int \Phi\mathbf{E}\cdot\hat{\mathbf{e}}_{n_1} dS_1 + \frac{\epsilon_0}{2}\int E^2 d\tau$$

(over the surfaces of the conductor)

where $\hat{\mathbf{e}}_n$ and $\hat{\mathbf{e}}_{n_1}$ are unit vectors normal to the respective surfaces.

Since Φ varies as r^{-1} and E as r^{-2} while dS increases as r^2, the first integral is proportional to $1/r$ and, hence, vanishes when r is very large i.e. on the surface S_0.

In the second integral, since the integration is over the surfaces of the medium, $\hat{\mathbf{e}}_{n_1} dS_1$ is the vector drawn outwards from the medium and, hence, into the conducting surface as shown in the figure. Since $E = -\sigma/\epsilon_0$ in this direction

$$\tfrac{1}{2}\int \rho\Phi d\tau = -\tfrac{1}{2}\int \sigma\Phi dS_1 + \frac{\epsilon_0}{2}\int E^2 d\tau$$

and the total energy is

$$W = -\tfrac{1}{2}\int \sigma\Phi dS_1 + \frac{\epsilon_0}{2}\int E^2 d\tau + \tfrac{1}{2}\int \sigma\Phi dS_1 = \frac{\epsilon_0}{2}\int E^2 d\tau \qquad (2.13)$$

Since $E=0$ within a conductor, we may regard this energy as distributed throughout the surrounding medium with density $\frac{\epsilon_0 E^2}{2}$.

This formula (2.13), however, has certain limitations. Suppose we wish to compute the electrostatic energy of a point charge q using (2.13). The energy will be

$$W=\frac{\epsilon_0}{2}\int\frac{q^2}{(4\pi\epsilon_0)^2 r^4}\,4\pi r^2 dr$$

$$=\frac{q^2}{8\pi\epsilon_0}\int_0^\infty\frac{dr}{r^2}$$

$$=\frac{q^2}{8\pi\epsilon_0}\left[-\frac{1}{r}\right]_0^\infty$$

$$=\infty$$

Self-energy of a point charge is, therefore, infinite, a result physically absurd. The concept of energy density, therefore, is not compatible with the concept of point charge.

2.6 Electric Response of a Non-conducting Medium to an Electric Field

In the preceding sections we considered the behaviour of conducting materials in an electric field. We shall now discuss the properties of the materials of the second group mentioned in Sec. 2.1 — dielectrics — which do not conduct electricity.

Cavendish, and later Faraday, observed that the capacity of a conductor depends not only on the shape and size of the conducting plates but also on the nature of the dielectric material by which they are separated. Capacitance of a capacitor increases on inserting a slab of dielectric between its plates. The capacity is given by $C=Q/V$. Since this increases on inserting a dielectric (Q remaining the same), as observed by Cavendish and Faraday, it follows that the potential difference V falls, i.e. the field is weakened. If the capacity is increased by a factor ϵ_r, the field is weakened by the factor $\frac{1}{\epsilon_r}$. The field in a dielectric medium, therefore, is given by

$$\mathbf{E}=\frac{q\mathbf{r}}{4\pi\epsilon_0\epsilon_r\,|\mathbf{r}|^3} \tag{2.14}$$

The parameter ϵ_r which depends only on the nature of the dielectric medium is called the **dielectric constant** of the medium. For vacuum $\epsilon_r=1$; for air it is 1.00057; for most solid substances it varies from 1 to 10, for water it is 81.

2.7 Polarization

To find how the dielectric constant ϵ_r is related to the intrinsic properties of the materials, let us consider what happens inside a dielectric when an electric field is applied to it.

Consider an isolated atom with atomic number Z. In the absence of an electric field (Fig. 2.7a) the nucleus is in a position of stable equilibrium since the "centre of gravity" of its positive charge coincides with that of the negative charge. Suppose now the atom is placed in an external electric field. What happens? The field will push the positively charged nucleus slightly in the direction of the field and the negatively charged electrons in the opposite direction (Fig. 2.7b). As soon as the centre of the electron cloud moves away from the nucleus a strong electrostatic force of attraction is generated between the nucleus and the electron cloud which tends to restore the displaced charges to their original positions. How far will they be displaced? Clearly, until the internal restoring force balances the external field. The centres of positive and negative charges no longer coincide and the atom behaves like a dipole. The amount of dipole moment induced will be proportional to the field because larger field will displace the charges more than the smaller field. We say that the atom is polarized under the influence of the external field. If the separation between the "centres" is $\boldsymbol{\delta}$, the dipole moment induced is

$$\mathbf{p} = Ze\boldsymbol{\delta} \tag{2.15}$$

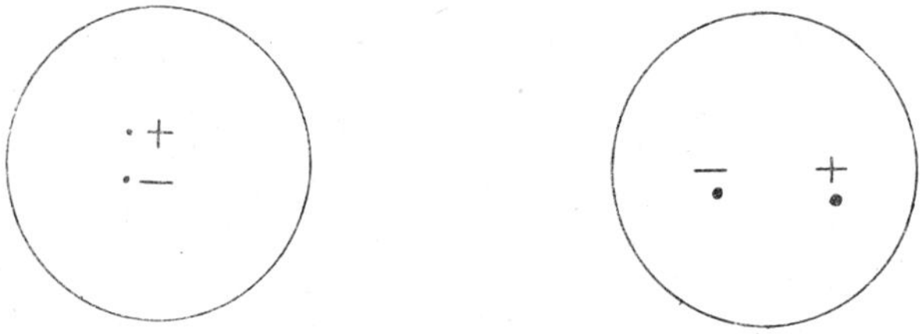

(a) Unpolarized atom (b) Polarized atom

Fig. 2.7

When an insulator is placed in an electric field, all its atoms become polarized and, hence, the medium as a whole is polarized (electronic polarization) (Fig. 2.8).

Insulating materials, in general, can be classified into two broad groups. In one, the molecules do not behave like dipoles unless they are subjected to an external electric field. In the other, the molecules do behave like dipoles even in the absence of an external electric field i.e. they have permanent dipoles, but owing to the thermal agitation the dipole moments are randomly oriented (Fig. 2.9a) and as a consequence the material is left unpolarized. With the application of the field the dipoles tend to orient themselves in the direction of the external field (Fig. 2.9b). The

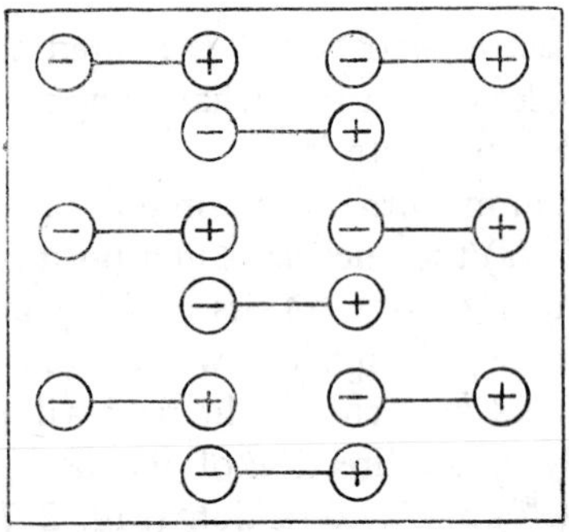

Fig. 2.8

molecules of the first group are known as non-polar molecules (e.g. H_2O_2, CO_2); while those of the other as polar molecules (e.g. NaCl, H_2O). Polarization, therefore, can result under the influence of the field from the alignment of molecules which have a natural asymmetry in their charge distribution (polar molecules) or from induced asymmetry in naturally symmetric molecules.

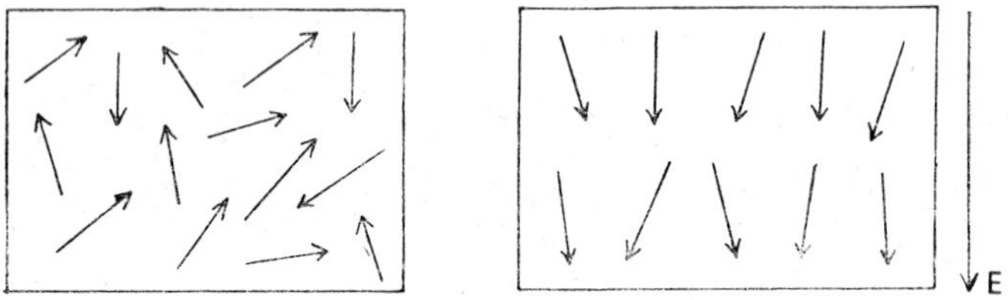

(a) (b) Oriental polarization

Fig. 2.9

When we are considering large systems, we are not primarily concerned with forces exerted by an atom or molecule on another, but with the average effect of large number of atoms or molecules. This is what is known as macroscopic description of the field in which the details of the atomic and inter-atomic effects are neglected. What is considered is the value of the field averaged over a region much bigger than a single atom. This is the field that would exist if the matter were continuous and had no atomic structure.

The macroscopic polarization density is defined as the electric dipole moment per unit volume. It is found by taking the vector sum of the microscopic dipole moments in unit volume. It is a vector in the direction of the individual dipole moments i.e. in the direction of charge separation $\boldsymbol{\delta}$ and is denoted by the symbol **P**. If there are N molecules per unit volume and in each molecule the positive and negative charges $\pm q$ are separated by a distance δ, the polarization density **P** is given by

$$\mathbf{P} = Nq\boldsymbol{\delta} \tag{2.16}$$

The dimensions of P are coul/m^2.

If the polarization of a dielectric slab placed in an electric field (Fig. 2.10) is uniform, there will be no net charge density within the slab as all the individual dipoles are aligned parallel to the field and, hence, negative charge of one dipole is next to the positive charge of the next dipole. Thus the average density of the positive and negative charges displaced under the external field is the same, the net charge being zero. It may be noted that even when the polarization is uniform the neutrality at the surface of the dielectric is not preserved. There will be positive charge on one surface and the negative on the other. These are called **fictitious** charges or **bound** charges, but more generally **polarization** charges.

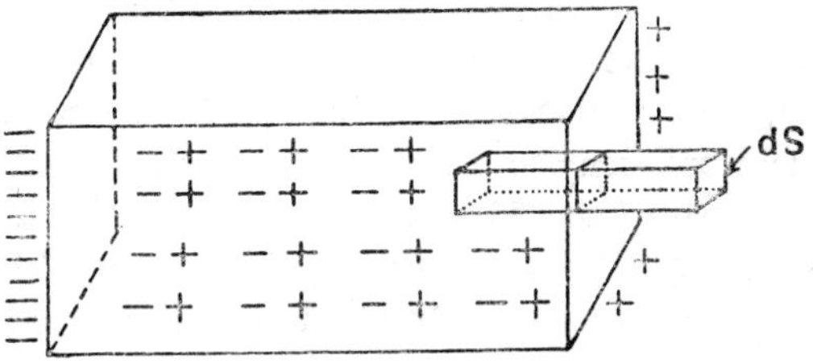

Fig. 2.10

If the polarization is not uniform we may expect some charge within the volume of the slab as the amount of positive charge displaced may not be the same as that of the displaced negative charge.

Consider a small area dS on one of the surfaces of the slab (Fig. 2.10). If σ_P is the surface density of the polarization charge, the total charge on dS is $\sigma_P\, dS$. This charge is also equal to the net charge in a box with cross-section dS and length δ equal to the separation of positive and negative charges.

$$\therefore \quad \sigma_P\, dS = Nq\delta dS$$

i.e.

$$\sigma_P = Nq\delta = P \tag{2.17}$$

where P is the magnitude of the polarization vector **P**. Thus, surface density of charge is equal to the polarization inside the material. Of course, Eq. (2.17) is true if the polarization vector is normal to the surface. In a general case

$$\sigma_P = \mathbf{P} \cdot \hat{\mathbf{e}}_n \tag{2.18}$$

Let us now examine how the increase in capacitance comes about when a dielectric slab is inserted between the plates of a capacitor.

Consider a parallel plate capacitor (Fig. 2.11) with a negative charge on the top and positive charge on the bottom plate. Let A be the area of each plate, σ the surface charge density and d their separation.

$$\therefore \quad E = \frac{\sigma}{\epsilon_0}, V = \frac{\sigma}{\epsilon_0} d, \ C = \frac{Q}{V} = \frac{\epsilon_0 A}{\delta} \tag{2.19}$$

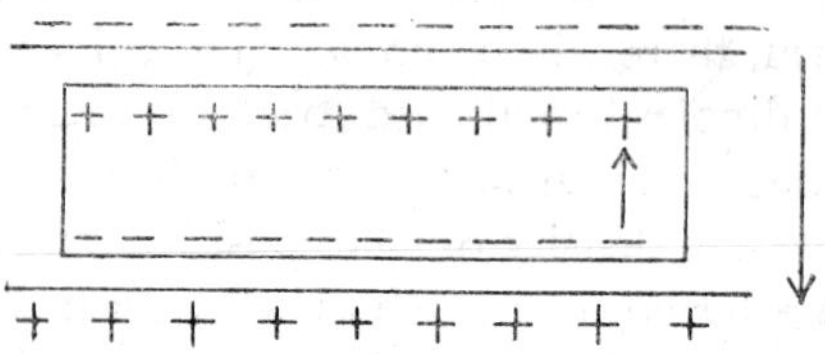

Fig. 2.11

Suppose a dielectric slab is introduced between the plates of the capacitor as shown in Fig. 2.11. We have shown a gap between the plates and the walls of the dielectric. Such a gap will always be present though its width may be only of molecular dimensions. A positive charge will be induced in the top and negative on the bottom side of the slab. This will produce a macroscopic field inside the material which is opposed to the field between the plates, thus, effectively reducing it. This macroscopic field is called the **"depolarizing field"**. We have seen above that the surface density of polarization charges is P. Therefore, the effective field inside the capacitor containing the dielectric is

$$E = \frac{(\sigma - P)}{\epsilon_0} \tag{2.20}$$

Now the strength of the polarization P is proportional to E i.e.

$$\mathbf{P} = \text{constant} \times \mathbf{E}$$

$$= \epsilon_0 \chi \mathbf{E} \tag{2.21}$$

The constant of proportionality is usually written as $\epsilon_0 \chi$. Substituting in (2.20)

$$E = \frac{\sigma}{\epsilon_0} - \chi E$$

$$\therefore \quad E = \frac{\sigma}{\epsilon_0} \frac{1}{1+\chi}$$

We see that the field is reduced by a factor $\frac{1}{1+\chi}$. The dielectric constant defined in Sec. 2.6, therefore, is

$$\epsilon_r = 1 + \chi \tag{2.22}$$

$$\therefore \quad V = Ed = \frac{\sigma d}{\epsilon_0(1+\chi)} \tag{2.23}$$

and

$$C = Q/V = \frac{\sigma A}{\sigma d/\epsilon_0(1+\chi)}$$

$$= \frac{\epsilon_0 A}{d} (1 + \chi)$$

$$= \frac{\epsilon_0 \epsilon_r A}{d} \tag{2.24}$$

Thus the capacity is increased by a factor ϵ_r. The values of ϵ_r at room temperature of some materials are given in Table 2.1.

Table 2.1

Material	
Carbon tetrachloride	2.24
Paraffin wax	2.25
Polythelene	2.30
Nylon	3.50
Porcelain	6.00
Mica	7.00
Water	81.00

2.8 Laws of Electrostatic Field in the Presence of Dielectrics

We have seen (Eq. 2.14) that the field in a dielectric medium is not the same as would be expected from Coulomb's law. The law, therefore, is not of universal validity. It has been proved experimentally for air and is found not to be true for dielectrics. Hence, the laws of electrostatics so far established require reconsideration in dielectrics.

We have, so far, restricted our attention to uniform polarization. In practice, however, dielectrics are often found in situations where the polarization is not uniform. This may be due either to dielectric being non-uniform or due to the variation of the field even inside a uniform dielectric. Let us consider a situation in which the polarization **P** is not the same everywhere.

When an unpolarized slab of a material is placed in an electric field, a certain amount of net charge will pass through a given element of area dS due to polarization process. The amount of charge will be equal to the product of the component of **P** normal to dS and the magnitude of dS i.e. $\mathbf{P}\cdot\hat{e}_n\, dS$. As mentioned earlier, the displacement of the charges can result in a volume charge density if P is not uniform. The total charge displaced out of any volume V by the polarization is

$$\int_S \mathbf{P}\cdot\hat{e}_n dS$$

where S is the surface that bounds the volume. An equal excess charge of the opposite sign is left behind. If this is denoted by q'

$$q' = -\int_S \mathbf{P}\cdot\hat{\mathbf{e}}_n dS \tag{2.25}$$

Let ρ' be the density of the polarization charges within the volume

$$\therefore \quad q' = \int_V \rho' d\tau$$

Hence,

$$\int_V \rho' d\tau = -\int_S \mathbf{P}\cdot\hat{\mathbf{e}}_n dS \tag{2.26}$$

Using Divergence theorem, we have

$$\int_V \rho' d\tau = -\int_V \nabla\cdot\mathbf{P} d\tau \tag{2.27}$$

$$\therefore \quad \rho' = -\nabla\cdot\mathbf{P} \tag{2.28}$$

Thus the polarization charge density ρ' (in coul/m^3) is given by the negative divergence of the polarization density **P**.

Gauss' law in free space is

$$\nabla\cdot\mathbf{E} = \rho/\epsilon_0$$

An equivalent law for dielectric medium can be written by taking inton account the fact that besides free charges there are also polarization charges present. Thus

$$\nabla\cdot\mathbf{E} = \frac{\rho+\rho'}{\epsilon_0} = \frac{\rho}{\epsilon_0} - \frac{\nabla\cdot\mathbf{P}}{\epsilon_0}$$

$$\therefore \quad \nabla\cdot\left(\mathbf{E} + \frac{\mathbf{P}}{\epsilon_0}\right) = \frac{\rho}{\epsilon_0}$$

i.e.
$$\nabla\cdot(\epsilon_0\mathbf{E} + \mathbf{P}) = \rho. \tag{2.29}$$

We now introduce a new vector field

$$\mathbf{D} = \epsilon_0\mathbf{E} + \mathbf{P} \tag{2.30}$$

The equation (2.29) can now be written in terms of D as

$$\nabla\cdot\mathbf{D} = \rho \tag{2.31}$$

This is the modified form of Gauss' law which also includes the effects of polarization charges. Thus flux of D out of a closed surface $S =$ total free charge enclosed by S.

The Gauss' law in this form is very convenient for computation.

The vector **D** has the same dimensions as **P** and is called the **electric displacement.**

Since $\mathbf{P} = \epsilon_0 \chi \mathbf{E}$

by (2·30) we have

$$\mathbf{D} = \epsilon_0 \mathbf{E} + \epsilon_0 X \mathbf{E}$$
$$= (1 + X)\epsilon_0 \mathbf{E} = \epsilon_0 \epsilon_r \mathbf{E}$$
$$= \epsilon \mathbf{E} \quad (2.32)$$

where $\epsilon = \epsilon_0 \, \epsilon_r$ is called the "permittivity" of the medium, ϵ_0 is the "permittivity" of the free space and ϵ_r is the relative permittivity of the medium.

Hence, the Gauss' law (2.31) can also be written as

$$\nabla \cdot \epsilon \mathbf{E} = \rho \quad (2.33)$$

Note that the other equation in electrostatics viz. $\nabla \times \mathbf{E} = 0$ remains unchanged in dielectrics.

EXAMPLE 2.1. An artificial dielectric consists of a large number of metal spheres of macroscopic size arranged in a three dimensional (lattice) structure (Fig. 2.12). Find the permittivity of the dielectric.

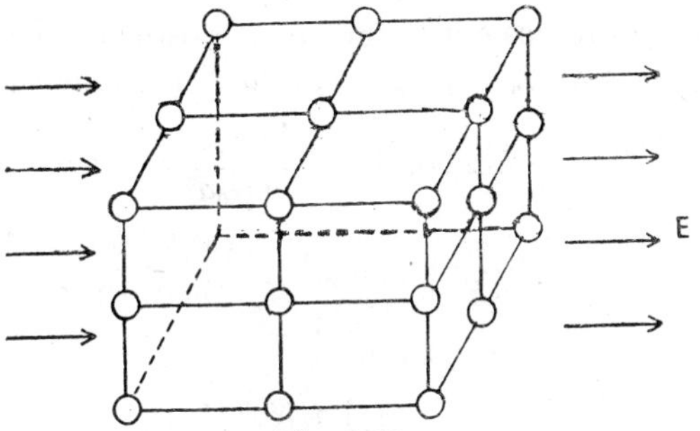

Fig. 2.12

Let a uniform field **E** be applied to the dielectric. This will induce charges on the individual spheres (Fig.2.13a). Each sphere becomes analogous to a polarized atom and may be represented by a dipole moment $\mathbf{p} = q\mathbf{l}$ (Fig.2.13b). If N is the number of spheres per unit volume, the polarization P is given by

$$\mathbf{P} = Nq\mathbf{l}$$

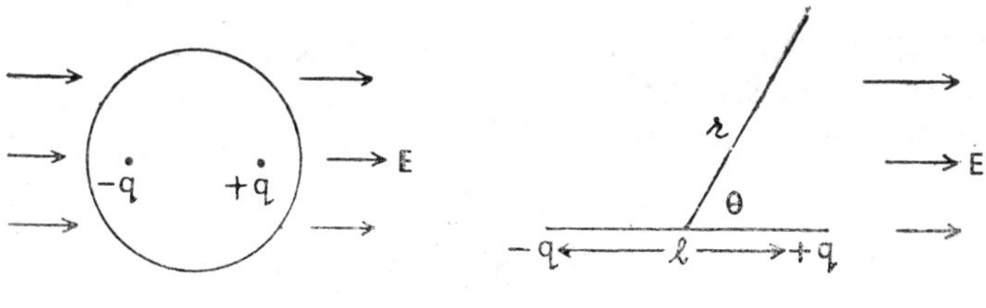

(a) (b)

Fig. 2.13

Since $\quad \mathbf{D} = \epsilon\mathbf{E} = \epsilon_0\mathbf{E} + \mathbf{P}$ (by 2.30 and 2.32)

$$\therefore \quad \epsilon = \epsilon_0 + \frac{P}{E} = \epsilon_0 + \frac{Nql}{E} \tag{2.34}$$

Now the potential at a point due to the uniform field is

$$V_0 = -\int_0^r E \cos\theta \, dr = -Er\cos\theta \tag{2.35}$$

where r is the distance of the point from the origin which, for convenience, is taken at the centre of the dipole and θ is the angle between the axis of the dipole and the radial line.

The potential due to the dipole is

$$V_d = \frac{p\cos\theta}{4\pi\epsilon_0 r^2} = \frac{ql\cos\theta}{4\pi\epsilon_0 r^2} \tag{2.36}$$

The total potential at the point is

$$V = V_0 + V_d = -Er\cos\theta + \frac{ql\cos\theta}{4\pi\epsilon_0 r^2} \tag{2.37}$$

Since the metal sphere has equal amount of positive and negative charges on it, its potential is zero. Therefore, if 'a' is the radius of the sphere

$$0 = -Ea\cos\theta + \frac{ql\cos\theta}{4\pi\epsilon_0 a^2} \tag{2.38}$$

$$\therefore \quad \frac{ql}{E} = 4\pi\epsilon_0 a^3 \tag{2.39}$$

Substituting this in Eq. (2.34), we have

$$\epsilon = \epsilon_0 + 4\pi\epsilon_0 N a^3$$

or

$$\epsilon_r = \frac{\epsilon}{\epsilon_0} = 1 + 4\pi N a^3 \tag{2.40}$$

i.e.

$$\epsilon_r = 1 + 3VN \tag{2.41}$$

where V is the volume of the sphere $= (4/3)\pi a^3$. This shows that the permittivity of the dielectric under consideration depends on both the number of the spheres and their sizes.

2.9 Energy of the Field in the Presence of a Dielectric

We have seen in Sec. 2.5 that when the charges are distributed with the volume density ρ and surface density σ, the potential energy is given by

$$W = \tfrac{1}{2}\int_V \rho\Phi d\tau + \tfrac{1}{2}\int_S \sigma\Phi dS_1$$

This result is not affected by the presence of a dielectric. By Gauss' theorem $\nabla\cdot\mathbf{D} = \rho$. Hence,

$$\tfrac{1}{2}\int_V \rho\Phi d\tau = \tfrac{1}{2}\int_V \Phi(\nabla\cdot\mathbf{D})d\tau = \tfrac{1}{2}\int_V \left\{\nabla\cdot(\Phi\mathbf{D}) - \mathbf{D}\cdot\nabla\Phi\right\}d\tau$$

$$= \tfrac{1}{2}\int_S \Phi\mathbf{D}\cdot\hat{\mathbf{e}}_n dS + \tfrac{1}{2}\int_V \mathbf{D}\cdot\mathbf{E}d\tau \qquad (2.42)$$

where we have used the vector identity

$$\text{div}\,(\Phi\mathbf{D}) = \Phi\ \text{div}\,\mathbf{D} + \mathbf{D}\cdot\text{grad}\,\Phi$$

and the Divergence theorem. The first integral must be taken over a closed surface bounding the whole volume (which is zero) and also over the surfaces of the conductors. Since the integration is over the surface of the medium, the yector $\hat{\mathbf{e}}_n dS$ is drawn outwards from the medium i.e. into the conducting surface as shown in Fig. 2.6.

Further, one can easily show that the normal component of $\mathbf{D}$ i.e.

$$\mathbf{D}\cdot\hat{\mathbf{e}}_{n_1} dS_1 = -\sigma dS_1$$

$$\therefore\quad \tfrac{1}{2}\int_V \rho\Phi d\tau = -\tfrac{1}{2}\int_{S_1} \sigma\Phi dS_1 + \tfrac{1}{2}\int_V \mathbf{D}\cdot\mathbf{E}d\tau$$

and the total energy W

$$W = \tfrac{1}{2}\int_V \rho\Phi d\tau + \tfrac{1}{2}\int_{S_1} \sigma\Phi dS_1$$

$$= \tfrac{1}{2}\int_V \mathbf{D}\cdot\mathbf{E}d\tau \qquad (2.43)$$

This expression shows that we may regard the energy as being distributed throughout all space, the energy density being $\tfrac{1}{2}\mathbf{D}\cdot\mathbf{E}$.

2.10 Boundary Conditions

For solving any electrostatic problem involving more than one dielectric medium, it is important to know what happens at the boundary surfaces separating the different materials.

Imagine a disc enclosing a part of the boundary surface between two media and whose axis is normal to the boundary (Fig.2.14).

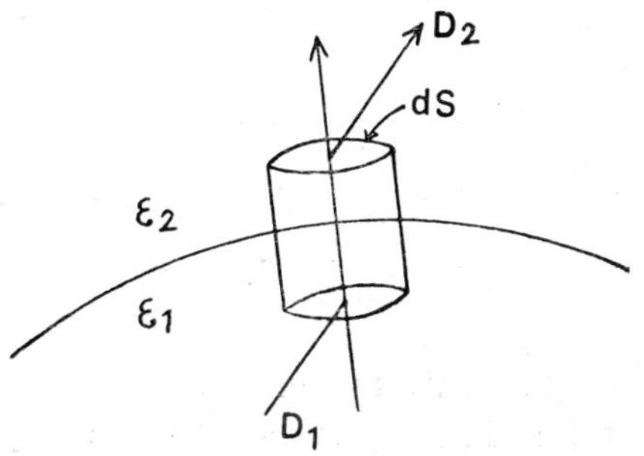

Fig. 2.14

Let ϵ_1, ϵ_2 be the relative permittivities of the two media. Let us assume that the thickness of the disc is very small. The only contribution to the outward flux from the disc comes from its flat surfaces. By Gauss' theorem

$$\nabla \cdot \mathbf{D} = \rho$$

$$\therefore \quad \int_V \nabla \cdot \mathbf{D} d\tau = \int_V \rho d\tau \tag{2.44}$$

where ρ is the density of free charges. Since there is no free charge on the surface

$$\int_V \nabla \cdot \mathbf{D} d\tau = 0$$

By Divergence theorem

$$\int_V \nabla \cdot \mathbf{D} d\tau = \int_S \mathbf{D} \cdot \hat{\mathbf{e}}_n dS \tag{2.45}$$

$$\therefore \quad \int_S \mathbf{D} \cdot \hat{\mathbf{e}}_n dS = \mathbf{D}_1 \cdot \hat{\mathbf{e}}_{n_1} dS + \mathbf{D}_2 \cdot \hat{\mathbf{e}}_{n_2} dS = 0$$

where D_1, D_2 are the values of D in the two media and $\hat{\mathbf{e}}_{n1}$, $\hat{\mathbf{e}}_{n2}$ are the unit vectors.

Further, $\qquad \hat{\mathbf{e}}_{n_1} = -\hat{\mathbf{e}}_{n_2} = \hat{\mathbf{e}}_n$

$$\therefore \quad (\mathbf{D}_1 \cdot \hat{\mathbf{e}}_n - \mathbf{D}_2 \cdot \hat{\mathbf{e}}_n)\, dS = 0 \quad \text{or} \quad \mathbf{D}_1 \cdot \hat{\mathbf{e}}_n = \mathbf{D}_2 \cdot \hat{\mathbf{e}}_n \tag{2.46)}$$

Therefore, the normal components of **D** in the two media are equal. That is, **D** is continuous.

Let us now construct at the boundary a rectangular path $ABCD$ (Fig. 2.15) with sides $AB = CD = dl$ parallel to the surfaces and $BC = DA = dt$ perpendicular to it.

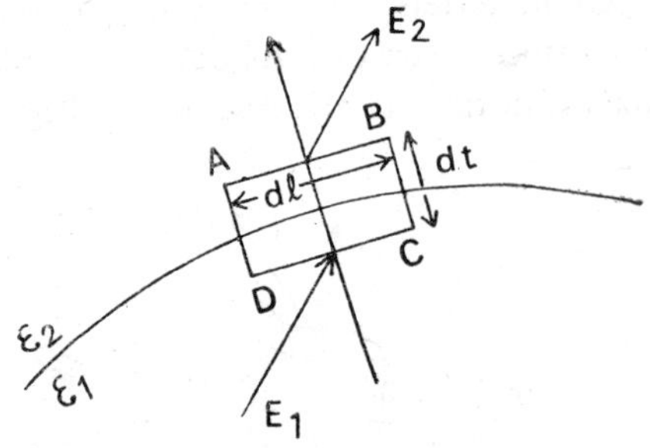

Fig. 2.15

Let $\mathbf{E}_1$, $\mathbf{E}_2$ be the fields in the two media. Since electric field is conservative no net work is done by taking a unit charge round the path $ABCDA$.

$$\therefore \quad \oint \mathbf{E}\cdot d\mathbf{l} = E_{\|(1)}dl - E_{\|(2)}dl + \text{contribution from } BC \text{ and } DA$$

$$= 0 \tag{2.47}$$

Where $E_{\|(1)}$ and $E_{\|(2)}$ are the components of $\mathbf{E}$ in the two media parallel to the surface. Assuming that BC, DA i.e. dt is vanishingly small, we have

$$E_{\|(1)}\, dl - E_{\|(2)}dl = 0$$

$$\because \quad E_{\|(1)} = E_{\|(2)} \tag{2.48}$$

That is, the tangential component of $\mathbf{E}$, $E_{\|}$ is continuous across the boundary.

Equations (2.46) and (2.48) give the two conditions to be satisfied by the displacement vector $\mathbf{D}$ and electric field $\mathbf{E}$ at the boundary.

These conditions can also be expressed in terms of potential. If we have two media in contact with potentials Φ_1 and Φ_2 respectively, at all connecting points of the two media

$$\Phi_1 = \Phi_2$$

and

$$\epsilon_1 \frac{\partial \Phi_1}{\partial n} = \epsilon_2 \frac{\partial \Phi_2}{\partial n} \tag{2.49}$$

Note that the condition (2.46) for the displacement vector was arrived at on the assumption that there is no free charge on the boundary. If there is any charge on the boundary, say with surface density σ, equation (2.46) changes to

$$\mathbf{D}_1 \cdot \hat{\mathbf{e}}_n - \mathbf{D}_2 \cdot \hat{\mathbf{e}}_n = \sigma \tag{2.50}$$

That is the normal component of D abruptly changes by σ

EXAMPLE 2.2. Show that at a change of dielectric medium field lines are refracted according to the law $\epsilon_1 \cot \theta_1 = \epsilon_2 \cot \theta_2$, where θ_1 and θ_2 are the angles between the directions of the field and the common normal to the boundary (Fig. 2.16) and ϵ_1 and ϵ_2 are the relative permittivities of the two media.

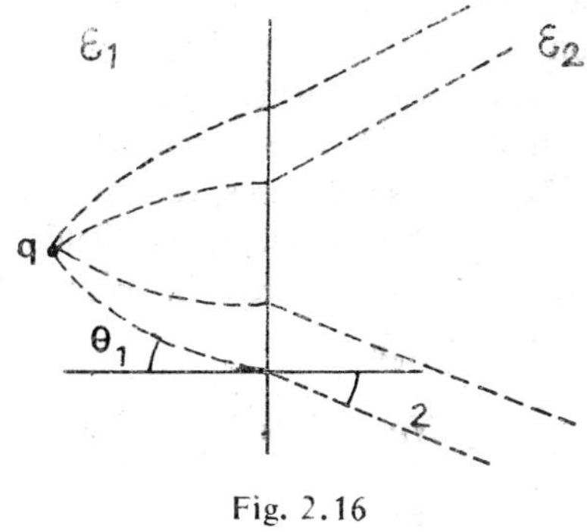

Fig. 2.16

By the first boundary condition (2.46)

$$\mathbf{D}_1 \cdot \hat{\mathbf{e}}_n = \mathbf{D}_2 \cdot \hat{\mathbf{e}}_n$$

i.e.

$$D_1 \cos \theta_1 = D_2 \cos \theta_2$$

or

$$\epsilon_1 E_1 \cos \theta_1 = \epsilon_2 E_2 \cos \theta_2 \qquad (2.51)$$

The second condition gives

$$E_1 \sin \theta_1 = E_2 \sin \theta_2 \qquad (2.52)$$

Combining (2.51) and (2.52), we have

$$\epsilon_1 \cot \theta_1 = \epsilon_2 \cot \theta_2$$

2.11 Gaseous Non-polar Dielectrics

In electromagnetic theory, a dielectric is treated as a continuous medium and the macroscopic quantities such as charge density and electric polarization are defined as continuous functions. Since a dielectric is really an assembly of positive and negative charges, it is necessary to relate the electrical properties of the constituent discrcte particles and the corresponding properties of the continuous medium. In this and the next sections we shall proceed to establish this relation.

If the field is not too large, the strength of the dipole induced in an atom, ion or molecule is proportional to the actual field acting on the particle. Let us call it $\mathbf{E}_{local}$

$$\therefore \quad \mathbf{p} = \alpha\epsilon_0 \mathbf{E}_{local} \qquad (2.53)$$

where $\alpha\epsilon_0$ is the constant of proportionality and $\mathbf{p}$, the dipole moment induced.

It must be noted that the local field $\mathbf{E}_{local}$ acting on a particular particle is not the same as the macroscopic field, which is the field averaged over a region large enough to contain many particles, and calculated by taking the average of the vector sum of the electric fields set up by all the dipoles in the medium including that of the particle for which we require the vector $\mathbf{E}_{local}$. The local field $\mathbf{E}_{local}$ is equal to the macroscopic field $\mathbf{E}$ due to the dipoles more than a few atomic diameters away which could be treated as a continuous medium with a polarization $\mathbf{P}$ plus the contribution from the nearby dipoles excluding the field of the dipole at which we need the value of $\mathbf{E}_{local}$, since the field due to this dipole does not act on itself. This is particularly true in the case of liquids and solids. In the case of gases, the molecules, for most of the time, are far apart, so that the short range fields are negligible and $\mathbf{E}_{local}$ is the same as the macroscopic field $\mathbf{E}$. Hence, in the case of gases

$$\mathbf{p} = \alpha\epsilon_0\mathbf{E}$$

and

$$\mathbf{P} = N\mathbf{p} = N\alpha\epsilon_0\mathbf{E} = \epsilon_0 \chi \mathbf{E} \text{ (by 2.21)} \qquad (2.54)$$

$$\therefore \quad \alpha = \frac{\chi}{N} \qquad (2.55)$$

where α is called the polarizability of the atom, ion or molecule. It measures the resistance of the particle to the displacement of its electron cloud. The scalar X appearing in equations (2.21) and (2.54) is called the **electric susceptibilty** of the medium.

EXAMPLE 2.3. Assuming a simple classical model for an atom, derive an expression for the induced dipole moment, and, hence, for its polarizability.

A simple classical model of the atom is a point positive charge (Ze) surrounded by a spherically symmetric cloud of negative charge ($-Ze$) in which the density is uniform up to the atomic radius r_0, and zero at larger radii.

If the atom is placed in an electric field **E**, the nucleus will be displaced in the direction of **E** by a distance, say 'd'. The force acting on the nucleus in the direction of **E** is Ze **E**. An electrostatic force between the nucleus and the charge cloud will tend to restore the initial configuration. The negative charge attracting the nucleus is, by Gauss' law, the part of the cloud within the sphere of radius 'd'. This charge is

$$\tfrac{4}{3}\pi d^3\rho = \tfrac{4}{3}\pi d^3 \frac{Ze}{\tfrac{4}{3}\pi r_0^3} = \frac{Zed^3}{r_0^3}$$

where ρ is the density of the charge.

$$\therefore \quad \frac{(Ze)(Zed^3/r_0^3)}{4\pi\epsilon_0 d^2} = Ze\mathbf{E}$$

i.e.

$$Zed = 4\pi\epsilon_0 r_0^3\mathbf{E}$$

But the atomic dipole **p** is

$$\mathbf{p} = Zed = 4\pi\epsilon_0 r_0^3\,\mathbf{E} = \alpha\epsilon_0\,\mathbf{E} \tag{2.56}$$

$$\therefore \quad \alpha = 4\pi r_0^3 \tag{2.57}$$

We see that the polarizability is independent of the field. The dielectrics in which this condition is satisfied is said to be **a linear** dielectric.

EXAMPLE. 2.4. Calculate the individual dipole moment **p** of a molecule of carbon tetrachloride given the following data. Find also the average electron displacement.

Relative permittivity $\epsilon_r = 2.24$
Density $= 1.60$ gm/cm^3
Molecular weight $= 156$
Field $= 10^7$ volts/metre

$$\text{Molecular density} = N = \frac{\text{Avogadro's number}}{\text{Molecular weight}} \times \text{density}$$

$$= \frac{6.02 \times 10^{23}}{156} \times 1.60$$

$$= 6.17 \times 10^{21} \text{ molecules/cm}^3$$

$$\text{i.e. } 6.17 \times 10^{27} \text{ molecules/m}^3$$

The dipole moment of a single molecule p is

$$p = \frac{P}{N} = \frac{\epsilon_0 \chi E}{N} = \frac{\epsilon_0(\epsilon_r - 1) E}{N} \quad (\because \quad \epsilon_r = 1 + \chi)$$

$$= \frac{8.85 \times 10^{-12} \times 1.24 \times 10^7}{6.17 \times 10^{27}}$$

$$= 1.77 \times 10^{-32} \text{ coulomb metres}$$

Let the average electron displacement be 'd'. Since there are 74 electrons in each CCl_4 molecule,

$$p = 74\, de \text{ and}$$

$$d = \frac{p}{74e}$$

$$= \frac{1.77 \times 10^{-32}}{74 \times 1.6 \times 10^{-19}}$$

$$= 1.5 \times 10^{-15} \text{m}$$

2.12 Gaseous Polar Dielectrics

Let us now consider the dielectric behaviour of molecules which have a permanent dipole moment. As shown above the local field acting on a molecule in a gas is the same as the external field **E**. This field induces an extra dipole moment in the molecule giving it the same kind of polarizability as in non-polar molecules. Besides this, the field also tends to line up the individual dipoles. This, however, does not happen effectively owing to the thermal motion of the molecules. We know that the potential energy of a dipole when placed in an electric field is given by

$$W = -\mathbf{p} \cdot \mathbf{E} = -p\, E \cos\theta \text{ (see Eq. 1.66)}$$

The energy varies with the orientation of the dipole, being minimum when $\theta = 0$ i.e. when the dipole is aligned along the field. In order to have an idea about the order of magnitude of this energy let us calculate the energy required to reverse the dipole, so as to make it point in the direction opposite to that of the field. This is obviously $2p\,E$. Let us assume the atomic spacing to be 10^{-10}m and the field $E = 10^6$ volts/m the maximum that can be achieved in a gaseous dielectric

$$\therefore \quad p = e \times \text{atomic spacing}$$

$$= 1.6 \times 10^{-19} \times 10^{-10}$$

$$= 1.6 \times 10^{-29} \text{ coulomb m}$$

and

$$2pE = 2\times 1.6\times 10^{-29}\times 10^{6} \text{ joules}$$
$$= 2\times 10^{-4} \text{ eV} \quad (2.58)$$

This is much less than the average kinetic energy of a molecule at room temperature which is of the order of 0.04 eV. Hence, the random thermal motion will not be affected much by the presence of the field **E** and there will still be molecules with their dipoles pointing in all directions. However, the dipoles will, in general, tend to line up in a direction in which their potential energy is minimum i.e. in the direction of the field, there, thus, being a slight excess of molecules pointing in this direction. We shall now calculate the net polarization caused by this alignment. This can be done using the methods of statistical mechanics.

Let N be the number of molecules per unit volume. Since, in the absence of an electric field, all orientations of molecules are equally probable, the number of molecules per unit volume having their dipoles within a solid angle $d\Omega$ will be equal to $N\frac{d\Omega}{4\pi}$. The solid angle in the range of orientation between θ and $\theta + d\theta$ (Fig. 2.17) is

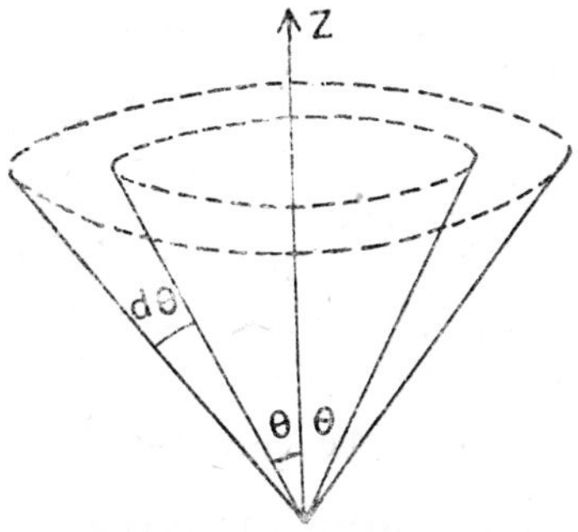

Fig. 2.17

$$d\Omega = \frac{2\pi r \sin\theta \, r\, d\theta}{r^2} = 2\pi \sin\theta \, d\theta \quad (2.59)$$

$$\therefore \quad N\frac{d\Omega}{4\pi} = \tfrac{1}{2} N \sin\theta \, d\theta \quad (2.60)$$

If now a field **E** is applied along the z-axis, the dipoles making angle θ with the field acquire an additional potential energy $W = -p\,E\cos\theta$ and the probability distribution will depend upon the Boltzmann factor $e^{-W/kT}$. The number of molecules per unit volume with their dipoles in the range between θ and $\theta + d\theta$ is

$$N(\theta)\, d\theta = Ce^{-W/kT} d\Omega \quad (2.61)$$

Since W/kT is very small, we may write this as

$$N(\theta)\, d\theta = C\left(1 - \frac{W}{kT}\right) d\Omega$$
$$= C\, 2\pi \sin\theta \left(1 + \frac{pE\cos\theta}{kT}\right) d\theta \quad (2.62)$$

The total number of molecules per unit volume is

$$N = \int_0^{\pi} N(\theta)\, d\theta = 2\pi C \int_0^{\pi} \sin\theta \left(1 + \frac{pE\cos\theta}{kT}\right) d\theta$$

$$= 4\pi C$$

$$\therefore \quad C = \frac{N}{4\pi}$$

and

$$N(\theta) = \frac{N}{2} \sin\theta \left(1 + \frac{pE\cos\theta}{kT}\right) \tag{2.63}$$

Equation (2.63) shows that there will be more molecules in the field direction ($\cos\theta = 1$) than in the opposite direction ($\cos\theta = -1$). This confirms our observation made earlier that there will be a net alignment when the field **E** is applied.

The component of a dipole moment pointing in the direction θ, along the field is $p \cos\theta$. Hence, the net dipole moment per unit volume along the the field is

$$P = \int_0^{\pi} N(\theta)\, p \cos\theta\, d\theta$$

$$= \frac{N}{2}\int_0^{\pi} \sin\theta \left(1 + \frac{pE\cos\theta}{kT}\right) p\cos\theta d\theta$$

$$= \frac{Np^2E}{3kT} \tag{2.64}$$

This is the contribution of the permanent dipole moments to the polarization. The quantity $\frac{p^2}{3\,kT}$ is often termed as **orientational polarizability**. Besides this, the molecules will also acquire an induced dipole moment along the field irrespective of the direction of the permanent dipole moment, leading to an additional polarization $N\alpha_0\epsilon_0\mathbf{E}$ where α_0 is what might be called as **deformation polarizability**. Therefore, the net polarization **P** is given by

$$\mathbf{P} = \left(N\alpha_0\epsilon_0 + \frac{Np^2}{3kT}\right)\mathbf{E} \tag{2.65}$$

and the susceptibility

$$\chi = N\left(\alpha_0 + \frac{p^2}{3kT\epsilon_0}\right) \tag{2.66}$$

Equation (2.66) shows that the measurement of susceptibilities at different temperatures will yield information about permanent and induced dipole moments. The plot of χ against $1/T$ must be a straight line. The slope will give the permanent dipole moment and intercept at $1/T = 0$ will give the susceptibility due to the induced polarization. One can also learn something about the shape of molecules from the knowledge of susceptibilities and the dipole moments. For example, the molecule CS_2 and H_2O both contain two identical atoms bound to a common partner. There is electron transfer across C—S bond as well as across O—H bond and dipole moments associated with each bond. Yet the dipole moment of H_2O is 6.2×10^{-30} coulomb metres, while that of CS_2 is zero. We conclude that since dipole moments of CS_2 cancel each other, it must be a linear molecule: S — C — S; while H_2O is not a linear molecule (see Fig. 1.33).

2.13 Non-polar Liquids

In solid and liquid dielectrics, the molecules are so close together that the effects of the short-range fields cannot be overlooked. The behaviour becomes more complicated when the polar molecules are present, because the fields generated by the permanent dipoles are sometimes much larger than any macroscopic field. In a non-polar liquid, however, the short-range fields are not so strong and $\mathbf{E}_{\text{local}}$ can be deduced to a good approximation by adopting the following procedure.

Imagine a small spherical cavity carved out in a uniformly polarized material. Since the medium is polarized, polarization charges will appear on any surface of the medium. The field at a point A inside the cavity, $\mathbf{E}_{\text{local}}$ is the sum of the macroscopic field $\mathbf{E}$ and the fields generated by the polarization charges on the surface of the cavity (Fig. 2.18).

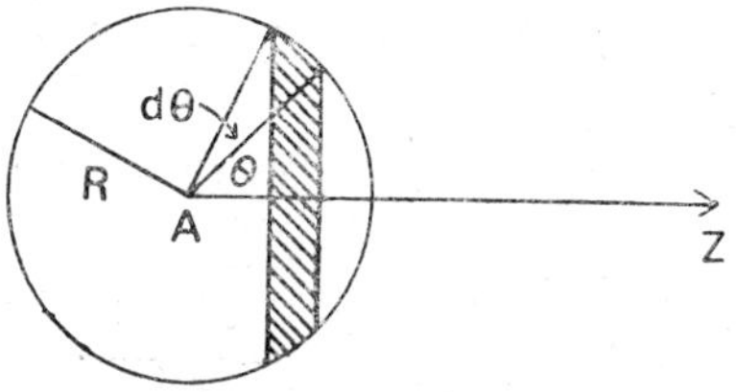

Fig. 2.18

Let the macroscopic field act in the z-direction. The polarization charge on a small element of area dS, on the surface of the cavity, is $\sigma_P\, dS$, where σ_P is the surface density of the polarization charge.

$$\sigma_P dS = \mathbf{P}\cdot\hat{\mathbf{e}}_n dS = \epsilon_0 \chi \mathbf{E}\cdot\hat{\mathbf{e}}_n dS$$

$$= -\ \epsilon_0 \chi E \cos\theta\ dS \tag{2.67}$$

if θ is the angle between the field and the normal to dS. The negative sign appears because the outward normal to the dielectric is the inward normal to the cavity. The net field at A, the centre of the cavity, generated by the polarization charges is, by symmetry, directed along the z-axis. The magnitude of the z-component of the field at A due to this charge is

$$\frac{\epsilon_0 \chi E \cos\theta\ dS \cos\theta}{4\pi\epsilon_0 R^2}$$

where R is the radius of the cavity. The field due to the entire band of area $dS = 2\pi\ R \sin\theta R d\theta$ is

$$\frac{\epsilon_0 \chi E \cos^2\theta}{4\pi\epsilon_0 R^2} \times 2\pi R^2 \sin\theta\ d\theta$$

and the total field at A due to all the polarization charges on the surface is

$$\tfrac{1}{2}\chi E \int_0^{\pi} \cos^2\theta \sin\theta\ d\theta = \tfrac{1}{3}\chi E$$

Hence, the effective field at A

$$\mathbf{E}_{\text{local}} = \mathbf{E} + (1/3)\chi\mathbf{E}$$
$$= \mathbf{E}\{1 + (1/3)\chi\} \qquad (2.68)$$

$\therefore$
$$\mathbf{P} = N\alpha\epsilon_0\mathbf{E}_{\text{local}}$$
$$= N\alpha\epsilon_0\mathbf{E}\{1+(1/3)\chi\} \qquad (2.69)$$

$\therefore$
$$\epsilon_0\chi\mathbf{E} = N\alpha\epsilon_0\mathbf{E}(1+\tfrac{1}{3}\chi) \qquad (2.70)$$

i.e.

$$\chi = \frac{N\alpha}{1 - N\alpha/3} \qquad (2.71)$$

This can also be written in a different form. Remembering that $\chi = \epsilon_r - 1$ and $P = \varepsilon_0(\epsilon_r - 1)\mathbf{E}$, we can write (2.70) as

$$\epsilon_0(\epsilon_r - 1)\mathbf{E} = N\alpha\epsilon_0\mathbf{E} + \frac{N\alpha\epsilon_0\mathbf{E}}{3}(\epsilon_r - 1)$$

or

$$\epsilon_0(\epsilon_r - 1) = N\alpha\epsilon_0\left\{\frac{2+\epsilon_r}{3}\right\}$$

Hence,

$$\frac{\epsilon_r - 1}{\epsilon_r + 2} = \frac{N\alpha}{3} \qquad (2.72)$$

This was first derived by Clausius and Mossoti and is generally known as **Clausius-Mossoti equation**. The Clausius-Mossoti equation shows that the local field is larger than the macroscopic field. It also gives the size of the increase. However, it must be borne in mind that the formula is not very accurate since it depends on a crude approximation.

In Table 2.2 we give some experimental data.

Table 2.2 Relative permittivity of some substances at N.T.P.

Substance	*Gas*		*Liquid*	
	ϵ_r	ρ (*density*)	ϵ_r	ρ (*density*)
CS_2	1.0029	0.0034	2.64	1.293
CCl_4	1.0030	0.0030	2.24	1.590
O_2	1.0005	0.0014	1.51	1.190
CO_2	1.0010	0.0019	1.61	0.975

2.14 Solid Dielectrics-Electrets

It is possible to prepare materials in the solid state having a permanent polarization, which exists even when a field is not applied. For example, if some wax is melted and a strong electric field is applied to it when it is still in the liquid form, the dipoles get partly lined up and remain that way when the liquid freezes. The solid material thus formed has, like a magnet, a permanent dipole moment, and is called an **electret.** It, however, soon gets discharged as it attracts free charges from the air.

A permanent internal polarization is also found in some complicated crystals, but we do not notice it because the external fields are discharged, as in electrets.

A theory of the dielectric constant can be worked out for crystals that do not have permanent dipole moment, on the same lines as for non-polar liquids.

2.15 Electric Field Stresses

We shall discuss in this section a method for the calculation of forces on charged and polarized bodies in an electrostatic field—the method of **field stresses.**

Faraday, while studying the curvature of lines of force in electrostatic fields, had noticed an apparent tendency of adjacent lines to repel each other as if each tube of force were inherently disposed to distend laterally. In addition to this repellent force in the transverse direction, he believed that there exists an attractive force in the direction of lines of force. Maxwell was fully persuaded of the existence of these **pressures** and **tensions** in electrostatic fields.

Let us see how the concept of stresses which is used in ordinary mechanics can be applied to a system of electric forces. There should be no difficulty if we consider stresses in a material body. But we know that electric forces also act through empty space. How can we talk about a stress in the absence of any material? The idea of a stress in empty space seems to be meaningless; or else, we have to accept the view that space is filled with an elastic substance—ether—which transmits stress. The *ether-theory* was in vogue until the theory of relativity led physicists to abandon the idea of all-pervading ether. What is, then, the way out? Well, we do not discuss how charges act on each other across empty space; we merely observe that they do, and seek to construct a conceptual framework involving stress and strain which enables us to calculate such interactions.

Let us now obtain analytical expressions for the stresses and pressures in electric fields.

The tension along the lines of force must be supposed to maintain the ponderomotive force which acts on the conductor on which the lines of force terminate. This force urges the element of area away from the conductor and is equal to $\frac{\sigma^2}{2\epsilon}=\frac{1}{2}\sigma E$ per unit area, where σ is the surface charge density and E is the electric field strength. If the tubes are in a state of such a tension, it follows that they must also exert a lateral pressure upon neighbouring tubes. If this were not the case, the tubes would shrink until they become straight lines. We can find the expression for this pressure by considering a special case.

Consider a spherical condenser formed of spheres of radii '*a*' and '*b*'. Suppose this condenser is cut into two equal halves by a plane through its centre. The two halves will repel each other. We must ascribe this action to stresses in the medium across the plane of section. Since the lines of force are radial, one can see that the stress is perpendicular to the lines of force and, hence, is a pressure, say $\mathcal{P}$. The area over which the pressure acts is $\pi(b^2-a^2)$. The total repulsion between the two halves, therefore, is $-\mathcal{P}\pi(b^2-a^2)$.

The force per unit area over each hemisphere is $\frac{1}{2}\sigma E$. The total force acting on the inner hemisphere is $\frac{1}{2}\sigma E\pi a^2$ and that on the outer hemisphere is $\frac{1}{2}\sigma E\pi b^2$. The resultant repulsion on the complete half of the condenser is $\frac{1}{2}\sigma E\pi(b^2-a^2)$

$$\therefore \qquad \tfrac{1}{2}\sigma E\pi(b^2-a^2) = -\mathcal{P}\pi(b^2-a^2)$$

i.e.

$$\mathcal{P} = -\tfrac{1}{2}\sigma E \tag{2.73}$$

If these stresses exist, they will account for the observed mechanical action on the conductors. The concept of field stresses as applied to electrostatic field is one of the examples of fruitful relationship between mechanics and the theory of electricity.

PROBLEMS

2.1 Two large parallel plates each of area A, are separated by a distance 'd', their potentials being 0 and V respectively. A third plate carrying a charge 'q' is placed midway between them. Find the potential of this plate.

2.2 Find the diminution of energy in a system consisting of two conductors of capacities C_1, C_2 and potentials V_1, V_2 respectively when the conductors are connected with one another.

2.3 Show that the capacity of a parallel plate conductor is increased by a factor $\frac{t}{t-d}$, if a conducting sheet of thickness 'd' is introduced between the plates, the separation between the plates being 't'.

2.4 A parallel plate conductor is filled with two dielectrics of equal size but unequal ϵ_r. Find the capacity of this conductor.

2.5 A sphere of dielectric material is uniformly polarized. Find the electric field inside and outside this sphere.

2.6 Find the dipole moment induced in eaeh atom when a gas of helium is placed in an electric field, given that the dielectric constant of helium at 0°C and 1 atmosphere pressure is 1.000074.

2.7 A spherical condenser consists of two concentric spheres of radii 'a' and 'd' ($a<d$). A spherical shell of permittivity ϵ, bounded by the spheres of radii 'b' and 'c' lies between them, with its centre coinciding with that of the spheres. Show that the capacity C of the condenser is given by

$$\frac{4\pi\epsilon_0}{C} = \frac{1}{a} - \frac{1}{d} + \frac{(\epsilon_0 - \epsilon)}{\epsilon}\left(\frac{1}{b} - \frac{1}{c}\right)$$

2.8 The centre of a conducting sphere of radius 'r' and carrying a charge 'q' is on the flat boundary between two dielectrics of permittivities ϵ_1 and ϵ_2. Find the potential of this system and the charge distribution on the surfaces of the sphere.

2.9 A sphere of mass M and radius R floats on a liquid of density μ and permittivity ϵ. One quarter of it is immersed in the liquid when it is not charged. What should be the charge on the sphere so that it sinks to a depth exactly half its diameter?

2.10 The relative permittivity of NH_3 measured at 273°K and 373°K at one atmosphere is 1.00834 and 1.00487 respectively. Calculate the permanent dipole moment for the gas. Find also the radius of the molecule, assuming the polarizability to be the same as that of a conducting sphere.

2.11 Show that the energy expanded in polarizing a dieletric is $\frac{(\epsilon_r - \epsilon_0)E^2}{2}$.

2.12 Two long, coaxial. cylindrical, conducting surfaces are lowered vertically in a liquid dielectric. When a potential difference V is established between the surfaces, the liquid is found to rise to a height 'h' between the electrodes. Find the susceptibility of the liquid.

Chapter 3

Boundary Value Problems in Electrostatic Fields

We outline in this chapter some of the techniques generally used in solving electrostatic problems. We shall discuss a few problems where the techniques can be fruitfully used. The examples chosen will also illustrate important properties of the electric field.

3.1 Poisson and Laplace Equations

We have seen that an electric field can be completely described in terms of its potential.

$$\mathbf{E} = -\nabla\Phi$$

Taking the divergence we get

$$\nabla\cdot\mathbf{E} = -\nabla\cdot\nabla\Phi = -\nabla^2\Phi$$

We also know by Gauss' law that

$$\nabla\cdot\mathbf{E} = \rho/\epsilon_0$$

$$\therefore \qquad \nabla^2\Phi = -\rho/\epsilon_0 \tag{3.1}$$

This is known as **Poisson's equation.**

If there are no charges in the region, $\rho = 0$ and

$$\nabla^2\Phi = 0 \tag{3.2}$$

This is **Laplace's equation** and is valid only in the region free of charges.

The Laplacian ∇^2 is a scalar operator and has the following form in the three coordinate systems generally used:

(i) Cartesian system : $\dfrac{\partial^2}{\partial x^2}+\dfrac{\partial^2}{\partial y^2}+\dfrac{\partial^2}{\partial z^2}$ (3.3)

(ii) Spherical polar coordinate system:

$$\frac{1}{r^2}\frac{\partial}{\partial r}\left(r^2\frac{\partial}{\partial r}\right)+\frac{1}{r^2\sin\theta}\frac{\partial}{\partial\theta}\left(\sin\theta\frac{\partial}{\partial\theta}\right)+\frac{1}{r^2\sin^2\theta}\frac{\partial^2}{\partial\phi^2} \tag{3.4}$$

(iii) Cylindrical polar coordinates:

$$\frac{1}{r}\frac{\partial}{\partial r}\left(r\frac{\partial}{\partial r}\right)+\frac{1}{r^2}\frac{\partial^2}{\partial\phi^2}+\frac{\partial^6}{\partial z^2} \tag{3.5}$$

The solution of an electrostatic problem is obtained when the potential is determined at all points. In other words, the problems can be solved directly by solving Poisson or Laplace equation, subject to the condition that Φ, thus obtained, satisfies certain boundary conditions appropriate to the configurations of electrodes. Such problems are called **boundary value problems.** The boundary value problems usually present formidable mathematical difficulties and can be solved approximately using numerical techniques. There are a few problems, however, which, because of the high degree of symmetry involved, can be solved exactly. These may be handled by the use of a series of known functions. Before we take up the discussion of such problems we shall first consider one important theorem that follows from Poisson's and Laplace's equation and also the boundary conditions to be satisfied by the potential.

3.2 Earnshaw's Theorem

The theorem states that "a charged body cannot rest in stable equilibrium under the influence of electrostatic forces alone".

For any extremum value of potential Φ it must satisfy the condition $\nabla^2\Phi \neq 0$. For Φ to be maximum $\nabla^2\Phi$ must be negative and for Φ to be minimum $\nabla^2\Phi$ must be positive. In the region where there are no charges $\nabla^2\Phi = 0$ (Laplace's equation) and, hence, the potential cannot have a maximum or a minimum value. For a positive charge to be stable, it must be at a point of minimum potential i.e. $\nabla^2\Phi$ must be positive. However, from (3.1), we find that $\nabla^2\Phi$ is negative indicating that Φ is maximum, and, hence, the charge is not stable. In the case of a negative charge, the condition for stability demands that $\nabla^2\Phi$ is negative, while Poisson's equation gives $\nabla^2\Phi = \rho/\epsilon_0$. This implies that there can be no points of stable equilibrium in an electrostatic field.

This does not mean that it is not possible to balance a charge by electrical forces. For example, if there are four negative charges at the four corners of a square, the net force on a positive charge at the centre of the square will be zero. But the equilibrium is not stable. i.e., if the charge is displaced slightly, it will not return to its equilibrium position.

One could ask "How do you account, then, for the stability of atoms and molecules which consist of positive and negative charges?". If the only forces acting within atoms were electrostatic, the charges would coalesce and neutralize each other. The very fact that they are stable indicates that besides electrostatic forces, forces of different nature also exist in atoms and molecules.

3.3 Boundary Conditions and Uniqueness Theorem

The problems in electrostatics, generally, involve finite bounded regions of space, and it is necessary that the physically reasonable solution of Poisson's or Laplace's equation pertaining to a problem is the correct and only answer. We shall now show that if Φ is real with its first and second derivative continuous in a region and

(a) the value of the potential Φ on the closed surface is specified (Dirichlet condition) or

(b) the normal derivative of Φ, $\frac{\partial \Phi}{\partial n}$, i.e. the field vector is specified everywhere on the closed surface (Neumann condition), then the solution is unique.

The theorem we are going to prove is known as **Uniqueness theorem** and is a fundamental theorem of potential theory.

Suppose the solution is not unique and that there exist two solutions Φ_1 and Φ_2 of Laplace's equation within the bounded region but satisfying the boundary conditions

$$\Phi_1 = \Phi_2$$

and
$$\frac{\partial \Phi_1}{\partial n} = \frac{\partial \Phi_2}{\partial n} \text{ at the boundary.} \tag{3.6}$$

Let
$$\Phi = \Phi_1 - \Phi_2$$

Hence,
$$\Phi = 0 \text{ and } \frac{\partial \Phi}{\partial n} = 0 \text{ at the boundary.} \tag{3.7}$$

Since Φ_1 and Φ_2 are both solution of Laplace's equation.

$$\nabla^2 \Phi_1 = \nabla^2 \Phi_2 = 0$$

$$\therefore \quad \nabla^2 (\Phi_1 - \Phi_2) = \nabla^2 \Phi = 0 \tag{3.8}$$

i.e. Φ is also a solution of Laplace's equation. Let us use the well-known Green's theorem, viz.

$$\int_S \psi \frac{\partial \phi}{\partial n} dS = \int_V [\psi \nabla^2 \phi + \nabla \phi \cdot \nabla \psi] d\tau \tag{3.9}$$

where ψ and ϕ are arbitrary scalar functions, $\frac{\partial \phi}{\partial n}$ is the normal derivative of ϕ on the surface and dS an $d\tau$ are the surface and volume elements respectively.

Putting $\psi = \phi = \Phi$, we have

$$\int_S \Phi \frac{\partial \Phi}{\partial n} \, dS = \int_V \left[\Phi \nabla^2 \Phi + \left| \nabla \Phi \right|^2 \right] d\tau \tag{3.10}$$

For any of the boundary condition (3.6), the left hand side vanishes. Further $\nabla^2 \Phi = 0$ by (3.8)

$$\therefore \quad \int_V \left| \nabla \Phi \right|^2 d\tau = 0 \tag{3.11}$$

Since the integrand is a positive definite quantity

$$\nabla \Phi = 0 \tag{3.12}$$

Consequently, $\Phi = \Phi_1 - \Phi_2 = C$ is a constant throughout the volume. If Dirichlet condition is satisfied on the surface, $C = 0$ on the surface (by 3.6). Hence. it is zero throughout the region and $\Phi_1 = \Phi_2$ i.e. the solution is unique. If, on the other hand, Neumann condition is satisfied on the surface

$$\frac{\partial \Phi}{\partial n} = 0 \quad \text{i.e.} \quad \Phi = \Phi_1 - \Phi_2 = \text{a constant}$$

Since the constant is arbitrary we can take it to be zero and the solution is again unique.

$$\Phi_1 = \Phi_2$$

Notice that if Dirichlet condition is satisfied over a part of the surface S and Neumann condition over the remaining part, the left-hand side of (3.10) still vanishes and the solution is unique.

3.4 Solution of Laplace's Equation in Rectangular Coordinates

We now discuss some of the methods for solving problems in which there are no free charges. We are, therefore, concerned with Laplace's equation which when expressed in rectangular coordinates becomes

$$\frac{\partial^2 \Phi}{\partial x^2} + \frac{\partial^2 \Phi}{\partial y^2} + \frac{\partial^2 \Phi}{\partial z^2} = 0 \tag{3.13}$$

Let us assume that the solution of this equation, $\Phi\,(x, y, z)$, can be represented as the product of three functions: $X(x)$, $Y(y)$ and $Z(z)$, each one of which depends on one coordinate only. Thus, let

$$\Phi(x, y, z) = X(x)\,Y(y)\,Z(z) \tag{3.14}$$

Substituting this in (3.13) and dividing throughout by $X(x)\,Y(y)\,Z(z)$, we get

$$\frac{1}{X(x)}\frac{d^2X}{dx^2} + \frac{1}{Y(y)}\frac{d^2Y}{dy^2} + \frac{1}{Z(z)}\frac{d^2Z}{dz^2} = 0 \tag{3.15}$$

Notice that we have replaced partial derivatives by total derivatives. This is permissible in the present case, since each term involves a function of one coordinate only.

For the equation (3.15) to be valid for arbitrary values of independent coordinates, each term on the left must be separately equal to a constant. Thus,

$$\frac{1}{X(x)}\frac{d^2X}{dx^2} = k_1^2; \quad \frac{1}{Y(y)}\frac{d^2Y}{dy^2} = k_2^2; \quad \frac{1}{Z(z)}\frac{d^2Z}{dz^2} = k_3^2 \tag{3.16}$$

subject to the condition that

$$k_1^2 + k_2^2 + k_3^2 = 0 \tag{3.17}$$

You can see that not all these constants can be real nor all imaginary. At least one of them must be real and one imaginary. The third may be real or imaginary. Which constants are real and which are imaginary will be decided by the physical situation of the problem we are dealing with.

Suppose the constants k_1^2, k_2^2, k_3^2 are

$$k_1^2 = -\alpha^2,\ k_2^2 = -\beta^2, \text{ and } k_3^2 = \gamma^2$$

Equation (3.16) can now be written as

$$\frac{1}{X(x)}\frac{d^2X}{dx^2} = -\alpha^2, \quad \frac{1}{Y(y)}\frac{d^2Y}{dy^2} = -\beta^2$$
$$\frac{1}{Z(z)}\frac{d^2Z}{dz^2} = \gamma^2 \tag{3.18}$$

You can verify that the solutions of these equations are:

$$\begin{aligned} X(x) &= A_1 e^{i\alpha x} + A_2 e^{-i\alpha x} \\ Y(y) &= B_1 e^{i\beta y} + B_2 e^{-i\beta y} \\ Z(z) &= C_1 e^{\gamma z} + C_2 e^{-\gamma z} \end{aligned} \tag{3.19}$$

Clearly, the first two solutions are oscillatory while the last is exponential. The potential Φ is the product of these functions. The constants involved can be found from the boundary conditions to be satisfied by the potential.

It may be noted that (3.19) is one particular solution of equation (3.15). There may be other values of α, β, γ which could give valid solutions. The general solution, therefore, can be written as

$$\Phi(x, y, z) = \sum_{r,s,t}^{\infty} (A_1^r e^{i\alpha_r x} + A_1^r e^{-i\alpha_r x})$$
$$\times (B_1^s e^{i\beta_s y} + B_2^s e^{-i\beta_s y})(C_1^t e^{\gamma_t z} + C_2^t e^{-\gamma_t z}) \tag{3.20}$$

We shall now illustrate the use of this method by considering couple of examples.

EXAMPLE 3.1. You are given a set of wires parallel to the y-axis in the plane $z = 0$. The wires are equally spaced at $x = \pm n\left(\frac{a}{2}\right)$, where $n = 0,$

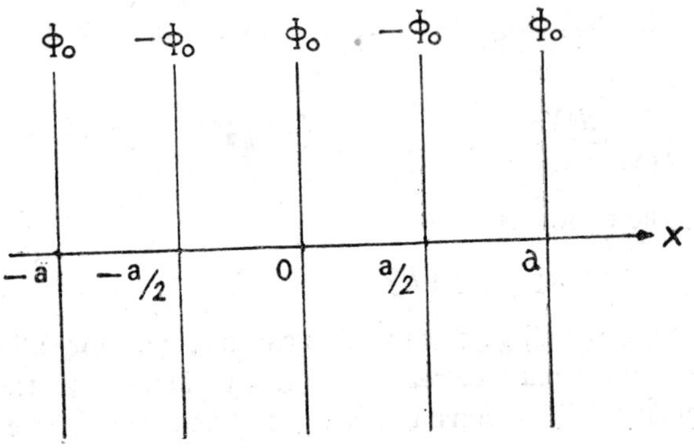

Fig. 3.1

1, 2, 3, ... Wires at odd positions (i.e. when n is odd) are at potential $-\Phi_0$; while those at even positions are at potential Φ_0. Find the potential at all points in space.

The potential Φ is independent of y and varies periodically with x. Since it is an even function of x, we represent it by a cosine series $\left(\cos\frac{2\pi kx}{a}\right)$. The functional dependence of Φ on z is not known; but from what we have said above, we expect it to be exponential. Thus, Φ is of the general form.

$$\Phi = \sum_{k=-\infty}^{\infty} f_k(z) \cos\frac{2\pi kx}{a}. \tag{3.21}$$

The form of $f(z)$ is yet to be determined. If this Φ is to be a valid potential, it must satisfy Laplace's equation $\frac{\partial^2\Phi}{\partial x^2} + \frac{\partial^2\Phi}{\partial z^2} = 0$ in the region above the wires where there are no charges.

$$\therefore \quad \sum_{k=-\infty}^{\infty} \left\{-\frac{4\pi^2k^2}{a^2} f_k(z)\cos\frac{2\pi kx}{a} + \frac{\partial^2 f_k(z)}{\partial z^2}\cos\frac{2\pi kx}{a}\right\} = 0.$$

This must be true for every value of k.

$$\therefore \quad -\frac{4\pi^2k^2}{a^2} f_k(z) + \frac{\partial^2 f_k(z)}{\partial z^2} = 0$$

Solving this equation, we have

$$f_k(z) = A_k e^{\pm\frac{2\pi k}{a}z}$$

where A_k is a constant.

Physics of the problem demands that for $z > 0$, we retain only the negative sign of the exponent.

$$\therefore \quad f_k(z) = A_k e^{-\frac{2\pi k}{a}z}$$

Hence,

$$\Phi(x, z) = \sum_k A_k e^{-\frac{2\pi k}{a}z} \cos\frac{2\pi kx}{a} \quad \text{for } z > 0 \tag{3.22}$$

At $z = 0$, Φ, as given by (3.22), must coincide with the prescribed potential and this requirement determines the values of A_k.

EXAMPLE 3.2 Consider a rectangular box as shown in Fig. 3.2. The dimensions of the box in the x, y, z directions are a, b, c respectively. The shaded surface is at potential Φ_0. All other surfaces are at zero potential. Find the potential at a point inside the box.

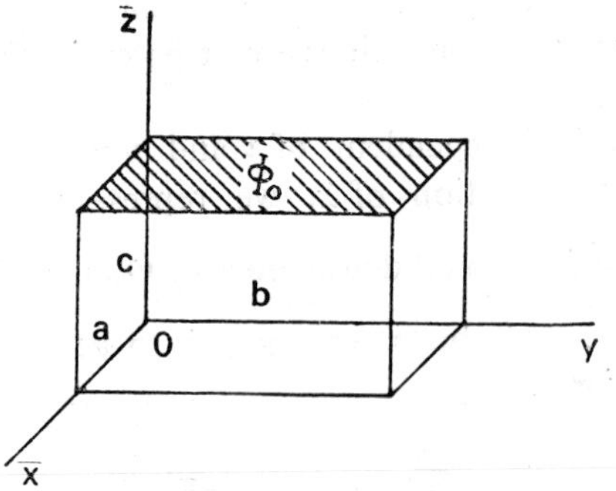

Fig. 3.2

Since the potential is zero at $x = 0$, $x = a$ and also at $y = 0$, $y = b$, the functions X, Y must be of the form

$$X = \sin \alpha x, \quad Y = \sin \beta y$$

$$\text{where } \alpha = \frac{m\pi}{a} \quad \text{and} \quad \beta = \frac{n\pi}{b} \quad \begin{cases} m = 1, 2, 3, \ldots \\ n = 1, 2, 3, \ldots \end{cases}$$

Further, because X and Y are oscillatory Z part of the potential must be exponential. The potential, therefore, must be of the form

$$\Phi = \sin \frac{m\pi}{a} x \sin \frac{n\pi}{b} y \sinh \gamma_{mn} z$$

$$\text{where } \gamma_{mn} = \sqrt{\alpha^2 + \beta^2} = \pi \sqrt{\frac{m^2}{a^2} + \frac{n^2}{b^2}} \quad \text{(see 3.17).}$$

The exponential factor has been expressed in terms of hyperbolic sine function.

$$\{\text{Note: } e^{\pm \gamma_{mn} z} = \cosh \gamma_{mn} z \pm \sinh \gamma_{mn} z\}$$

Hence, the general solution is,

$$\Phi = \sum_{n, m}^{\infty} A_{mn} \sin \frac{m\pi}{a} x \sin \frac{n\pi}{b} y \sinh \gamma_{mn} z \tag{3.23}$$

3.5 Laplace's Equation in Spherical Polar Coordinates

When the problem has axial symmetry, it is generally convenient to use spherical polar coordinates r, θ, ϕ, and to take the axis of symmetry as the polar axis $\theta = 0$.

Laplace's equation in spherical polar coordinates takes the form

$$\frac{1}{r^2} \frac{\partial}{\partial r} \left(r^2 \frac{\partial V}{\partial r} \right) + \frac{1}{r^2 \sin \theta} \frac{\partial}{\partial \theta} \left(\sin \theta \frac{\partial V}{\partial \theta} \right) + \frac{1}{r^2 \sin^2 \theta} \frac{\partial^2 V}{\partial \phi^2} = 0 \tag{3.24}$$

We have used symbol V for the potential in order to avoid confusion with the coordinate ϕ.

As in the case of rectangular coordinates we will use the method of separation of variables and search for a solution of the form

$$V = RS \tag{3.25}$$

where R is function of r only and S is a function of the angular coordinates θ and ϕ only. Substituting in (3.24) we have

$$\frac{S}{r^2}\frac{\partial}{\partial r}\left(r^2\frac{\partial R}{\partial r}\right)+\frac{R}{r^2\sin\theta}\frac{\partial}{\partial\theta}\left(\sin\theta\frac{\partial S}{\partial\theta}\right)+\frac{R}{r^2\sin^2\theta}\frac{\partial^2 S}{\partial\phi^2}=0 \tag{3.26}$$

Dividing throughout by RS/r^2 we get

$$\frac{1}{R}\frac{\partial}{\partial r}\left(r^2\frac{\partial R}{\partial r}\right)+\frac{1}{S\sin\theta}\frac{\partial}{\partial\theta}\left(\sin\theta\frac{\partial S}{\partial\theta}\right)+\frac{1}{S\sin^2\theta}\frac{\partial^2 S}{\partial\phi^2}=0 \tag{3.27}$$

We see that the first term is a function of r only; while the remaining two terms are independent of r. The equation is satisfied if we take

$$\frac{1}{R}\frac{\partial}{\partial r}\left(r^2\frac{\partial R}{\partial r}\right)=K \tag{3.28}$$

and

$$\frac{1}{S\sin\theta}\frac{\partial}{\partial\theta}\left(\sin\theta\frac{\partial S}{\partial\theta}\right)+\frac{1}{S\sin^2\theta}\frac{\partial^2 S}{\partial\phi^2}=-K \tag{3.29}$$

where K is a constant. The solutions of these equations take simpler forms, if we write the constant K as $l\,(l+1)$ where the constant l is still arbitrary. One can readily verify that the solution of (3.28) is

$$R = Ar^l + \frac{B}{r^{l+1}} \tag{3.30}$$

where A, B are arbitrary constants.

Equation (3.29) can now be written as

$$\frac{1}{\sin\theta}\frac{\partial}{\partial\theta}\left(\sin\theta\frac{\partial S}{\partial\theta}\right)+\frac{1}{\sin^2\theta}\frac{\partial^2 S}{\partial\phi^2}=-l(l+1)S \tag{3.31}$$

Any solution of this equation S_l is a function of θ and ϕ and is called a **surface harmonic** of degree 'l'. The solution of Laplace's equation (3.24), therefore, is

$$V = RS = \left(Ar^l + \frac{B}{r^{l+1}}\right)S_l \tag{3.32}$$

Any solution of Laplace's equation is called a **spherical harmonic.**

To solve (3.31) we adopt the same technique as used in solving (3.24) and write

$$S = P\,(\theta)\,Q\,(\phi) \tag{3.33}$$

where $P(\theta)$ is a function of θ only and $Q(\phi)$ is a function of ϕ only. Substituting in (3.31), we have,

$$\frac{Q}{\sin\theta}\frac{d}{d\theta}\left(\sin\theta\frac{dP}{d\theta}\right)+\frac{P}{\sin^2\theta}\frac{d^2Q}{d\phi^2}+l(l+1)\,PQ=0 \qquad (3.34)$$

Dividing throughout by $PQ/\sin^2\theta$ we get

$$\frac{\sin\theta}{P}\frac{d}{d\theta}\left(\sin\theta\frac{dP}{d\theta}\right)+l(l+1)\sin^2\theta+\frac{1}{Q}\frac{d^2Q}{d\phi^2}=0 \qquad (3.35)$$

We see that the variables are again separable. The first two terms are function of θ only and the last term is a function of ϕ only.

Let us put

$$\frac{1}{Q}\frac{d^2Q}{d\phi^2}=-m^2 \qquad (3.36)$$

where m^2 is a constant. The solution of this equation is

$$Q_m=Ce^{\pm im\phi} \qquad (3.37)$$

where C is a constant.

In order that the potential be single valued, it is necessary that

$$e^{\pm im\phi}=e^{\pm im(\phi+2\pi)} \qquad (3.38)$$

This is assured only if m is an integer.

The functions Q_m are normalized by choosing the constant C in such a way as to give $\int_0^{2\pi} Q_m^* Q_m d\phi = 1$. For this to be satisfied, C must be equal to $1/\sqrt{2\pi}$. The functions Q_m are also orthogonal

i.e. $$\int_0^{2\pi} Q_m^* Q_n d\phi=0 \qquad \text{if } m\neq n$$

$$\therefore \quad \int_0^{2\pi} Q_m^* Q_n d\phi=\delta_{mn} \qquad (3.39)$$

3.6 Legendre's Equation

Using (3.36), the θ part of the equation (3.35) becomes

$$\frac{\sin\theta}{P}\frac{d}{d\theta}\left(\sin\theta\frac{dP}{d\theta}\right)+l(l+1)\sin^2\theta=m^2$$

or dividing throughout by $\sin^2\theta/P$

$$\frac{1}{\sin\theta}\frac{d}{d\theta}\left(\sin\theta\frac{dP}{d\theta}\right)+l(l+1)\,P-\frac{m^2}{\sin^2\theta}\,P=0 \qquad (3.40)$$

We transform this equation by putting $x=\cos\theta$.

$$\left(\therefore \quad \frac{d}{d\theta}=\frac{dx}{d\theta}\frac{d}{dx}=-\sin\theta\frac{d}{dx}\right)$$

The equation changes to

$$\frac{d}{dx}\left[(1-x^2)\frac{dP}{dx}\right]+\left[l(l+1)-\frac{m^2}{1-x^2}\right]P=0 \qquad (3.41)$$

This is known as the generalized Legendre's equation. If we put $m=0$ we get the ordinary Legendre equation, viz.

$$\frac{d}{dx}\left[(1-x^2)\frac{dP}{dx}\right]+l(l+1)\,P=0$$

or

$$(1-x^2)\,P''-2x\,P'+l\,(l+1)\,P=0 \qquad (3.42)$$

We shall first consider the solution of (3.42). We shall try to obtain it by series integration. Let us assume the solution to be.

$$P\,(x)=a_0x^\lambda+a_1x^{1+\lambda}+a_2x^{2+\lambda}+\ldots=\sum_{k=0}^{\infty}a_kx^{k+\lambda} \qquad (3.43)$$

where the constants a_k and λ are yet to be determined.

Substituting this in (3.42), we have

$$\sum_{k=0}^{\infty}a_k\,(k+\lambda)\,(k+\lambda-1)\;x^{k+\lambda-2}-\sum_{k=0}^{\infty}a_k\Big[(k+\lambda)\,(k+\lambda-1)$$

$$+\,2\,(k+\lambda)-l(l+1)\Big]x^{k+\lambda}=0 \qquad (3.44)$$

This equation must hold for every value of x. Hence, the coefficient of every power of x must vanish. The lowest power of x is found from (3.44) (by putting $k=0$) to be $\lambda-2$. The coefficient of $x^{\lambda-2}$, therefore, must be, zero, i.e.

$$a_0\,\lambda\,(\lambda-1)=0 \qquad (3.45)$$

This equation is known as the **indicial equation**. Since $a_0\neq 0$ ($\because$ it is assumed to be the coefficient of lowest power), either

$$\lambda=0 \quad \text{or} \quad \lambda=1$$

The coefficient of $x^{j+\lambda}$ where j is any value of k, can be found from (3.44) by putting $k=j+2$ in the first term and $k=j$ in the second. This is

$$a_{j+2}\,(j+\lambda+2)\,(j+\lambda+1)-a_j\,[(j+\lambda)\,(j+\lambda+1)-l(l+1)]$$

and it must be zero.

$$\therefore\quad a_{j+2}=\frac{(j+\lambda)\,(j+\lambda+1)-l\,(l+1)}{(j+\lambda+2)\,(j+\lambda+1)}\,a_j \qquad (3.46)$$

Thus, if a_j is known, a_{j+2} can be found. Starting from an arbitrary value of a_0, we can compute a_2, a_4, ... etc. On the other hand, if we assign any arbitrary value to a_1, we can compute a_3, a_5 ... etc.

For this computation we take $\lambda = 0$, since this provides all the necessary independent solutions.

$$\therefore \quad a_{j+2} = \frac{j(j+1) - l(l+1)}{(j+2)(j+1)} a_j \tag{3.47}$$

$$\left.\begin{aligned}
a_2 &= -\frac{l(l+1)}{2!} a_0 \\
a_4 &= \frac{6 - l(l+1)}{4.3} a_2 = -\frac{6 - l(l+1)}{4.3} \frac{l(l+1)}{2!} a_0 \\
&= \frac{(l-2)l(l+1)(l+3)}{4!} a_0 \\
a_6 &= -\frac{(l-4)(l-2)l(l+1)(l+3)(l+5)}{6!} a_0 \\
&\cdots\cdots\cdots\cdots\cdots\cdots \\
a_3 &= \frac{2 - l(l+1)}{3.2} a_1 = -\frac{(l-1)(l+2)}{3!} a_1 \\
a_5 &= \frac{(l-3)(l-1)(l+2)(l+4)}{5!} a_1 \\
&\cdots\cdots\cdots\cdots\cdots\cdots
\end{aligned}\right\} \tag{3.48}$$

Hence,

$$P_1 = a_0 \left\{ 1 - \frac{l(l+1)}{2!} x^2 + \frac{(l-2)l(l+1)(l+3)}{4!} x^4 \ldots\ldots \right\} \tag{3.49}$$

and

$$P_2 = a_1 \left\{ x - \frac{(l-1)(l+2)}{3!} x^3 + \frac{(l-3)(l-1)(l+2)(l+4)}{5!} x^5 \ldots\ldots\ldots \right\} \tag{3.50}$$

are both solutions of the Legendre's equation (3.42). The general solution being

$$W = AP_1 + BP_2 \tag{3.51}$$

We have found the solutions of Legendre's equation, but whether the solutions are of any interest depends on their converging properties. A series converges if the ratio of two successive terms, i.e. $\frac{a_{j+2}}{a_j} x^2$ is smaller than unity for large j.

We see from (3.47) that the ratio

$$\frac{a_{j+2}}{a_j} = 1 \text{ as } j \to \infty$$

Hence, the series will converge if $x^2 < 1$, i.e. if the values of x lie in the range $-1 < x < 1$. Since x in our equation stands for $\cos\theta$, its values range from $+1$ to -1. However, if $x = \pm 1$, the series diverges and the solution becomes unacceptable unless the series terminates and becomes a polynomial. Inspection of (3.49) and (3.50) will show that P_1 terminates for even values of 1, and P_2 for odd values of 1. The solutions then are just polynomials with x^l as the highest power. It is customary to normalize these polynomials to have the value unity at $x = 1$ and are then called Legendre polynomials of order 1. These polynomials are represented by the symbol $P_l(x)$. Normalization is brought about by taking

$$a_0 = \frac{1}{1 - \frac{l(l+1)}{2\,!} + \ldots \text{up to the cofficient of the highest power of } x} \tag{3.52}$$

$$a_1 = \frac{1}{1 - \frac{(l-1)(l+2)}{3\,!} + \ldots \text{up to the coefficient of the highest power of } x} \tag{3.53}$$

Thus, from (3.49) and (3.50), we have

$$\left.\begin{aligned}
&\text{for } l = 0,\ a_0 = 1 \text{ and } P_0(x) = 1\\
&\text{for } l = 1,\ a_1 = 1 \text{ and } P_1(x) = x\\
&\text{for } l = 2,\ a_0 = -\tfrac{1}{2} \text{ and } P_2(x) = \tfrac{1}{2}(3x^2 - 1)\\
&\text{for } l = 3,\ a_1 = -\tfrac{3}{2} \text{ and } P_3(x) = \tfrac{1}{2}(5x^3 - 3x)
\end{aligned}\right\} \tag{3.54}$$

and so on.

The general formula for $P_l(x)$ is

$$P_l(x) = \sum_{r=0}^{N} \frac{(-1)^r (2l - 2r)\,!}{2^l r!\,(l - r)!(l - 2r)!}\, x^{l-2r} \tag{3.55}$$

where $N = \frac{l}{2}$ if l is even and $N = \frac{l-1}{2}$ if l is odd.

A simpler representation of Legendre polynomials is given by Rodrigue's formula, viz.

$$P_l(x) = \frac{1}{2^l l\,!} \frac{d^l}{dx^l}(x^2 - 1)^l \tag{3.56}$$

You can satisfy yourself that this formula yields the same expression for $P_l(x)$ as the one obtained from (3.55) for a given value of l.

There is yet another way of expressing the Legendre polynomials. If we expand $(1-2xs+s^2)^{-1/2}$ by Maclaurin's theorem, we find that the coefficients of various powers of s are Legendre polynomials, i.e.

$$(1-2xs+s^2)^{-1/2} = \sum_{l=0}^{\infty} P_l(x)\, s^l \tag{3.57}$$

The function $(1-2xs+s^2)^{-1/2}$ is called **the generating function** of Legendre polynomials.

We will now show that the Legendre polynomials form a complete orthogonal set of functions.

We write the Legendre equation in the following form

$$\frac{d}{dx}\left[(1-x^2)P_l'(x)\right] + l(l+1)\, P_l(x) = 0. \tag{3.58}$$

Multiplying by $P_q(x)$ and integrating over the interval $(1, -1)$, we have,

$$\int_{-1}^{1} P_q(x)\frac{d}{dx}\left[(1-x^2)\, P_l'(x)\right]dx + l(l+1)\int_{-1}^{1} P_q(x)\, P_l(x)dx = 0$$

$$\therefore \quad \left[P_q(x)\left\{(1-x^2)\, P_l'(x)\right\}\right]_{-1}^{1} - \int_{-1}^{1} (1-x^2)\, P_l'(x)\, P_q'(x)\, dx$$

$$+\, l(l+1)\int_{-1}^{1} P_q(x)P_l(x)dx = 0$$

The first term vanishes at both limits

$$\therefore \quad -\int_{-1}^{1}(1-x^2)P_l'(x)P_q'(x)dx + l(l+1)\int_{-1}^{1} P_q(x)P_l(x)dx = 0 \tag{3.59}$$

Interchanging l and q we obtain

$$-\int_{-1}^{1}(1-x^2)\, P_q'(x)P_l'(x)dx + q(q+1)\int_{-1}^{1} P_l(x)P_q(x)dx = 0 \tag{3.60}$$

Subtracting (3.60) from (3.59) we have

$$\left\{l(l+1) - q(q+1)\right\}\int_{-1}^{1} P_l(x)P_q(x)\, dx = 0 \tag{3.61}$$

If $l \neq q$,

$$\int_{-1}^{1} P_l(x)P_q(x)dx = 0 \tag{3.62}$$

showing that the Legendre polynomials of different order are orthogonal.

If $l = q$, the integral is finite. The value of $\int_{-1}^{1}[P_l(x)]^2dx$ can be found easily by using the generating function $(1-2xs+s^2)^{-1/2}$.

Thus,

$$(1-2xs+s^2)^{-1} = \left[\sum_{l=0}^{\infty} P_l(x)s^l\right]^2 \tag{3.63}$$

$$\therefore \quad \int_{-1}^{1} (1 - 2xs + s^2)^{-1}\, dx = \int_{-1}^{1} \left[\sum_{l=0}^{\infty} P_l(x) s^l\right]^2 dx.$$

Integrating the left-hand side

$$\frac{1}{s} \ln \frac{1+s}{1-s} = \int_{-1}^{1} \left[\sum_{l=0}^{\infty} s^l P_l(x)\right]^2 dx$$

Since the product terms in the summation on the right hand side vanish because of the orthogonality condition

$$\frac{1}{s} \ln \frac{1+s}{1-s} = \sum_{l=0} s^{2l} \int_{-1}^{1} [P_l(x)]^2\, dx \tag{3.64}$$

Expanding the left-hand side and equating the coefficients of the power s^{2l} we have

$$\int_{-1}^{1} \left[P_l(x)\right]^2 dx = \frac{2}{2l+1}, \quad l = 0, 1, 2, \ldots \tag{3.65}$$

The orthogonality condition, therefore, can be written as

$$\int_{-1}^{1} P_l(x)\, P_q(x)\, dx = \frac{2}{2l+1}\, \delta_{ql} \tag{3.66}$$

and orthonormal Legendre's functions are

$$\sqrt{\frac{2l+1}{2}}\; P_l(x)$$

Hence, when $m = 0$, i.e. in the problems in which there is azimuthal symmetry , the solution of Laplace's equation is

$$\begin{aligned} V &= R(r)\, S(\theta) \\ &= \sum_{l=0}^{\infty} \left(A_l r^l + B_l r^{-(l+1)}\right) \sqrt{\frac{2l+1}{2}}\; P_l(\cos\theta) \end{aligned} \tag{3.67}$$

Since the Legendre polynomials form a complete set of orthogonal functions, any function $f(x)$ in the interval $-1 \leqslant x \leqslant 1$ can be expressed in terms of Legendre polynomials. That is,

$$\begin{aligned} f(x) &= C_0 + C_1 P_1(x) + C_2 P_2(x) + \ldots.. \\ &= \sum_{l=0}^{\infty} C_l P_l(x) \end{aligned} \tag{3.68}$$

In order to find the coefficients C_l, we multiply both sides by $P_m(x)$ and integrate

$$\int_{-1}^{1} f(x) P_m(x) dx = \sum_{l=0}^{\infty} C_l \int_{-1}^{1} P_l(x) P_m(x) dx = \frac{2C_m}{2m+1}$$

$$\therefore \quad C_m = \frac{2m+1}{2}\int_{-1}^{1} f(x)\, P_m(x)dx \tag{3.69}$$

This property has been found to be useful in solving some problems.

3.7 Associated Legendre Functions

Let us now discuss the solution of Laplace's equation when the potential problem has azimuthal variation i.e. $m \neq 0$.

We know that $y = P_l(x)$ is a solution of the equation

$$(1 - x^2)\frac{d^2y}{dx^2} - 2x\frac{dy}{dx} + l(l+1)\,y = 0 \tag{3.70}$$

Differentiating m times, we have

$$(1 - x^2)\frac{d^{m+2}y}{dx^{m+2}} - 2x(m+1)\frac{d^{m+1}y}{dx^{m+1}} + (l-m)(l+m+1)\frac{d^m y}{dx^m} = 0 \tag{3.71}$$

Put

$$v = \frac{d^m y}{dx^m} = \frac{d^m P_l(x)}{dx^m}. \tag{3.72}$$

The equation (3.71) changes to

$$(1 - x^2)\frac{d^2v}{dx^2} - 2x\,(m+1)\frac{dv}{dx} + (l-m)(l+m+1)\,v = 0 \tag{3.73}$$

This equation is clearly satisfied by

$$v = \frac{d^m P_l(x)}{dx^m}$$

If we make a further substitution

$$w = v(1 - x^2)^{m/2} \tag{3.74}$$

we obtain,

$$(1 - x)^2\frac{d^2w}{dx^2} - 2x\frac{dw}{dx} + \left[l(l+1) - \frac{m^2}{1 - x^2}\right]w = 0 \tag{3.75}$$

which is the same as our generalized Legendre equation (3.41) and its solution is

$$w = v\,(1 - x^2)^{m/2} = (1 - x^2)^{m/2}\frac{d^m P_l(x)}{dx^m} = P_l^m(x). \tag{3.76}$$

We represent it by the symbol $P_l^m(x)$.

The functions $P_l^m(x)$ defined as above are known as **associated Legendre polynomials.**

The orthogonality condition for associated Legendre polynomials is

$$\int_{-1}^{1} P_l^m(x)\, P_q^m(x)dx = \frac{2}{2l+1}\frac{(l+m)!}{(l-m)!}\,\delta_{ql} \tag{3.77}$$

The normalized surface harmonics, therefore, are given by

$$S_{lm}(\theta, \varphi) = \sqrt{\frac{(2l+1)(l-m)!}{4\pi(l+m)!}}\, P_l^m (\cos\theta)\, e^{im\varphi} \qquad (3.78)$$

The general solution of Laplace's equation can, now, be written as

$$V(r, \theta, \varphi) = \sum_{l=0}^{\infty} \sum_{m=-l}^{l} \left[A_{lm} r^l + B_{lm} r^{-(l+1)} \right] S_{lm}(\theta, \varphi) \qquad (3.79)$$

Some spherical harmonic functions are given in Table 3.1.

Table 3.1

$S_{1,0}$	$= +\left(\frac{3}{4\pi}\right)^{\frac{1}{2}} \cos\theta$
$S_{1,1}$	$= -\left(\frac{3}{8\pi}\right)^{\frac{1}{2}} \sin\theta\, e^{i\varphi}$
$S_{1,-1}$	$= +\left(\frac{3}{8\pi}\right)^{\frac{1}{2}} \sin\theta\, e^{-i\varphi}$
$S_{2,0}$	$= +\left(\frac{5}{16\pi}\right)^{\frac{1}{2}} (3\cos^2\theta - 1)$
$S_{2,1}$	$= -\left(\frac{15}{8\pi}\right)^{\frac{1}{2}} \cos\theta \sin\theta\, e^{-i\varphi}$

Are you bored? You, perhaps, think the treatment given in these few sections to be a mere hocus-focus, which has little to do with physics, which is otherwise so interesting. But do not give up! Take heart! Let us solve a few examples and you will see how the formulae given above are handy and readily yield the solutions of some complicated problems in physics. Before we take up these examples you have to note the following traits of the trade.

(i) If the region of interest includes the entire space inside some sphere, the terms that go to infinity at the origin must be removed i.e. all B_{lm} must be taken to be zero. This is because we are interested in a finite solution.

(ii) If the region does not include the origin, B_{lm} must be retained.

(iii) If we are interested in the spherically symmetrical boundary conditions, only terms with $m = 0$ need be considered.

EXAMPLE 3.3. Find the potential of a spherical cap of angle α charged to a uniform surface density σ_0 (Fig. 3.3).

The spherical cap may be considered as the surface cut from a sphere by a right circular cone of semivertical angle α.

In order to obtain the solution of this problem, we regard the entire sphere as charged to a surface density σ where

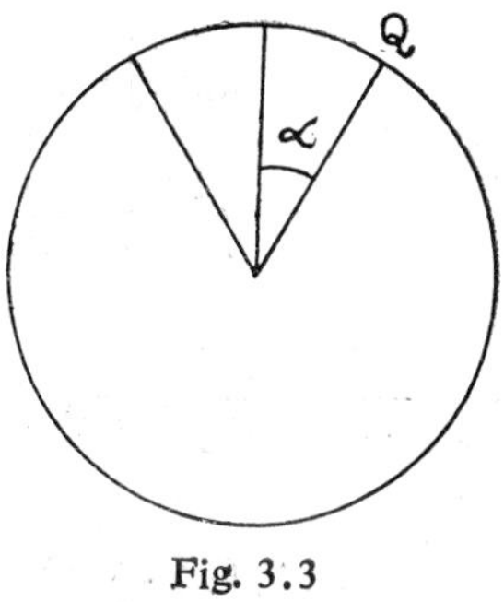

Fig. 3.3

$$\sigma = \sigma_0 \quad \text{from } \theta = 0 \quad \text{to} \quad \theta = \alpha$$
$$= 0 \quad \text{from } \theta = \alpha \quad \text{to} \quad \theta = \pi$$

Since the problem has azimuthal symmetry, we express σ in terms of harmonics

$$\sigma = a_0 + a_1 P_1 (\cos \theta) + a_2 P_2 (\cos \theta) + \dots \text{(see 3.68)} \tag{3.80}$$

where

$$a_l = \frac{2l+1}{2} \int_{\theta=\pi}^{\theta=0} \sigma P_l (\cos \theta)\, d(\cos \theta)$$

$$= \frac{2l+1}{2} \int_{\theta=\pi}^{\theta=\alpha} \sigma P_l (\cos \theta)\, d(\cos \theta) + \frac{2l+1}{2} \int_{\theta=\alpha}^{\theta=0} \sigma P_l (\cos \theta) d(\cos \theta)$$

$$= \frac{2l+1}{2} \sigma_0 \int_{\theta=\alpha}^{\theta=0} P_l(\cos \theta)\, d(\cos \theta) \quad (\because \sigma = 0 \text{ from } \theta = \pi \text{ to } \theta = \alpha)$$

$$\therefore \quad a_0 = \tfrac{1}{2} \sigma_0 \int_{\theta=\alpha}^{\theta=0} d(\cos \theta) = \frac{1}{2} \sigma_0 (1 - \cos \alpha)$$

It can be shown that

$$\int_{\theta=\alpha}^{\theta=0} P_l (\cos \theta)\, d(\cos \theta)$$

$$= \left[\frac{P_{l+1} (\cos \theta) - P_{l-1} (\cos \theta)}{2l+1} \right]_{\theta=\alpha}^{\theta=0} \quad \text{(Problem 3.1)}$$

$$\therefore \quad \sigma = \tfrac{1}{2} \sigma_0 \Big[(1 - \cos \alpha)$$
$$+ \sum_{l=1}^{\infty} \Big\{ P_{l-1} (\cos \alpha) - P_{l+1} (\cos \alpha) \Big\} P_l (\cos \theta) \Big] \tag{3.81}$$

Now the potential at a point P on the polar axis, distant r from the centre is given by

$$V_P = \frac{1}{4\pi\epsilon_0} \iint \frac{\sigma dS}{PQ} = \frac{1}{4\pi\epsilon_0} \iint \frac{\sigma dS}{(a^2 + r^2 - 2ar \cos \theta)^{1/2}} \tag{3.82}$$

where dS is an element of area at Q on the surface of the sphere and 'a' is the radius of the sphere. If $r < a$

$$V_P = \frac{1}{4\pi\epsilon_0} \iint \frac{\sigma}{a} \left(1 + \frac{r^2}{a^2} - \frac{2r}{a} \cos \theta \right)^{-1/2} dS$$

$$= \frac{1}{4\pi\epsilon_0} \iint \frac{\sigma}{a} \sum_{n=0}^{\infty} P_n (\cos \theta) \left(\frac{r}{a} \right)^n dS \quad \text{by (3.57)}$$

$$= \frac{1}{4\pi\epsilon_0} \int \frac{\sigma_0}{2a} \Big[(1 - \cos \alpha) + \sum_{l=1}^{\infty} \Big\{ P_{l-1} (\cos \alpha) - P_{l+1} (\cos \alpha) \Big\} P_l (\cos \theta) \Big]$$
$$\times \left[1 + P_1 (\cos \theta) \left(\frac{r}{a} \right) + P_2 (\cos \theta) \left(\frac{r}{a} \right)^2 + \dots \right] 2\pi a^2 \sin \theta \, d\theta$$

$$= \frac{a\sigma_0}{4\epsilon_0}\int_{-1}^{1}\left[(1-\cos\alpha)+\sum_{l=1}^{\infty}\left\{P_{l-1}(\cos\alpha)-P_{l+1}(\cos\alpha)\right\}P_l(\cos\theta)\right]$$

$$\times\left[1+P_1(\cos\theta)\left(\frac{r}{a}\right)+P_2(\cos\theta)\left(\frac{r}{a}\right)^2+\ldots\right]d(\cos\theta)$$

$$= \frac{a\sigma_0}{2\epsilon_0}\left[(1-\cos\alpha)+\sum_{n=1}^{\infty}\frac{P_{n-1}(\cos\alpha)-P_{n+1}(\cos\alpha)}{2n+1}\left(\frac{r}{a}\right)^n\right]$$

where we have used (3.66). Hence, the potential at a point (r, θ) is given by

$$V(r,\theta) = \frac{a\sigma_0}{2\epsilon_0}\left[(1-\cos\alpha)+\sum_{n=1}^{\infty}\frac{P_{n-1}(\cos\alpha)-P_{n+1}(\cos\alpha)}{2n+1}\left(\frac{r}{a}\right)^n \times P_n(\cos\theta)\right] \quad (3.83)$$

If $r > a$

$$V_P = \frac{a\sigma_0}{2\epsilon_0}\left[(1-\cos\alpha)\left(\frac{a}{r}\right)+\sum_{n=1}^{\infty}\frac{P_{n-1}(\cos\alpha)-P_{n+1}(\cos\alpha)}{2n+1}\left(\frac{a}{r}\right)^{n+1}P_n(\cos\theta)\right] \quad (3.84)$$

EXAMPLE 3.4. A conducting sphere of radius 'a' is made up of two hemispheres separated by a small insulating ring. The potential on the surface of the upper hemisphere is $+V$; while that on the surface of the lower hemisphere is $-V$. Find the potential at a point inside the sphere.

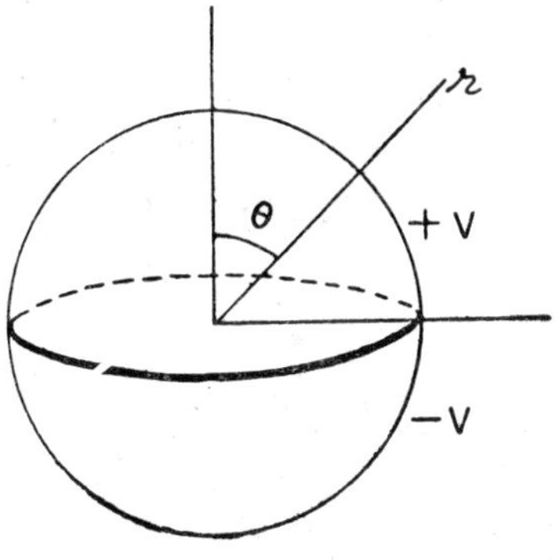

Fig. 3.4

Since the problem has azimuthal symmetry, $m = 0$ and the potential is given by

$$V(r,\theta) = \sum_{l=0}^{\infty}\left[A_l r^l + \frac{B_l}{r^{l+1}}\right]P_l(\cos\theta) \quad (3.85)$$

The potential must be finite at the origin, if there are no charges there. Therefore, the second term in the bracket, which gives infinite potential at the origin, must be removed i.e. $B_l = 0$ for all l. Hence

$$V(r,\theta) = \sum_{l=0}^{\infty} A_l r^l P_l(\cos\theta) \quad (3.86)$$

The potential on the surface of the sphere $(r = a)$ is

$$V(a,\theta) = \sum_{l=0}^{\infty} A_l a^l P_l(\cos\theta) \quad (3.87)$$

The coefficients A_l can be found using (3.69)

$$A_l a^l = \frac{2l+1}{2}\int_0^{\pi} V(a, \theta)\, P_l\,(\cos\theta)\sin\theta\, d\theta \tag{3.88}$$

$$= \frac{2l+1}{2}\left[\int_0^{\pi/2} V(a,\theta)P_l(\cos\theta)\sin\theta\, d\theta + \int_{\pi/2}^{\pi} V(a,\theta)P_l(\cos\theta)\sin\theta\, d\theta\right].$$

Now $V(\theta)$ is given by

$$V(\theta) = \begin{cases} +V & \text{for} \quad 0 \leqslant \theta \leqslant \pi/2 \\ -V & \text{for} \quad \pi/2 \leqslant \theta \leqslant \pi \end{cases} \tag{3.89}$$

Substituting x for $\cos\theta$, we have

$$A_l a^l = \frac{2l+1}{2}\left[-\int_{-1}^{0} V\, P_l(x)dx + \int_0^1 VP_l(x)dx\right] \tag{3.90}$$

$$\therefore \quad A_0 = 0$$

$$A_1 = \frac{3}{2a}\left\{-V\left(\frac{x^2}{2}\right)_{-1}^{0} + V\left(\frac{x^2}{2}\right)_0^1\right\} = \frac{3V}{2a} \tag{3.91}$$

$$A_2 = 0$$

$$A_3 = \frac{7}{2a^3}\left\{-V\left(\frac{5x^4}{8} - \frac{3x^2}{4}\right)_{-1}^{0} + V\left(\frac{5x^4}{8} - \frac{3x^2}{4}\right)_0^1\right\} = -\frac{7V}{8a^3}$$

etc.

Notice that all even numbered coefficients vanish.

The potential inside the sphere is

$$V(r, \theta) = A_1 r P_1\,(\cos\theta) + A_3 r^3 P_3\,(\cos\theta) + \ldots$$

$$= \frac{3Vr}{2a} P_1\,(\cos\theta) - \frac{7V}{8a^3} r^3 P_3\,(\cos\theta) + \ldots$$

EXAMPLE 3.5. Find the potential at all points in space exterior to a conducting sphere of radius 'a' placed in a uniform electric field E_0.

If we choose our axis in such a way that the polar axis coincides with the direction of the field, the problem has azimuthal symmetry, the symmetry axis being, say z-axis, with $\theta = 0$. The potential on the symmetry axis is given by

$$V(r, \theta) = \sum_{l=0}^{\infty}\left[A_l r^l + \frac{B_l}{r^{l+1}}\right] \qquad (\because \quad P_l\,(\cos\theta) = 1) \tag{3.92}$$

with $r = z$. If this potential is evaluated by some means, at an arbitrary point and the potential function is expanded in a power series as in (3.92), then the potential at any point in space is obtained by multiplying each power of r^l and of $\dfrac{1}{r^{l+1}}$ by $P_l\,(\cos\theta)$.

A uniform field E_0 in the z-direction is related to its potential by

$$E_0 = -\nabla\Phi_0$$

$$\therefore \qquad \Phi_0 = -E_0 z = -E_0 r\cos\theta = -E_0 r\, P_1\,(\cos\theta) \tag{3.93}$$

You will, perhaps, object to this definition of Φ_0 since it does not satisfy the usual condition that $\Phi = 0$ as $r \to \infty$. Note, however, that we are assuming uniform field of infinite extent, the source of which must lie at infinity and, hence, Φ will not vanish as $r \to \infty$.

The field immediately around the sphere will be distorted owing to the induced charges on the surface; but at large distances it will be equal to E (Fig. 3.5). The total potential at a point $V(r, \theta)$ is

$V(r_1 \theta) = \Phi_0$ + the potential due to the charges induced in the conducting sphere (3.94).

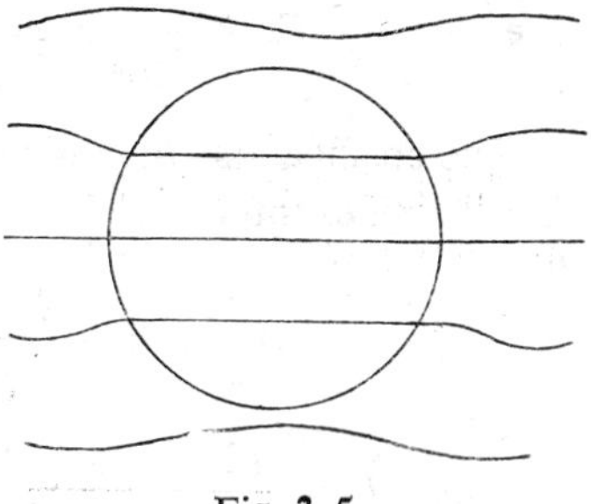

Fig. 3.5

At very large values of r, however, the potential due to the induced charges may be insignificant and

$$V(r, \theta) = \Phi_0.$$

Now

$$V(r, \theta) = \sum_{l=0}^{\infty} \left(A_l r^l + \frac{B_l}{r^{l+1}} \right) P_l (\cos \theta) \tag{3.95}$$

At large values of r, the terms in (3.95) involving B_l may be neglected and, hence,

$$V(r, \theta) = \sum_{l=0}^{\infty} A_l r^l P_l (\cos \theta)$$

$$= A_0 P_0 (\cos \theta) + A_1 r P_1 (\cos \theta) + A_2 r^2 P_2 (\cos \theta) + \ldots \tag{3.96}$$

Since this must be identical with Φ_0 as given by (3.93)

$$A_1 r P_1 (\cos \theta) = - E_0 r P_1 (\cos \theta)$$

Hence, $A_1 = - E_0$ and all other A_l vanish. Therefore, from (3.95)

$$V(r, \theta) = A_1 r P_1 (\cos \theta) + \sum_{l=0}^{\infty} \frac{B_l}{r^{l+1}} P_l (\cos \theta)$$

$$= - E_0 r P_1 (\cos \theta) + \sum_{l=0}^{\infty} \frac{B_l}{r^{l+1}} P_l (\cos \theta) \tag{3.97}$$

The sphere placed in the uniform field being a conductor, the induced charges generate an electric field within it, which cancels the external field giving zero field within the space and, hence, the potential on the surface $(r = a)$ is zero.

$$\therefore \quad V(a, \theta) = -E_0 a P_1(\cos\theta) + \sum_{l=0}^{\infty} B_l a^{-(l+1)} P_l(\cos\theta) = 0$$

i.e.
$$E_0 a P_1(\cos\theta) = \sum_{l=0}^{\infty} B_l a^{-(l+1)} P_l(\cos\theta) \tag{3.98}$$

The coefficients B_l can be found by the procedure adopted in the preceding example

$$B_l a^{-(l+1)} = \frac{2l+1}{2} \int_{-1}^{1} E_0 a P_1(x) P_l(x)\, dx$$

$$\therefore \quad B_l = \frac{(2l+1)a^{l+2}}{2} E_0 \int_{-1}^{1} P_l(x) P_1(x)\, dx$$

$$= \frac{(2l+1)\, a^{l+2}}{2} E_0 \frac{2}{2l+1} \delta_{1l} \text{ (by 3.66)}$$

$$\therefore \quad B_1 = E_0 a^3;\ B_2 = B_3 \ldots = 0 \tag{3.99}$$

Hence,

$$V(r, \theta) = -E_0 r P_1(\cos\theta) + \frac{E_0 a^3}{r^2} P_1(\cos\theta)$$

$$= -E_0 r \cos\theta \left(1 - \frac{a^3}{r^3}\right) \tag{3.100}$$

The field components can be found from this using the relations

$$E_r = -\frac{\partial V}{\partial r},\ E_\theta = -\frac{1}{r}\frac{\partial V}{\partial \theta}$$

EXAMPLE 3.6. Find the potential at any point in space due to a charge q uniformly distributed around a circular ring of radius a, its axis being the z-axis and its centres at $z = b$.

The potential at a point P on the axis of symmetry at $z = r$ is

$$V(z = r) = \frac{q}{AP} = \frac{q}{(r^2 + c^2 - 2rc\cos\alpha)^{1/2}} \tag{3.101}$$

where A is any point on the ring; its distance from the origin O is c, and α is the angle AO makes with the z-axis.

If $r > c$

$$(r^2 - 2rc\cos\alpha + c^2)^{-1/2} = r^{-1}\left(1 - \frac{2c\cos\alpha}{r} + \frac{c^2}{r^2}\right)^{-1/2}$$

$$= r^{-1} \sum_{l=0}^{\infty} P_l(\cos\alpha) \left(\frac{c}{r}\right)^l \quad \text{(by 3.57)}$$

$$= \sum_{l=0}^{\infty} \frac{c^l}{r^{l+1}} P_l(\cos\alpha)$$

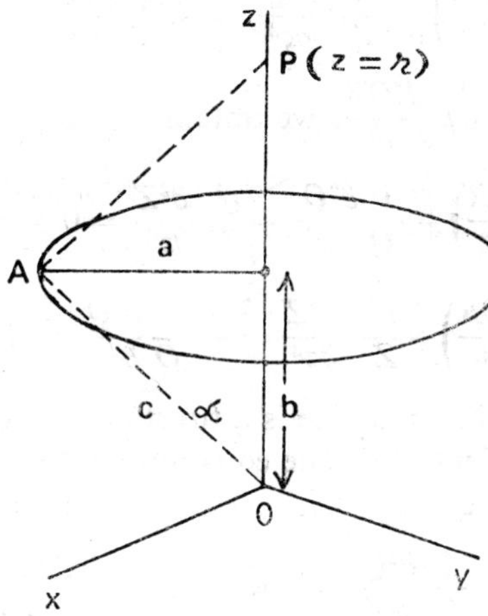

Fig. 3.6

Hence,

$$V(z = r) = q \sum_{l=0}^{\infty} \frac{c^l}{r^{l+1}} P_l (\cos \alpha) \tag{3.102}$$

and potential at any point in space

$$V(r, \theta) = q \sum_{l=0}^{\infty} \frac{c^l}{r^{l+1}} P_l (\cos \alpha) P_l (\cos \theta) \tag{3.103}$$

If $r < c$,

$$V(r, \theta) = q \sum_{l=0}^{\infty} \frac{r^l}{c^{l+1}} P_l (\cos \alpha) P_l (\cos \theta) \tag{3.104}$$

3.8 Laplace's Equation in Cylindrical Coordinates

Laplace's equation takes the following form in cylindrical coordinates

$$\frac{1}{r}\frac{\partial}{\partial r}\left(r\frac{\partial V}{\partial r}\right) + \frac{1}{r^2}\frac{\partial^2 V}{\partial \phi^2} + \frac{\partial^2 V}{\partial z^2} = 0 \tag{3.105}$$

We use once again the method of separation of variables and write

$$V(r, \phi, z) = R(r)\, Q(\phi)\, Z(z) \tag{3.106}$$

where R is a function of r only

Q is a function of ϕ only

and Z is a function of z only

Substituting in (3.105) we get

$$\frac{QZ}{r}\frac{\partial}{\partial r}\left(r\frac{\partial R}{\partial r}\right)+\frac{RZ}{r^2}\frac{\partial^2 Q}{\partial \phi^2}+RQ\frac{\partial^2 Z}{\partial z^2}=0 \qquad (3.107)$$

Dividing throughout by $\dfrac{RQZ}{r^2}$, we obtain

$$\frac{r}{R}\frac{\partial}{\partial r}\left(r\frac{\partial R}{\partial r}\right)+\frac{1}{Q}\frac{\partial^2 Q}{\partial \phi^2}+\frac{r^2}{Z}\frac{\partial^2 Z}{\partial z^2}=0$$

i.e.

$$\frac{r}{R}\frac{\partial}{\partial r}\left(r\frac{\partial R}{\partial r}\right)+\frac{r^2}{Z}\frac{\partial^2 Z}{\partial z^2}=-\frac{1}{Q}\frac{\partial^2 Q}{\partial \phi^2} \qquad (3.108)$$

We see that the right-hand side is a function of ϕ only; while the left-hand side is independent of ϕ. The equation is satisfied if each side is equal to a constant. Let

$$\frac{1}{Q}\frac{\partial^2 Q}{\partial \phi^2}=-\nu^2, \qquad (3.109)$$

where ν is a constant. The solution of this equation is

$$Q(\phi)=e^{\pm i\nu\phi} \qquad (3.110)$$

To ensure that $Q(\phi)$ is single valued, ν must take only integral values (see 3.38).

The left-hand side of equation (3.108) now becomes

$$\frac{r}{R}\frac{\partial}{\partial r}\left(r\frac{\partial R}{\partial r}\right)+\frac{r^2}{Z}\frac{\partial^2 Z}{\partial z^2}=\nu^2.$$

Dividing throughout by r^2 and transposing, we have

$$\frac{1}{rR}\frac{\partial}{\partial r}\left(r\frac{\partial R}{\partial r}\right)-\frac{\nu^2}{r^2}=-\frac{1}{Z}\frac{\partial^2 Z}{\partial z^2}. \qquad (3.111)$$

As before we put

$$\frac{1}{Z}\frac{\partial^2 Z}{\partial z^2}=k^2 \qquad (3.112)$$

where k is a constant.

Notice, however, that while in equation (3.109) the constant is prefixed with a negative sign, in (3.112) it is prefixed with a positive sign. You need not be disturbed by this; it is only a matter of convenience. The solution of (3.112) is

$$Z(z)=e^{\pm kz}. \qquad (3.113)$$

The 'r' part of the equation now becomes,

$$\frac{1}{rR}\frac{\partial}{\partial r}\left(r\frac{\partial R}{\partial r}\right)-\frac{\nu^2}{r^2}=-k^2 \qquad (3.114)$$

or

$$\frac{r}{R}\frac{\partial}{\partial r}\left(r\frac{\partial R}{\partial r}\right)+(k^2r^2-\nu^2)=0 \qquad (3.115)$$

If we put $x = kr$, the equation changes to

$$\frac{\partial^2 R}{\partial x^2} + \frac{1}{x}\frac{\partial R}{\partial x} + \left(1 - \frac{\nu^2}{x^2}\right) R = 0 \tag{3.116}$$

This is known as **Bessel's equation.** As in the case of Legendre equation, we will try to find its solution in the form

$$R(x) = \sum_{\lambda=0}^{\infty} a_\lambda\, x^{k+\lambda}. \tag{3.117}$$

Substituting in (3.116), we have

$$\left[\sum_{\lambda=0}^{\infty} a_\lambda\,(k+\lambda)(k+\lambda-1) + \sum_{\lambda=0}^{\infty} a_\lambda\,(k+\lambda) - \nu^2 \sum_{\lambda=0}^{\infty} a_\lambda \right] x^{k+\lambda-2}$$

$$+ \sum_{\lambda=0}^{\infty} a_\lambda x^{k+\lambda} = 0$$

i.e.

$$\sum_{\lambda=0}^{\infty} a_\lambda\,\{(k+\lambda)^2 - \nu^2\}\, x^{k+\lambda-2} + \sum_{\lambda=0}^{\infty} a_\lambda\, x^{k+\lambda} = 0 \tag{3.118}$$

The coefficient of the lowest power $\lambda = 0$ gives the indicial equation

$$a_0\,(k^2 - \nu^2) = 0 \qquad \therefore \quad k = \pm\nu \tag{3.119}$$

Consider the next highest power $\lambda = 1$. Its coefficient is $a_1\{(k+1)^2 - \nu^2\}$ and it must be zero.

$$\therefore \qquad a_1\,\{(k+1)^2 - \nu^2\} = 0$$

Because $k = \pm\,\nu$, we have

$$a_1\,(2\nu \pm 1) = 0$$

and since ν is an integer $(2\nu \pm 1) \neq 0$. Hence,

$$a_1 = 0. \tag{3.120}$$

The coefficient of x^{k+j} is

$$a_{j+2}\,\{(k+j+2)^2 - \nu^2\} + a_j = 0$$

i.e.

$$a_{j+2} = -\frac{a_j}{(k+j+2)^2 - \nu^2}\,. \tag{3.121}$$

This gives the relation between the coefficients of alternate terms. Because $a_1 = 0$, all odd number terms vanish.

If $k = \nu$, the even numbered coefficients are

$$a_0, -\frac{a_0}{2^2(\nu+1)}, \frac{1}{2^4(\nu+1)(\nu+2)}\frac{a_0}{2!}, \ldots$$

$$\ldots \frac{(-1)^s}{2^{2s}(\nu+1)(\nu+2)\ldots(\nu+s)}\frac{a_0}{s!}, \ldots$$

and

$$R(r) = a_0\left[x^\nu - \frac{1}{2^2(\nu+1)}x^{\nu+2} + \frac{1}{2^4(\nu+1)(\nu+2)\,2!}x^{\nu+4} + \ldots\right.$$

$$\left.\ldots + \frac{(-1)^s}{2^{2s}(\nu+1)\ldots(\nu+s)\,s!}x^{\nu+2s}\ldots\right]$$

$$= \sum_{s=0}^{\infty} a_0 \frac{(-1)^s x^{\nu+2s}}{2^{2s}\,s!\,(\nu+1)\ldots(\nu+s)} \qquad (3.122)$$

It is customary to define a_0 as

$$a_0 = \frac{1}{2^\nu\Gamma(\nu+1)} \qquad (3.123)$$

Hence,

$$R(r) = J_\nu(x) = \sum_{s=0}^{\infty} \frac{(-1)^s x^{\nu+2s}}{2^\nu\,\Gamma(\nu+1)\,2^{2s}s!\,(\nu+1)\ldots(\nu+s)}$$

$$= \sum_{s=0}^{\infty} \frac{(-1)^s}{\Gamma(s+1)\,\Gamma(\nu+s+1)}\left(\frac{x}{2}\right)^{\nu+2s} \qquad (3.124)$$

This is known as the **Bessel's tunction** of the first kind of order ν and is represented by the symbol $J_\nu(x)$. The ratio test shows that the function converges for any value of x.

If $k = -\nu$

$$R(r) = J_{-\nu}(x) = \sum_{s=0}^{\infty} \frac{(-1)^s}{\Gamma(s+1)\,\Gamma(s-\nu+1)}\left(\frac{x}{2}\right)^{2s-\nu} \qquad (3.125)$$

The general solution of the Bessel's equation, therefore, should be written as

$$R = A\,J_\nu(x) + B\,J_{-\nu}(x) \qquad (3.126)$$

where A and B are constants. This is indeed true when ν is not an integer in which case $J_\nu(x)$ and $J_{-\nu}(x)$ as defined by (3.124) and (3.125) are linearly independent solutions. However, if ν is an integer (as it is in the present case), it can be shown that the two solutions are linearly dependent. If ν is an integer, the first ν terms in the denominator of (3.125) for which $s = 0, 1, 2, \ldots\ldots\ldots (\nu - 1)$ vanish, because $\frac{1}{\Gamma(s-\nu+1)} = 0$. Hence, (3.125) can be written as

$$J_{-\nu}(x) = \sum_{s=\nu}^{\infty} \frac{(-1)^s}{\Gamma(s+1)\Gamma(s-\nu+1)} \left(\frac{x}{2}\right)^{2s-\nu} \tag{3.127}$$

Put $p = s - \nu$

$$\therefore \quad J_{-\nu}(x) = \sum_{p=0}^{\infty} \frac{(-1)^{p+\nu}}{\Gamma(p+1)\Gamma(p+\nu+1)} \left(\frac{x}{2}\right)^{2p+\nu}$$

$$= (-1)^\nu \sum_{p=0}^{\infty} \frac{(-1)^p}{\Gamma(p+1)\Gamma(p+\nu+1)} \left(\frac{x}{2}\right)^{2p+\nu} = (-1)^\nu J_\nu(x) \tag{3.128}$$

Hence, when ν is an integer, (3.126) is not a general solution of the second order Bessel's equation, and it is necessary to find another linearly independent solution. Conventionally, this solution is taken to be $N_\nu(x)$, the *Neumann function* or what is also known as *Bessel function of the second kind*, defined by

$$N_\nu(x) = \frac{J_\nu(x) \cos \nu\pi - J_{-\nu}(x)}{\sin \nu\pi} . \tag{3.129}$$

It is easy to verfy that this satisfies Bessel's equation. It can also be shown that $N_\nu(x)$ is independent of $J_\nu(x)$.

The complete solution of Bessel's equation in cylindrical coordinates is

$$V(r, \phi, z) = \sum_{m,\nu}^{\infty} \left[A_{m\nu} J_\nu(k_m r) + B_{m\nu} N_\nu(k_m r) \right] e^{\pm i\nu\phi} e^{\pm kz} \tag{3.130}$$

where we have taken account of the fact that the various values of k may give acceptable solutions.

We enlist below some of the useful properties of Bessel's functions:

1. In a manner similar to that adopted in the case of Legendre polynomials we can show that the Bessel's functions satisfy the following orthogonality condition:

If $x = k_m\rho$ is the mth root of $J_\nu(x)$ i.e $J_\nu(k_m\rho) = 0$, then, in the interval $0 \leqslant r \leqslant \rho$

$$\int_0^\rho J_\nu(k_m r) J_\nu(k_{m'} r)\, r dr = \frac{\rho^2}{2} J^2_{\nu+1}(k_{m'}\rho)\, \delta_{mm'} \tag{3.131}$$

2. Since $J_\nu(k_m r)$ form a complete orthogonal set, any function $f(r)$ can be expanded in the interval $0 \leqslant r \leqslant \rho$ in terms of them, i.e.

$$f(r) = \sum_{m=1}^{\infty} D_{m\nu} J_\nu(k_m r) \tag{3.132}$$

where

$$D_{m\nu} = \frac{2}{\rho^2 J^2_{\nu+1}(k_m\rho)} \int_0^\rho f(r) J_\nu(k_m r)\, r dr \tag{3.133}$$

3. $J_0'(k_m r) = \frac{d J_0(k_m r)}{dr} = -k_m J_1(k_m r)$ (3.134)

4. $J_\nu'(k_m r) = \frac{\nu}{r} J_\nu(k_m r) - k_m J_{\nu+1}(k_m r)$ (3.135)

5. $\int J_1(k_m r)\, dr = -\frac{1}{k_m} J_0(k_m r)$ (3.136)

6. $\int (k_m r) J_0(k_m r)\, dr = r J_1(k_m r)$ (3.137)

EXAMPLE 3.7. In a coaxial cable the potential of the outer cylinder of radius 'b' is maintained at zero and that of the inner cylinder of radius 'a' is V_1. Find the expression for the potential at a point in the region between the two cylinders.

Since the coaxial cable has a cylindrical symmetry, there is no ϕ dependence and, hence, $\nu = 0$. If, further the cable is long, we may neglect the end effects i.e. there is no z dependence. Hence $k = 0$.

The Laplace's equation, under these circumstances, reduces to

$$\frac{1}{r}\frac{d}{dr}\left(r\frac{dV}{dr}\right) = 0. \tag{3.138}$$

Integrating $r\frac{dV}{dr} = A$ (a constant)

and $$V(r) = A \ln r + C \tag{3.139}$$

where C is another constant.

The boundary conditions give,

$$\text{at } r = b,\ V(b) = A \ln b + C = 0 \quad \therefore \quad C = -A \ln b$$

and $$\text{at } r = a,\ V(a) = A \ln a - A \ln b = V_1$$

$$\therefore \quad A = -\frac{V_1}{\ln(b/a)}$$

and $$\ln V(r) = -\frac{V_1}{\ln(b/a)}(\ln r - \ln b) = -\frac{V_1}{\ln(b/a)} \ln\frac{r}{b} \tag{3.140}$$

EXAMPLE 3.8. The cylindrical surface of a cylinder of infinite length and radius 'a' is maintained at zero potential. The axis of the cylinder coincides with the z-axis. The cylinder is closed at $z = 0$ with a plate held at a potential V_1. Calculate the potential at a point interior to the cylinder.

The solution is independent of ϕ $\therefore$ $\nu = 0$. At large values of $z, V \to 0$. Hence e^{+kz} is not admissible. Further V must be finite at the origin. However, the second term inside the bracket (see 3.130) will give infinite contribution to the potential at the origin unless $B_\nu = 0$. Hence, the potential is given by

$$V(r, \phi, z) = \sum_m A_{mo}\, J_0(k_m r)\, e^{-k_m z} \tag{3.141}$$

Boundary conditions give

$$V(a, \phi, z) = -\sum_m A_{mo}\, J_0(k_m a)\, e^{-k_m z} = 0.$$

and

$$V(r, \phi, 0) = \sum_m A_{mo}\, J_0(k_m r) = V_1 \tag{2.142}$$

From (3.133)

$$A_{mo} = \frac{2}{a^2 J_1^2(k_m a)} \int_0^a V_1\, J_0(k_m r)\, r\,dr$$

$$= \frac{2V_1}{a^2 J_1^2(k_m a)} \frac{a}{k_m} J_1(k_m a) \qquad \text{(by 3.137)}$$

$$= \frac{2V_1}{a k_m J_1(k_m a)}$$

$$\therefore \quad V(r, \phi, z) = \frac{2V_1}{a} \sum_m \frac{J_0(k_m r)}{k_m J_1(k_m a)} e^{-k_m z} \tag{3.143}$$

3.9 Solution of Poisson Equation Using Green Function

Having discussed the methods for solving Laplace's equation, we now proceed to set ou+ the method for solving Poisson equation.

We shall first i ustrate the use of Poisson equation by considering a simple example which has important practical applications: The problem of the potential distribution in a space-charge limited diode.

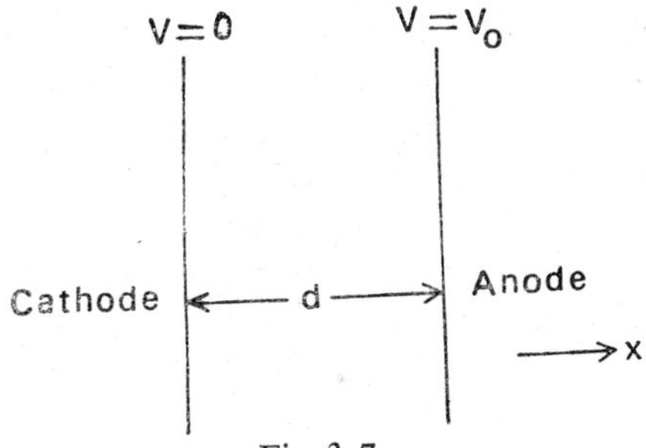

Fig. 3.7

Consider a parallel plate diode (Fig. 3.7). The surface area of the electrodes is assumed to be large enough so that the edge effects may be neglected. Under operating conditions electrons leave the cathode ($V=0$ at $x=0$) and are accelerated towards the anode ($V = V_0$ at $x = d$). During the process the space charge builds up in the region between the electrodes and limits the flow of the current. Since the region between the plates is vacuum, Poisson equation takes the form

$$\frac{d^2V}{dx^2} = -\rho/\epsilon_0 \tag{3.144}$$

where ρ is the charge density.

If J is current per unit area and 'u' is the electron velocity

$$\mathbf{J} = -\rho\mathbf{u} \tag{3.145}$$

Assuming that the emission velocity of electrons is negligible, we have from energy considerations

$$\frac{1}{2}mu^2 = eV \tag{3.146}$$

Using (3.145) and (3.146), we can write (3.144) as

$$\frac{d^2V}{dx^2} = \frac{J}{u\epsilon_0} = \frac{J}{\epsilon_0}\sqrt{\frac{m}{2eV}}$$

Integrating, we get

$$\frac{1}{2}\left(\frac{dV}{dx}\right)^2 = \frac{2J}{\epsilon_0}\left(\frac{m}{2e}\right)^{1/2} V^{1/2}$$

i.e.

$$\frac{dV}{dx} = \frac{2J^{1/2}}{\epsilon_0^{1/2}}\left(\frac{mV}{2e}\right)^{1/4}$$

Integrating once more and applying the boundary conditions

$$V^{3/4} = (3/2)\left(\frac{J}{\epsilon_0}\right)^{1/2}\left(\frac{m}{2e}\right)^{1/4} x.$$

Hence,

$$V = (3/2)^{4/3}\left(\frac{J}{\epsilon_0}\right)^{2/3}\left(\frac{m}{2e}\right)^{1/3} x^{4/3} \tag{3.147}$$

In general, because of the inhomogeneous nature of the equation, boundary value problems involving Poisson equation are somewhat more difficult. The solution of Poisson's equation (and also of Laplace's equation) can be obtained by means of what is called **Green's function.** The Green's function is of much theoretical importance because it enables a differential equation with suitable boundary conditions to be solved by quadrature. We shall first introduce Green's function and then describe, in brief, the method of solving Poisson equation using this function.

Let $\mathcal{L}$ be a differential operator and $g(x)$ a continuous function. Let us see how a function $f(x)$ which satisfies the inhomogeneous differential equation

$$\mathcal{L}f(x) = \frac{g(x)}{\epsilon_0} \tag{3.148}$$

and certain specified boundary conditions is found. If there exists a unique solution for each $g(x)$, there must exist an inverse operator $\mathcal{L}^{-1}$ such that for all $g(x)$ the formal solution of (3.148) is

$$f(x) = \mathcal{L}^{-1}\frac{g(x)}{\epsilon_0} \tag{3.149}$$

We assume that the operator $\mathcal{L}^{-1}$ does not only represent the operation which is inverse to that represented by $\mathcal{L}$, but also the application of the associated boundary conditions. For example, consider the equation

$$\frac{d}{dx}y = 2x$$

subject to the condition that $y = 1$ when $x = 0$.

The operation inverse to $\frac{d}{dx}$ is $\int dx$

$$\therefore \; y = \int 2x dx = x^2 + b$$

where b is a constant. The application of boundary condition gives $b = 1$. We, therefore, assume that the operator $\int dx$ when operated on $2x$ gives the unique solution

$$y = x^2 + 1$$

The solution (3.149) can also be expressed in terms of Dirac delta function. Thus,

$$f(x) = \int \mathcal{L}^{-1} \delta(x - x') \frac{g(x')}{\epsilon_0} dx' \tag{3.150}$$

By definition, the solution of the equation (3.148) when $g(x) = \delta(x - x')$ is called the Green's function $G(x - x')$, for the operator $\mathcal{L}$ and the given boundary conditions. The Green's function $G(x, x')$, thus, satisfies the equation

$$\mathcal{L}\, G(x, x') = \frac{\delta(x - x')}{\epsilon_0} \tag{3.151}$$

with the same boundary conditions as on the function $f(x)$

$$\therefore \quad G(x, x') = \mathcal{L}^{-1} \frac{\delta(x - x')}{\epsilon_0} \tag{3.152}$$

Hence, the solution $f(x)$ of the equation (3.148) is

$$fx = \int \mathcal{L}^{-1} \frac{\delta(x - x')}{\epsilon_0} g(x') dx'$$

$$= \int G(x, x') g(x') dx' \tag{3.153}$$

The problem now reduces to obtaining proper Green function for a given operator. Once this is found, the solution $f(x)$ is easily obtained by integration.

It may be noted that although for simplicity we have considered only one dimensional problem, the formula (3.153) is easily extended to three dimensions. that is,

$$G(\mathbf{r}, \mathbf{r}') = \mathcal{L}^{-1} \frac{\delta(\mathbf{r} - \mathbf{r}')}{\epsilon_0} \tag{3.154}$$

and

$$f(r) = \int G(\mathbf{r}, \mathbf{r}') g(r') dr' \tag{3.155}$$

We shall now show how the solution to Poisson equation with either Dirichlet or Neumann boundary condition is found by an approach using Green's function.

The Poisson equation is

$$\nabla^2\Phi = -\rho/\epsilon_0 \tag{3.156}$$

We have already obtained an expression for the scalar potential (1.41)

$$\Phi(r) = \int \frac{\rho(r)\,d\tau}{4\pi\epsilon_0 r} \tag{3.157}$$

Let us verify that this expression satisfies the Poisson equation (3.156). Operating with the Laplacian on both sides of (3.157) we have

$$\nabla^2\Phi = \nabla^2 \int \frac{\rho(r)d\tau}{4\pi\epsilon_0 r} = \int \frac{\rho}{4\pi\epsilon_0}\,\nabla^2\left(\frac{1}{r}\right) d\tau \tag{3.158}$$

We know by direct calculation that when $r \neq 0$,

$$\nabla^2\left(\frac{1}{r}\right) = \frac{1}{r}\frac{d^2}{dr^2}\left(r\cdot\frac{1}{r}\right) = 0 \tag{3.159}$$

At $r = 0$, it is undefined. We can find, however, the value of $\int \nabla^2\left(\frac{1}{r}\right) d\tau$ at $r = 0$, by following a limiting procedure. We assume

$$\int \nabla^2\left(\frac{1}{r}\right) d\tau = \lim_{\alpha\to 0} \int \nabla^2\left(\frac{1}{\sqrt{r^2+a^2}}\right) d\tau$$

Now $$\nabla^2\left(\frac{1}{\sqrt{r^2+a^2}}\right) = \frac{1}{r}\frac{d^2}{dr^2}\left(\frac{r}{\sqrt{r^2+a^2}}\right) = -\frac{3a^2}{(r^2+a^2)^{5/2}}$$

$$\therefore \lim_{\alpha\to 0}\int \nabla^2\left(\frac{1}{\sqrt{r^2+a^2}}\right) d\tau = \lim_{\alpha\to 0}\iiint\left\{-\frac{3a^2r^2}{(r^2+a^2)^{5/2}}\right\} \sin\theta\, d\theta\, d\phi\, dr$$

$$= \lim_{\alpha\to 0}\left[-12\pi\int_0^\infty \frac{a^2r^2}{(r^2+a^2)^{5/2}}\,dr\right]$$

Put $r = ap$

$$\therefore \int \nabla^2\left(\frac{1}{r}\right) d\tau = \lim_{\alpha\to 0}\int \nabla^2\left(\frac{1}{\sqrt{r^2+a^2}}\right) d\tau$$

$$= -12\pi\int_0^\infty \frac{p^2dp}{(p^2+1)^{5/2}} = -4\pi \tag{3.160}$$

Hence,

$$\nabla^2\Phi = \frac{\rho}{4\pi\epsilon_0}\cdot(-4\pi) = -\rho/\epsilon_0$$

The expression (3.157) for scalar potential, therefore, satisfies Poisson equation.

The results (3.159) and (3.160) can be expressed by a single relation

$$\nabla^2\left(\frac{1}{r}\right) = -4\pi\delta(r) \tag{3.161}$$

or more generally,

$$\nabla^2\left(\frac{1}{|\mathbf{r}-\mathbf{r}'|}\right) = -4\pi\delta(\mathbf{r}-\mathbf{r}') \tag{3.162}$$

Comparing this with the equation (3.151), we conclude that $\frac{1}{4\pi\epsilon_0|\mathbf{r}-\mathbf{r}'|}$ is a Green's function of the operator ∇^2 and is equal to the potential due to a unit point charge. This is, however, one of the Green functions of the operator ∇^2. The general form of the Green function is

$$G(\mathbf{r}, \mathbf{r}') = \frac{1}{4\pi\epsilon_0|\mathbf{r}-\mathbf{r}'|} + F(\mathbf{r}, \mathbf{r}') \tag{3.163}$$

provided F is a harmonic function, that is, it satisfies Laplace's equation

$$\nabla^2 F(\mathbf{r}, \mathbf{r}') = 0. \tag{3.164}$$

We may picture $F(\mathbf{r}, \mathbf{r}')$ as the potential of a system of charges external to the volume under consideration.

We know that the solution of the Poisson equation has to satisfy certain boundary conditions: Dirichlet condition in which Φ is specified or Neumann's condition in which $\frac{\partial\Phi}{\partial n}$ is specified. To see how to deal with boundary conditions, let us convert Poisson equation into integral form using Green's theorem, viz.

$$\int_V [\phi\nabla^2\psi - \psi\nabla^2\phi]\, d\tau = \int_S \left[\phi\frac{\partial\psi}{\partial n} - \psi\frac{\partial\phi}{\partial n}\right] dS \tag{3.165}$$

where ϕ and ψ are arbitrary scalar functions.

The Poisson equation for the potential can be converted into an integral equation if we put in (3.165) $\psi = \frac{1}{4\pi\epsilon_0|\mathbf{r}-\mathbf{r}'|}$ and Φ a scalar potential satisfying Poisson equation

$$\nabla^2\Phi = -\rho/\epsilon_0 \tag{3.166}$$

With these substitutions we can write (3.165) as

$$\int_V\left[-\Phi\frac{\partial(\mathbf{r}-\mathbf{r}')}{\epsilon_0} + \frac{\rho}{4\pi\epsilon_0^2|\mathbf{r}-\mathbf{r}'|}\right] d\tau$$

$$= \int_S\left[\Phi\frac{\partial}{\partial n'}\left(\frac{1}{4\pi\epsilon_0|\mathbf{r}-\mathbf{r}'|}\right) - \frac{1}{4\pi\epsilon_0|\mathbf{r}-\mathbf{r}'|}\frac{\partial\Phi}{\partial n'}\right] dS \tag{3.167}$$

where we have made use of (3.162).

If $\mathbf{r}$ is within the volume V, the integral of the first term on the left-hand side is $-\frac{\Phi(r)}{\epsilon_0}$.

Hence, it follows that

$$\Phi(r) = \int_V \frac{\rho(r')}{4\pi\epsilon_0|\mathbf{r}-\mathbf{r}'|}\, d\tau + \frac{1}{4\pi}\int_S\left\{\frac{1}{|\mathbf{r}-\mathbf{r}'|}\frac{\partial\Phi}{\partial n'} - \Phi\frac{\partial}{\partial n'}\left(\frac{1}{|\mathbf{r}-\mathbf{r}'|}\right)\right\} dS' \tag{3.168}$$

Note that this is not the solution of the Poisson equation, but simply the integral form of that equation.

One can easily obtain the generalization of this if we take as before Φ as the potential and $\psi = G(\mathbf{r}, \mathbf{r}')$ the Green function. We get in terms of $G(\mathbf{r}, \mathbf{r}')$

$$\frac{\Phi(r)}{\epsilon_0} = \int_V \frac{\rho G(\mathbf{r}, \mathbf{r}')}{\epsilon_0} d\tau + \int_S \left[G(\mathbf{r}, \mathbf{r}') \frac{\partial \Phi}{\partial n'} - \Phi \frac{\partial G(\mathbf{r}, \mathbf{r}')}{\partial n'} \right] dS' \tag{3.169}$$

Here $$G(\mathbf{r}, \mathbf{r}') = \frac{1}{4\pi\epsilon_0 |\mathbf{r} - \mathbf{r}'|} + F(\mathbf{r}, \mathbf{r}')$$

We now choose $F(\mathbf{r}, \mathbf{r}')$ so as to eliminate one or the other of the surface integrals in (3.169), i.e. we choose $F(\mathbf{r}, \mathbf{r}')$ so as to satisfy Dirichlet or Neumann condition. Thus, for Dirichlet boundary condition

$$G_D(\mathbf{r}, \mathbf{r}') = 0 \text{ for } r' \text{ on } S \tag{3.170}$$

$$\Phi(r) = \int_V \rho G(\mathbf{r}, \mathbf{r}') d\tau - \epsilon_0 \int_S \Phi \frac{\partial G(\mathbf{r}, \mathbf{r}')}{\partial n'} dS' \tag{3.171}$$

For Neumann condition we have to specify $\dfrac{\partial G_N}{\partial n'}$ on the surface.

Since $$\nabla^2 G(\mathbf{r}, \mathbf{r}') = -\frac{\delta(\mathbf{r} - \mathbf{r}')}{\epsilon_0},$$

$$\int_V \nabla^2 G(\mathbf{r}, \mathbf{r}') d\tau = -\int \frac{\delta(\mathbf{r} - \mathbf{r}')}{\epsilon_0} d\tau = -\frac{1}{\epsilon_0}.$$

Using Divergence theorem,

$$\int_S \frac{\partial G}{\partial n'} dS' = -\frac{1}{\epsilon_0} \tag{3.172}$$

Therefore, the simplest Neumann condition on G_N is

$$\frac{\partial G_N(\mathbf{r}, \mathbf{r}')}{\partial n'} = -\frac{1}{\epsilon_0 S} \text{ for } r' \text{ on } S \tag{3.173}$$

Hence,

$$\Phi(r) = \int_V \rho(r') G(\mathbf{r}, \mathbf{r}') d\tau + \int_S G(\mathbf{r}, \mathbf{r}') \frac{\partial \Phi}{\partial n'} dS' + \frac{\Phi_s}{\epsilon_0 S}$$

$$= \int_V \rho(r') G(\mathbf{r}, \mathbf{r}') d\tau + \int_S G(\mathbf{r}, \mathbf{r}') \frac{\partial \Phi}{\partial n'} dS' + \frac{\langle \Phi \rangle_s}{\epsilon_0} \tag{3.174}$$

where $\langle \Phi \rangle_s$ is the average value of the potential over the surface S. If surface is infinite $\langle \Phi \rangle_s = 0$.

3.10 The Multipole Expansion

In this section we show how the potential due to a static distribution of charges can be expressed as a sum over contributions from poles of different multiplicity viz, monopoles, dipoles, quadrupoles etc.

Assume that a certain charge distribution of density ρ_e is contained within a volume V (Fig. 3.8).

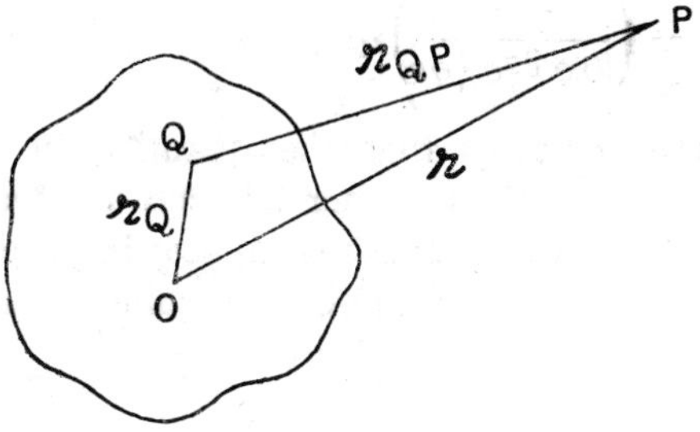

Fig. 3.8

The potential at a point P due to this charge is

$$V_P = \frac{1}{4\pi\epsilon_0}\int \frac{\rho_e\, d\tau}{r_{QP}} \tag{3.175}$$

where $d\tau$ is a volume element at Q; r_Q and r give the distances of P from Q and the origin O. If the point P is far away i.e. $r >> r_Q$, we can obtain the lowest-zero order-approximation by neglecting the positioning of the charges about the origin. That is, by assuming $r_Q = 0$ for each point Q and

$$r_{QP} = r \tag{3.176}$$

As a consequence, we have

$$V_P = \frac{1}{4\pi\epsilon_0}\int_V \frac{\rho_e\, d\tau}{r} = \frac{1}{4\pi\epsilon_0 r}\int \rho_e\, d\tau = \frac{Q}{4\pi\epsilon_0 r} \tag{3.177}$$

where $Q = \int \rho_e d\tau$ is the total charge contained in the distribution. The potential at P, therefore, is that due to a point charge Q placed at the origin.

To find a better approximation to the behaviour of (3.175) at P, we have to find a closer approximation to $\frac{1}{r_{QP}}$.

We have

$$r_{QP} = |\mathbf{r} - \mathbf{r}_Q| = \sqrt{(x-x_Q)^2 + (y-y_Q)^2 + (z-z_Q)^2} \tag{3.178}$$

where x, y, z and x_Q, y_Q, z_Q are the Cartesian coordinate of P and Q respectively.

Expressing $\frac{1}{r_{QP}}$ as a Maclaurin series. We have

$$\frac{1}{r_{QP}} = \left(\frac{1}{r_{QP}}\right)_O + \left[\left(\frac{\partial}{\partial x_Q}\frac{1}{r_{QP}}\right) x_Q + \left(\frac{\partial}{\partial y_Q}\frac{1}{r_{QP}}\right) y_Q\right.$$

$$+\left(\frac{\partial}{\partial z_Q}\frac{1}{r_{QP}}\right)z_Q\Big]_o+\Big[\frac{1}{2}\left(\frac{\partial^2}{\partial x_Q^2}\frac{1}{r_{QP}}\right)x_Q^2$$

$$+\frac{1}{2}\left(\frac{\partial^2}{\partial y_Q^2}\frac{1}{r_{QP}}\right)y_Q^2+\frac{1}{2}\left(\frac{\partial^2}{\partial z_Q^2}\frac{1}{r_{QP}}\right)z_Q^2$$

$$+\left(\frac{\partial^2}{\partial x_Q\,\partial y_Q}\frac{1}{r_{QP}}\right)x_Qy_Q+\left(\frac{\partial^2}{\partial x_Q\partial z_Q}\frac{1}{r_{QP}}\right)x_Qz_Q$$

$$+\left(\frac{\partial^2}{\partial y_Q\,\partial z_Q}\frac{1}{r_{QP}}\right)y_Qz_Q\Big]_o+\ \ldots.. \tag{3.179}$$

The subscript o indicates that the quantity in the parenthesis is evaluated at the origin. It is rather difficult to evaluate the various expansion coefficients of this series that appear in parenthesis. However, the low order terms that are of interest are quite manageable. Thus, the first term is

$$\left(\frac{1}{r_{QP}}\right)_o=\frac{1}{r} \tag{3.180}$$

as can be seen from (3.179). From an inspection of (3.178) we find that

$$\frac{\partial}{\partial x_Q}\frac{1}{r_{QP}}=-\left(\frac{\partial}{\partial x}\frac{1}{r_{QP}}\right)_o=-\frac{\partial}{\partial x}\frac{1}{r}$$

$$\left(\frac{\partial^2}{\partial x_Q\,\partial y_Q}\frac{1}{r_{QP}}\right)_o=\left(\frac{\partial^2}{\partial x\partial y}\frac{1}{r_{QP}}\right)_o=\frac{\partial^2}{\partial x\partial y}\frac{1}{r}$$

$$\therefore\qquad \frac{1}{r_{QP}}=\frac{1}{r}-\left(x_Q\frac{\partial}{\partial x}+y_Q\frac{\partial}{\partial y}+z_Q\frac{\partial}{\partial z}\right)\frac{1}{r}$$

$$+\frac{1}{2}\left(x_Q\frac{\partial}{\partial x}+y_Q\frac{\partial}{\partial y}+z_Q\frac{\partial}{\partial z_Q}\right)^2\frac{1}{r}+\ldots$$

$$=\frac{1}{r}-(\mathbf{r}_Q\cdot\nabla)\frac{1}{r}+\frac{1}{2}(\mathbf{r}_Q\cdot\nabla)^2\frac{1}{r}+\ldots \tag{3.181}$$

Hence,
$$V_P=\frac{1}{4\pi\epsilon_0}\int\frac{\rho_e d\tau}{r_{QP}}$$

$$=\frac{1}{4\pi\epsilon_0}\int\left[\frac{1}{r}-(\mathbf{r}_Q\cdot\nabla)\frac{1}{r}+\frac{1}{2}(\mathbf{r}_Q\cdot\nabla)^2\frac{1}{r}+\ldots\right]\int\rho_e d\tau$$

$$=V_0+V_1+V_2+\ldots \tag{3.182}$$

where each term of the series follows from the corresponding term in the $\frac{1}{r_{QP}}$ series.

The first term gives the potential at P due to a charge $Q=\int\rho_e d\tau$ at the origin. This may be called the **monopole potential**. The second term is

$$V_1=-\frac{1}{4\pi\epsilon_0}\int(\mathbf{r}_q\cdot\nabla)\frac{1}{r}\,\rho_e d\tau$$

$$= -\frac{1}{4\pi\epsilon_0}\nabla\left(\frac{1}{r}\right)\cdot\int \mathbf{r}_Q\,\rho_e\,d\tau = -\frac{1}{4\pi\epsilon_0}\nabla\left(\frac{1}{r}\right)\cdot\mathbf{p}$$

$$(\because\ \mathbf{p} = \int \mathbf{r}_Q\,\rho_e\,d\tau)$$

$$= \frac{\mathbf{p}\cdot\mathbf{r}}{4\pi\epsilon_0|\mathbf{r}|^3} \tag{3.183}$$

which is the potential of a dipole (see 1.61). In the same manner we can show that the next term V_2 corresponds to the potential due to a quadrupole.

At this stage it would be proper to warn the reader that expansion in Cartesian coordinates becomes very clumsy particularly when you are dealing with higher terms. It is convenient in these circumstances to expand the potential using spherical harmonics. We shall illustrate the method by considering a simple example.

Suppose we are interested in the potential due to a charge q_α (at Q) at a point P (Fig. 3.9).

Suppose the charge q_α, point P and the origin O are in the plane $\phi = 0$. The potential at P due to q_α is

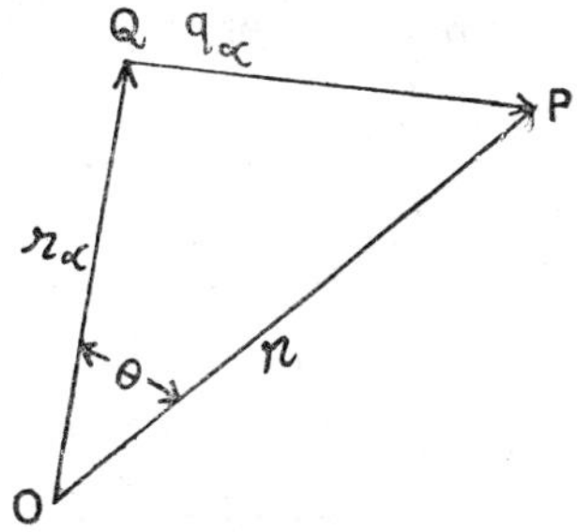

Fig. 3.9

$$V_\alpha = \frac{q_\alpha}{4\pi\epsilon_0|\mathbf{r}-\mathbf{r}_\alpha|} = \frac{q_\alpha}{4\pi\epsilon_0(r^2-2\mathbf{r}\cdot\mathbf{r}_\alpha+r_\alpha^2)^{1/2}}$$

$$= \frac{q_\alpha}{4\pi\epsilon_0(r^2-2rr_\alpha\cos\theta+r_\alpha^2)^{1/2}} = \frac{q_\alpha}{4\pi\epsilon_0}(r^2-2rr_\alpha\cos\theta+r_\alpha^2)^{-1/2}$$

$$= \frac{q_\alpha}{4\pi\epsilon_0 r}\left(1-\frac{2r_\alpha\cos\theta}{r}+\frac{r_\alpha^2}{r^2}\right)^{-1/2} \tag{3.184}$$

where r, r_α are the distances of P, Q from the origin and θ is the angle between them.

We know that $\left(1-\frac{2r_\alpha\cos\theta}{r}+\frac{r_\alpha^2}{r^2}\right)^{-1/2}$ is the generating function of Legendre polynomials, i.e.

$$\left(1-\frac{2r_\alpha\cos\theta}{r}+\frac{r_\alpha^2}{r^2}\right)^{-1/2} = 1+\frac{r_\alpha}{r}P_1(\cos\theta)+\frac{r_\alpha^2}{r^2}P_2(\cos\theta)+\ldots$$

$$\therefore\quad V_\alpha = \frac{q_\alpha}{4\pi\epsilon_0 r}+\frac{q_\alpha r_\alpha}{4\pi\epsilon_0 r^2}P_1(\cos\theta)+\frac{q_\alpha r_\alpha^2}{4\pi\epsilon_0 r^3}P_2(\cos\theta)+\ldots \tag{3.185}$$

Hence,

$$V = \Sigma V_\alpha = \sum\frac{q_\alpha}{4\pi\epsilon_0 r}+\sum\frac{q_\alpha r_\alpha}{4\pi\epsilon_0 r^2}P_1(\cos\theta)+\ldots \tag{3.186}$$

You can verify that the first term gives the contribution from a monopole, the second from a dipole etc.

3.11 Method of Electrostatic Images

We shall now discuss a general method for solving electrostatic problems, without specifically solving a differential equation. The method can be fruitfullly applied to some typical examples. We shall illustrate the method by a simple example.

(i) *Point charge and an infinite conducting plane*

Consider a point charge 'q' placed at a perpendicular distance 'd' from a conducting plate of infinite extent (Fig. 3.10a). The charge 'q' will induce charges in the plate. Since the plate is conducting, its potential must be constant at all points on it. Let us assume this potential to be zero. This means the induced charges produce at the plate a potential equal and opposite to that of the inducing charge. On the left hand side of the plate, we know, the field is zero. We want to find the potential and the field at a point in the region on the right hand side of the plate. We are also interested in the induced charge distribution on the conducting plate.

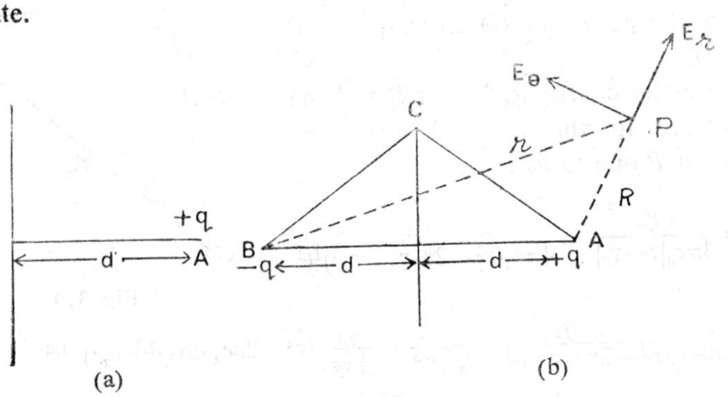

Fig. 3.10

Consider now another problem. Suppose, a charge $-q$ is placed exactly at a distance 'd' behind the plane of the plate (Fig. 3.10b). The potential at any point on the plane, say C, is zero $\left(\frac{q}{4\pi\epsilon_0 r} - \frac{q}{4\pi\epsilon_0 r} = 0\right)$. That is, if the conducting plate were removed and the charge $-q$ is placed at B, every point on the plane previously occupied by the conducting plate would still be at zero potential. Thus, the introduction of the charge $-q$ at B and removal of the plate, does not affect the flux on the right hand side of the plane. The conditions of the original problem, then, remain unaltered by the introduction of this 'device'. It is, indeed, true that the introduction of the charge $-q$ does not describe the situation compatible with the actual situation (i.e. the field on the left hand side of the plane is no longer zero). However, we need not take any cognizance of this fact, since we are interested only in the region on the right hand side of the plane. Since the condition of the problem remains unaltered by placing a charge $-q$ at B, by the principle of uniqueness, the field produced by the charge $+q$ at A and $-q$ at B is identical with that produced by the charge $+q$ at A and the induced charge on the plate. We can, therefore, solve the problem by ignoring the conducting plate and assuming a charge $-q$ at B. From optical analogy, the charge $-q$ at B is called the "image" of the charge $+q$ at A. It is important to realize that the charge $-q$ has no real existence, it is only a virtual image.

The potential at a point P (Fig. 3.10b) due to the charge q and the induced charges on the conductor is

$$V_P = \frac{q}{4\pi\epsilon_0 R} - \frac{q}{4\pi\epsilon_0\sqrt{R^2+4d^2+4Rd\cos\theta}} \tag{3.187}$$

whre θ is the angle between BA and AP and R is the distance of P from the charge.

The field components are

$$E_R = -\frac{\partial V_P}{\partial R} = \frac{q}{4\pi\epsilon_0 R^2} - \frac{q\,(R+2d\cos\theta)}{4\pi\epsilon_0\,(R^2+4d^2+4Rd\cos\theta)^{3/2}} \tag{3.188}$$

$$E_\theta = -\frac{1}{R}\frac{\partial V_P}{\partial \theta} = \frac{2qd\sin\theta}{4\pi\epsilon_0\,(R^2+4d^2+4Rd\cos\theta)^{3/2}} \tag{3.189}$$

In order to determine the surface charge density at a point on the conducting plate, we first calculate the field normal to the plate, E, at C.

Now E at $P = E_r\cos\theta - E_\theta\sin\theta$

$$= \frac{q}{4\pi\epsilon_0}\left[\frac{\cos\theta}{R^2} - \frac{R\cos\theta+2d}{(R^2+4d^2+4Rd\cos\theta)^{3/2}}\right]$$

$$\therefore \quad E \text{ at } C = -\frac{q}{4\pi\epsilon_0}\frac{2d}{R^3} = -\frac{qd}{2\pi\epsilon_0 R^3}$$

By Gauss' law $\qquad E = -\dfrac{\sigma}{\epsilon_0}$

$$\therefore \quad \sigma = \frac{qd}{2\pi R^3} \tag{3.190}$$

The surface density varies inversely as the cube of the distance of the position on the plate from the charge $+q$.

The force exerted on the charge '$+q$' by the induced charge on the plate is the same as that exerted by the image on $+q$, i.e.

$$F = -\frac{q^2}{16\pi\epsilon_0 d^2} \tag{3.191}$$

The negative sign shows that the charge is attracted towards the plate.

We, thus, conclude that the problem of charge distribution on the surface and the potential and the field produced by the system can be solved by introducing an image or a set of images suitably placed in a region external to the region under consideration and ignoring the actual surface of electrification. In other words by replacing boundary conditions by images". For the method to be successful one has to be able to guess a simple equivalent image distribution. That is the distribution that keeps the surface still an equipotential surface with requisite potential.

Let us apply the method to some simple cases:

(ii) *A point charge in the vicinity of a grounded sphere*

Consider a conducting sphere of radius 'a' maintained at zero potential and a charge 'q' at A at a distance 'd' from the centre of the sphere ($d > a$) (Fig. 3.11a).

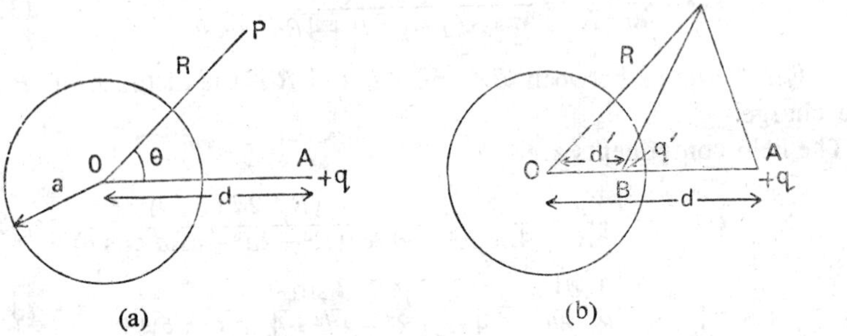

Fig. 3.11

Suppose the centre of the sphere coincides with the origin. Since 'q' is outside the sphere and our interest is in the region outside the sphere, the image must be within the sphere. By symmetry the image charge must be placed on the line joining A to the centre of the sphere. The situation, therefore, can be replaced by the equivalent problem (Fig. 3.11b) with the boundary conditions

$$V(R, \theta)\Big|_{|R|=a} = 0 \tag{3.192}$$

Let B be the position of the image charge, $OB = d'$. The potential at a point P is

$$V_P = \frac{q}{4\pi\epsilon_0 |\mathbf{R} - \mathbf{d}|} + \frac{q'}{4\pi\epsilon_0 |\mathbf{R} - \mathbf{d}'|}$$

$$= \frac{q}{4\pi\epsilon_0 \mid R\hat{\mathbf{e}}_R - d\hat{\mathbf{e}}_d \mid} + \frac{q'}{4\pi\epsilon_0 \mid R\hat{\mathbf{e}}_R - d'\hat{\mathbf{e}}_d \mid} \tag{3.193}$$

where $\hat{\mathbf{e}}_R$, $\hat{\mathbf{e}}_d$ are the unit vectors in the direction OP, OA respectively.

$$\therefore \quad V_P = \frac{q}{4\pi\epsilon_0 R \mid \hat{\mathbf{e}}_R - \frac{d}{R}\hat{\mathbf{e}}_d \mid} + \frac{q'}{4\pi\epsilon_0 d' \mid \hat{\mathbf{e}}_d - \frac{R}{d'}\hat{\mathbf{e}}_R \mid} \tag{3.194}$$

On the surfacc of the sphere $R = a$

$$V_P\Big|_{|R|=a} = \frac{q}{4\pi\epsilon_0 a \left| \hat{\mathbf{e}}_R - \frac{d}{a}\hat{\mathbf{e}}_d \right|} + \frac{q'}{4\pi\epsilon_0 d' \left| \hat{\mathbf{e}}_d - \frac{a}{d'}\hat{\mathbf{e}}_R \right|} = 0$$

This holds if

$$\frac{q}{a} = -\frac{q'}{d'} \text{ and } \frac{d}{a} = \frac{a}{d'} \text{ i.e. } d' = \frac{a^2}{d}$$

$$q' = -\frac{qd'}{a} = -\frac{qa}{d} \tag{3.195}$$

The points A and B are inverse points with respect to the sphere.

Alternatively, at $|R| = a$

$$\left|\mathbf{R} - \mathbf{d}\right|^{-1}_{|R|=a} = \left(R^2 + d^2 - 2Rd\cos\theta\right)^{-1/2}_{|R|=a}$$
$$= (a^2 + d^2 - 2ad\cos\theta)^{-1/2}$$
$$= d^{-1}(1 - 2h\cos\theta + h^2)^{-1/2} \tag{3.196}$$

and

$$\left|\mathbf{R} - \mathbf{d}'\right|^{-1}_{|R|=a} = a^{-1}(1 - 2h'\cos\theta + h'^2)^{-1/2} \tag{3.197}$$

where $h = \frac{a}{d}$ and $h' = \frac{d'}{a}$

Now $(1 - 2hx - h^2)^{-1/2} = \sum_{n=0}^{\infty} h^n P_{n(x)}$

where $|h| < 1$ and $|x| \leqslant |$

$$\therefore \quad \left|\mathbf{R} - \mathbf{d}\right|^{-1}_{|R|=a} = \sum_{n=0}^{\infty} a^n d^{-(n+1)} P_n(\cos\theta) \tag{3.198}$$

and

$$\left|\mathbf{R} - \mathbf{d}'\right|^{-1}_{|R|=a} = \sum_{n=0}^{\infty} d'^n a^{-(n+1)} P_n(\cos\theta) \tag{3.199}$$

Hence,

$$V(a,\theta) = \frac{1}{4\pi\epsilon_0}\sum_{n=0}^{\infty}(qa^n d^{-(n+1)} + q'd'^n a^{-(n+1)})P_n(\cos\theta) = 0 \tag{3.200}$$

Since the Legendre polynomials are linearly independent, coefficient of each Legendre polynomial in the expansion vanishes. Thus, for $n = 0$

$$qd^{-1} + q'a^{-1} = 0 \text{ or } q' = -\left(\frac{a}{d}\right)q \tag{3.195a}$$

and for $n = 1$

$$qad^{-2} + q'd'a^{-2} = 0$$

i.e.

$$qad^{-2} - \left(\frac{a}{d}\right)qd'a^{-2} = 0$$

or

$$d' = a^2/d \tag{3.195b}$$

For $n > 1$, the conditions (3.195 a,b) reduce all succeeding coefficients to zero.

Substitution of the values for q' and d' into (3.194) yield the desired expression for the potential.

Notice that as the charge q is brought nearer and nearer to the surface q' becomes larger and larger and moves away from the centre.

The charge density on the sphere can be found from relation

$$E_x = -\frac{\partial V}{\partial x} = -\frac{\sigma}{\epsilon_0}$$

and it can be shown by direct integration that the total induced charge on the sphere is equal to the magnitude of the image charge.

(iii) *The sphere is charged and insulated and a charge q is placed in its vicinity*

Suppose now that the conducting sphere is insulated and given a charge Q. The charge 'q' is at a distance d from the centre as before. The potential at a point due to this system can be found by the method of linear superposition. We first assume that the sphere is grounded; hence its potential is zero and the induced charge q' equal to the magnitude of the image charge of q is distributed over it. We now insulate the sphere by disconnecting it from the ground, and give it an additional amount of charge $Q - q'$. The total charge on the sphere now is Q. Since the forces due to the point charge q are balanced by the charge q', the added charge $Q - q'$ will be uniformly distributed over the surface of the sphere. The potential at a point outside the sphere is now given by the sum of the potentials due to q, q' and the added charge $Q - q'$. The contribution to the potential due to $Q - q'$ will be the same as that of a point charge of magnitude $Q - q'$ at the centre. Thus,

$$V_P = \frac{q}{4\pi\epsilon_0|\mathbf{R} - \mathbf{d}|} - \frac{qa}{4\pi\epsilon_0\, d\,\left|\mathbf{R} - \frac{a^2}{d}\,\hat{\mathbf{e}}_d\right|} + \frac{Q + \frac{qa}{d}}{4\pi\epsilon_0\,|R|} \tag{3.201}$$

(iv) *The sphere is maintained at a fixed potential*

Since the sphere is maintained at a potential V, it carries a charge equal to Va (a = capacity of the sphere).

$$\therefore\ V_P = \frac{q}{4\pi\epsilon_0\,|\mathbf{R} - \mathbf{d}|} - \frac{qa}{4\pi\epsilon_0 d\left|\mathbf{R} - \frac{a^2}{d}\,\hat{\mathbf{e}}_d\right|} + \frac{Va}{4\pi\epsilon_0|R|} \tag{3.202}$$

(v) *Conducting sphere in a uniform electric field*

We have already solved this problem (Ex. 3.5.) using spherical harmonics. We shall now show how it can be solved using the method of images.

A grounded conducting sphere of radius 'a' is placed in a uniform electric field E_0 with its centre at the origin. The field immediately around the sphere will become distorted owing to the induced charges on the surface of the sphere (Fig. 3.12).

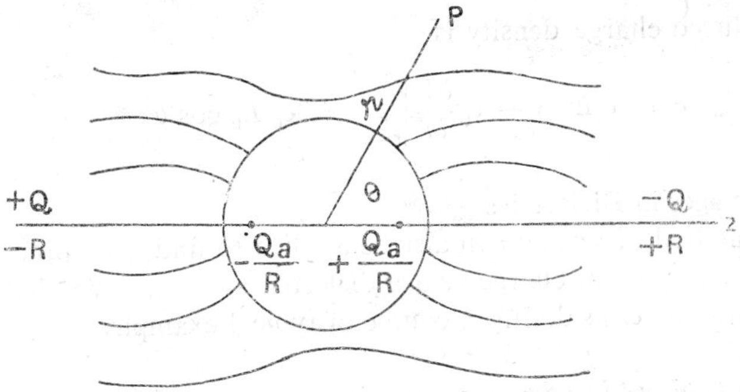

Fig. 3.12

A uniform field may be regarded as the field produced by the appropriate positive and negative charges at infinity. Consider, for example, a charge $-Q$ at $z = +R$ and a charge $+Q$ at $z = -R$. Then the field produced at a point near the origin is approximately $\frac{2Q}{4\pi\epsilon_0 R^2}$ parallel to the z-axis. In the limit when $R, Q \to \infty$ keeping $\frac{Q}{R^2}$ constant the field is uniform and equal to $E_0 = \frac{2Q}{4\pi\epsilon_0 R^2}$.

The potential at a point P distant 'r' from the origin due to the system consisting of charges $\pm Q$ at $z = \mp R$ and the induced charges on the sphere, is the same as that produced by the charge $+Q$ at $-R$, $-Q$ at $+R$, the image charge $-\frac{Qa}{R}$ at $-\frac{a^2}{R}$ and the image charge $\frac{Qa}{R}$ at $+\frac{a^2}{R}$.

$$\therefore \quad V_P = \frac{Q}{4\pi\epsilon_0 (r^2 + R^2 + 2rR\cos\theta)^{1/2}} - \frac{Q}{4\pi\epsilon_0 (r^2 + R^2 - 2rR\cos\theta)^{1/2}}$$

$+$ potentials due to the image charges.

Because $r \ll R$, the first two terms give the contribution (expanding and neglecting higher terms) $-\frac{2Q}{4\pi\epsilon_0 R^2} r\cos\theta$. We may regard the image charges $-\frac{Qa}{R}$ and $+\frac{Qa}{R}$ separated by a distance $2a^2/R$ as forming a dipole and, hence, their contribution to the potential at P is $\frac{p\cos\theta}{4\pi\epsilon_0 r^2}$, where $p = \frac{2Q\,a^3}{R^2}$.

$$\therefore \quad V_P = -\frac{2Q}{4\pi\epsilon_0 R^2} r\cos\theta + \frac{(2Qa^3/R^2)\cos\theta}{4\pi\epsilon_0 r^2}$$

$$= -E_0 r\cos\theta + \frac{E_0 a^3\cos\theta}{r^2} = -E_0\left(r - \frac{a^3}{r^2}\right)\cos\theta \qquad (3.203)$$

The induced charge density is

$$\sigma = \epsilon_0 E = -\epsilon_0 \frac{\partial V}{\partial r}\bigg|_{r=a} = 3\epsilon_0 E_0 \cos\theta \qquad (3.204)$$

3.12 Images in Dielectrics

The method of images can also be applied to find the potential difference due to a point charge near a dielectric surface. We shall illustrate the method by considering a couple of typical examples.

(i) *The field produced by a point charge 'q' placed at a perpendicular distance 'd' from a semi-infinite dielectric bounded by a plane surface* (Fig. 3.13)

Let the surface coincide with the plane $z = 0$; let O be the origin and A, the point on the z-axis where the charge 'q' is placed. The field of the point charge 'q' will polarize the dielectric which in turn will affect the field in vacuum on the right-hand side. We assume that the field produced by the polarization charges at a point on the right-hand side is the same as that produced by a charge 'q'', the image of q placed at A'. The question that we have to answer is "what is the value of 'q'' which would be compatible with our assumption?"

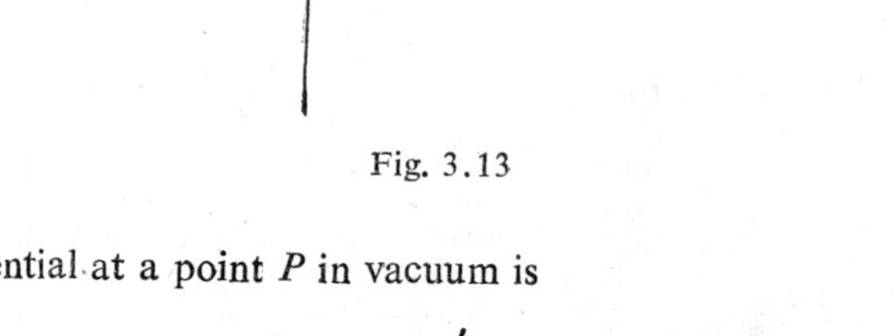

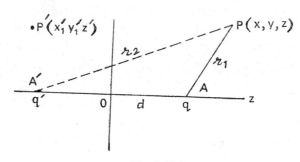

Fig. 3.13

The potential at a point P in vacuum is

$$V_P = \frac{q}{4\pi\epsilon_0 r_1} + \frac{q'}{4\pi\epsilon_0 r_2}$$

Or in Cartesian coordinates, the potential at $P(x, y, z)$

$$V_P = \frac{1}{4\pi\epsilon_0}\left[\frac{q}{\{x^2 + y^2 + (z-d)\}^{1/2}} + \frac{q'}{\{x^2 + y^2 + (z+d)^2\}^{1/2}}\right] \qquad (3.205)$$

What would be the potential at a point P' in the dielectric medium due to the point charge 'q'? It is certainly not the one that 'q' would produce at P' in vacuum. We assume that this potential is the same as that produced by a charge 'q''' at A if the space was vacuum and the dielectric was not there. This potential at P' (x', y', z') is

$$V_P' = \frac{q''}{4\pi\epsilon_0 \{x'^2 + y'^2 + (z' - d)^2\}^{1/2}} \tag{3.206}$$

The potential in vacuum at the point P when it is in the plane $z = 0$, is,

$$V_P \bigg|_{z=0} = \frac{1}{4\pi\epsilon_0} \left\{ \frac{q + q'}{(x^2+y^2+d^2)^{1/2}} \right\} \tag{3.207}$$

while that at P' when it is at the same point but in the dielectric

$$V_P' \bigg|_{z=0} = \frac{q''}{4\pi\epsilon_0 (x'^2+y'^2+d^2)^{1/2}} \tag{3.208}$$

On the boundary $V_P = V_P'$ which automatically satisfies the first boundary condition that the tangential components of E must be the same on either side of the boundary. This condition gives

$$q + q' = q'' \tag{3.209}$$

The second boundary condition is that the normal components of D must be continuous at the boundary. That is

$$\frac{\partial V_P}{\partial z} = \epsilon \frac{\partial V_P'}{\partial z} \quad \text{at} \quad z = 0 \tag{3.210}$$

where ϵ is the permittivity of the dielectric.

Now $$\frac{\partial V_P}{\partial z} = \frac{1}{4\pi\epsilon_0} \left[- \frac{q(z-d)}{\{x^2 + y^2 + (z-d)^2\}^{3/2}} - \frac{q'(z+d)}{\{x^2 + y^2 + (z+d)^2\}^{3/2}} \right]$$

$$= \frac{1}{4\pi\epsilon_0} \frac{d\,(q - q')}{(x^2 + y^2 + d^2)^{3/2}} \quad \text{at} \quad z = 0 \tag{3.211}$$

Similarly, $$\frac{\partial V_P'}{\partial z} = \frac{q''\,d}{4\pi\epsilon_0 (x^2 + y^2 + d^2)^{3/2}} \quad \text{at} \quad z' = 0 \tag{3.212}$$

Equation (3.210) gives

$$\frac{d(q - q')}{(x^2 + y^2 + d^2)^{3/2}} = \frac{\epsilon\, q''\, d}{(x^2 + y^2 + d^2)^{3/2}}$$

or

$$q - q' = \epsilon\, q'' \tag{3.213}$$

Solving (3.209) and (3.213)

$$q'' = \frac{2q}{\epsilon + 1}; q' = - \frac{q(\epsilon - 1)}{\epsilon + 1} \tag{3.214}$$

The force of attraction between the dielectric and the point charge is

$$F = \frac{qq'}{4\pi\epsilon_0 (2d)^2} = - \frac{q^2 (\epsilon - 1)}{16\pi\epsilon_0 d^2 (\epsilon + 1)} \tag{3.215}$$

(ii) *Dielectric sphere in a uniform field*

Consider a dielectric sphere of radius 'a' and dielectric constant ϵ_1 placed in a uniform dielectric field E_0 existing in a medium with dielectric constant ϵ_2. Since there are no free charges inside and outside the surface, the potential V (outside) and V (inside) the sphere must satisfy Laplace's equation viz.,

(i) $\nabla^2 V_1 = 0, \nabla^2 V_2 = 0$

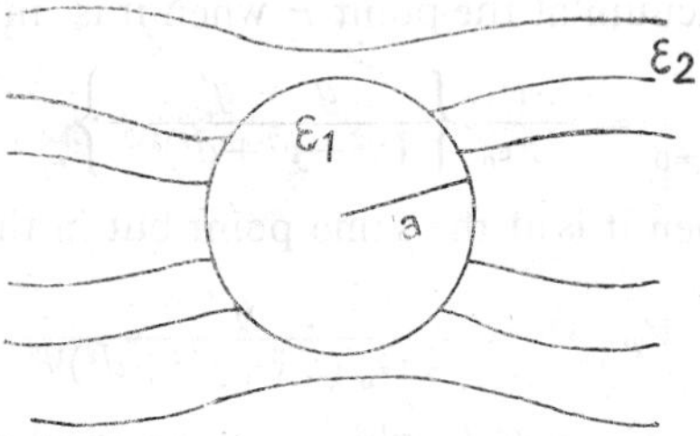

Fig. 3.14

Besides, the potentials must satisfy boundary conditions also, i.e.

(ii) $V(r, \theta)$ must be continuous on $r = a$ for all θ i.e. $V_1 = V_2$ on $r = a$

(iii) The normal components of D must be continuous on $r = a$ for all θ

i.e. $$\epsilon_1 \frac{\partial V_1}{\partial r} = \epsilon_2 \frac{\partial V_2}{\partial r} \quad \text{on} \quad r = a$$

and

(iv) $V_1(r, \theta)$ must be finite at $r = 0$

(v) $V_2(r, \theta) = V_0 = -E_0 z = -E_0 r \cos\theta$ as $r \to \infty$

The potential which is the solution of the Laplace's equation, can be expressed in the form

$$V = \sum_{l=0}^{\infty} \left[A_l r^l + B_l r^{-(l+1)} \right] P_l(\cos\theta) \tag{3.216}$$

Since the potential at the origin must be finite (condition iv), no term of the form $Br^{-(l+1)}$ should be included in the expression for V_1, as it gives infinite contribution to the potential at $r = 0$

$$\therefore \quad V_1 = \sum_{l=0}^{\infty} A_l r^l P_l(\cos\theta) \tag{3.217}$$

The potential outside V_2 is given by

$$V_2 = \sum_{l=0}^{\infty} \left[C_l r^l + D_l r^{-(l+1)} \right] P_l(\cos\theta) \tag{3.218}$$

The constants can be found from the boundary conditions.

The potential V_2 at infinity is

$$V_2 = C_0 P_0 (\cos\theta) + C_1 r P_1 (\cos\theta) + C_2 r^2 P_2 (\cos\theta) + \ldots \tag{3.219}$$

But this must be equal to $-E_0 r P_1 (\cos\theta)$ by condition (v).

All terms in (3.219) vanish except the second term and, hence,

$$l = 1; \quad C_1 = -E_0 \tag{3.220}$$

The boundary condition (ii) gives

$$V_1 = V_2 \text{ on } r = a \text{ for all } \theta$$

$$\therefore \quad A_1 a = -E_0 a + \frac{D_1}{a^2} \tag{3.221}$$

The condition (iii) gives

$$\epsilon_1 \frac{\partial V_1}{\partial r} = \epsilon_2 \frac{\partial V_2}{\partial r} \quad \text{on} \quad r = a \text{ for all } \theta$$

$$\therefore \quad \epsilon_1 A_1 = -\epsilon_2 \left(E_0 + \frac{2D_1}{a^3}\right) \tag{3.222}$$

Solving (3.221) and (3.222) we have

$$D_1 = \frac{\epsilon_1 - \epsilon_2}{\epsilon_1 + 2\epsilon_2} E_0 a^3; \; A_1 = -\frac{3\epsilon_2}{\epsilon_1 + 2\epsilon_2} E_0 \tag{3.223}$$

Hence, $$V_1 = -\frac{3\epsilon_2}{\epsilon_1 + 2\epsilon_2} E_0 r \cos\theta = \frac{3\epsilon_2}{\epsilon_1 + 2\epsilon_2} E_0 z \tag{3.224}$$

and $$V_2 = -E_0 r \cos\theta + \frac{\epsilon_1 - \epsilon_2}{\epsilon_1 + 2\epsilon_2} \frac{E_0 a^3 \cos\theta}{r^2} \tag{3.225}$$

The field inside the sphere is

$$E_{\text{int}} = -\frac{\partial V_1}{\partial z} = \frac{3\epsilon_2}{\epsilon_1 + 2\epsilon_2} E_0 \tag{3.226}$$

This field is constant, uniform and parallel to the field applied.

$$\text{If } \epsilon_1 > \epsilon_2, \quad E_{\text{int}} < E_0 \tag{3.227}$$

This, as explained in the preceding chapter, is due to the induced surface polarization charges which give rise to an opposing field.

The external field can be obtained from (3.225). This is equal to the field E_0 plus a dipole field of moment

$$p = \frac{\epsilon_1 - \epsilon_2}{\epsilon_1 + 2\epsilon_2} a^3 E_0 \tag{3.228}$$

The dielectric sphere in a uniform field, therefore, acts as a simple dipole.

(iii) *A spherical cavity in a dielectric medium*

The problem of a spherical cavity in a dielectric medium with dielectric constant ϵ can be solved in the same manner as in the case of the dielectric sphere.

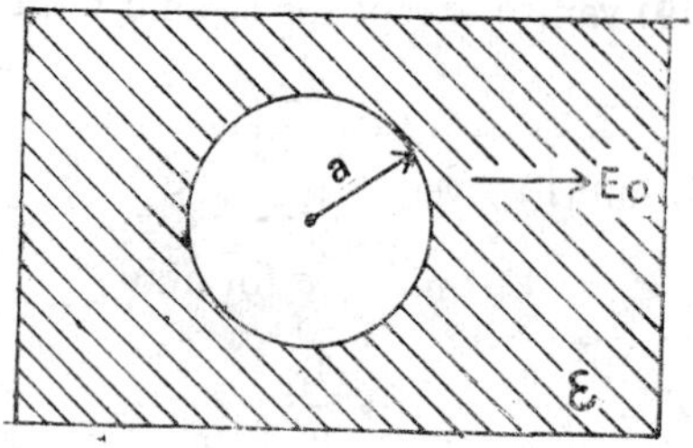

Fig. 3.15

Put $\epsilon_1 = 1$ and $\epsilon_2 = \epsilon$ in (3.226). We get

$$E_{\text{int}} = \frac{3\epsilon}{1 + 2\epsilon} E_0 \tag{3.229}$$

$$\text{If } \epsilon > 1, \qquad E_{\text{int}} > E_0$$

The field outside is equal to the applied field E_0 plus the field due to the dipole at the origin of moment

$$p = \frac{1 - \epsilon}{1 + 2\epsilon} a^3 E_0 = -\frac{(\epsilon - 1)}{1 + 2\epsilon} a^3 E_0 \tag{3.230}$$

which is oriented oppositely to the field applied.

PROBLEMS

3.1 Prove the following relation for Legendre's polynomials

$$\int P_l (\cos\theta)\, d(\cos\theta) = \frac{P_{l+1}(\cos\theta) - P_{l-1}(\cos\theta)}{2l + 1}$$

3.2 A linear quadrupole with charges q, $-2q$, q and separation 'a' is placed within a grounded conducting spherical shell of radius 'b' with the middle charge $-2q$ at the origin. Find the potential inside and outside the sphere. Show that in the limit as a $\to 0$

$$V(r, \theta, \phi) \sim \frac{2Q}{4\pi\varepsilon_0 r^3}\left(1 - \frac{r^5}{b^5}\right) P_2(\cos\theta)$$

where $Q = q a^2$.

3.3 A hemisphere of radius 'a' is charged uniformly with a charge Q. Assuming that the centre of the base of the hemisphere is at the origin, show that the potential at a point ($r \gg a$) is given by

$$V \sim \frac{Q}{4\pi\varepsilon_0 r}\left[1 + \frac{3a}{2r}\cos\theta\right]$$

3.4 A sphere is charged with a surface density σ everywhere except for a spherical segment near the pole, limited by a circle with $\theta = \alpha$. Find the potential inside and outside the sphere.

3.5 A grounded conducting sphere of radius 'a' is placed in a potential field $V(r, \theta, \phi) = \frac{r^2 \sin 2\theta}{4\pi\varepsilon_0} e^{i\phi}$ with its centre at the origin. Find the potential at a point outside the sphere.

3.6 An insulated spherical conducting uncharged shell placed in a uniform field E_0 is cut into two hemispheres by a plane perpendicular to the field. Find the force required to prevent the two hemispheres from separating.

3.7 A conducting circular cylinder of radius 'a' and infinite length is placed in a uniform field with its axis at right angles to the field. Find the potential at a point outside the cylinder if the cylinder is maintained at zero potential.

3.8 The axis of a hollow right circular cylinder of radius 'a' coincides with the z-axis, its end faces being at $z = 0$ and $z = b$. The potential on the cylindrical surface is $V(\phi, z)$ while the potential on the end faces is zero. Find, using cylindrical coordinates, a series solution for the potential at a point inside the cylinder.

3.9 Calculate the potential at a point due to a line charge 'q' per unit length placed parallel to an infinite plane conductor maintained at zero potential.

3.10 A conducting plane has a hemispherical boss of radius 'a', the centre of which lies on the plane. The plane is earthed and a point charge 'q' is placed on the axis of symmetry of the system at a point $b > a$ from the plane. Find the potential and the charge induced on the boss.

3.11 A conducting sphere of radius 'a' is kept at zero potential. An electric dipole of moment 'q' pointing away from the sphere is at a distance 'd' from the centre of the sphere. Show that the image is a dipole moment $\frac{pa^3}{d^3}$ and a charge pa/d^2 at the inverse point.

3.12 A spherical cavity of radius 'a' is cut out of a conducting block of metal maintained at zero potential. A charge 'q' is placed at a distance 'd' from the centre of gravity. Show that the force on it is $\frac{q^2 ad}{4\pi\varepsilon_0 (a^2-d^2)^2}$.

3.13 A spherical cavity of radius 'a' is cut out of an infinite medium of relative permittivity ε_r Show that the potential at a point in the dielectric medium due to a dipole 'p' situated at the centre of the cavity is the same as that produced by a dipole 'p'' inserted in a continuous dielectric where

$$\mathbf{p}' = \frac{3\varepsilon_r}{2\varepsilon_0 + 1}\mathbf{p}$$

3.14 An infinite circular cylinder of relative permittivity ε_r is placed in a uniform field $\mathbf{E}_0$ with its axis perpendicular to the field. Show that the field inside the cylinder is $\frac{2}{\varepsilon_0 + 1}\mathbf{E}_0$. Find the depolarizing field inside the cylinder.

3.15 A dielectric cylinder of radius 'a' in a two dimensional field has on its surface an induced charge distribution $q \sin n\theta$ coul/m^2. Find the potential at a point (i) inside and (ii) outside the cylinder.

Chapter 4

Magnetostatics

We have dealt so far with the electric charges that are stationary. Logically the next step would be to consider the motion of the charges and the forces associated with them, which we propose to do in the present chapter. You will perhaps, be surprised that although this chapter bears the title "Magnetostatics", we are still lingering in the realm of electricity. How is magnetism linked with the subject of electricity? You will find the answer to this question in the following sections.

4.1 Electric Current

We have seen in Chapter 2 that when a conductor is placed in an electric field, the electrons acquire an average velocity in the direction opposite to the field, This results into a net drift of charges and we say that there is an electric current flowing in the conductor. It may be noted that the conductor carrying the current is electrically neutral since the number of conducting electrons is balanced by the number of positively charged atoms. Electric current is not necessarily due to electrons. It may as well be due to other charges in motion as for example, in semiconductors or in plasmas where there is a current due to positive charges also. The treatment given below is quite general and applicable to charges in motion.

If there are N charges per unit volume at a point P, each carrying a charge 'q' and moving with a mean velocity $\mathbf{u}$, the amount of charge crossing unit area perpendicular to the mean drift velocity, per unit time, is

$$\mathbf{j} = Nq\mathbf{u} \tag{4.1}$$

We call '$\mathbf{j}$' *the current density.* It is a vector function of position, its direction being, by convention, that in which the positive charges move. If dS is an element of area in the material, the amount of charge flowing across that area in unit time is $\mathbf{j}\cdot\hat{\mathbf{e}}_n dS$, where $\hat{\mathbf{e}}_n$ is the unit vector normal to dS. This is the flux of $\mathbf{j}$ across dS.

The net charge that passes through the surface per unit time is called the *electric current* and is denoted by I. Thus,

$$I = \frac{dQ}{dt} = \int \mathbf{j}\cdot\hat{\mathbf{e}}_n\, dS \tag{4.2}$$

Consider now a closed surface S enclosing a volume V. If ρ is the volume density of charge the total charge within the volume is $\int\rho d\tau$. Since electric charge can neither be created nor destroyed (i.e. the charge is conserved), it follows that the net flow of the charge, if any, out of this volume must be equal to the rate of decrease of the total charge inside the volume. That is,

$$\int \mathbf{j}\cdot\hat{\mathbf{e}}_n\, dS = -\frac{d}{dt}\int \rho d\tau \tag{4.3}$$

Since the surface S is fixed in space, the rate of change of the charge within the volume is solely due to the time variation of ρ

$$\therefore \qquad \int_S \mathbf{j}.\hat{\mathbf{e}}_n\, dS = -\int \frac{\partial \rho}{\partial t}\, d\tau \tag{4.4}$$

On transforming the surface integral into a volume integral by Divergence theorem, we have

$$\int_S \mathbf{j}\cdot\hat{\mathbf{e}}_n\, dS = \int \operatorname{div}\mathbf{j}\, d\tau = -\int \frac{\partial \rho}{\partial t}\, d\tau$$

$$\therefore \qquad \int_V \left(\operatorname{div}\mathbf{j} + \frac{\partial \rho}{\partial t}\right) d\tau = 0 \tag{4.5}$$

This integral, must be zero for any arbitrary volume V. Hence, the integrand must vanish identically, i.e.

$$\operatorname{div}\mathbf{j} + \frac{\partial \rho}{\partial t} = 0 \tag{4.6}$$

This is known as the **equation of continuity** and is based, as stated above, on the law of conservation of charge.

In the steady state $\qquad \dfrac{\partial \rho}{\partial t} = 0$

$$\therefore \qquad \operatorname{div}\mathbf{j} = 0 \tag{4.7}$$

This equation is valid in the region which does not contain a source or a sink of current.

4.2 Ohm's Law—Electrical Conductivity

It is found experimentally that for many materials, the current density in the steady state is linearly proportional to the applied field **E**.

$$\therefore \qquad \mathbf{j} \propto \mathbf{E} \quad \text{or} \quad \mathbf{j} = \sigma\mathbf{E} \tag{4.8}$$

where σ, a constant for a given material at a given temperature, is called *electrical conduciivity* of the material.

A steady current I is given by

$$I = \int_S \mathbf{j}\cdot\hat{\mathbf{e}}_n\, dS = \sigma\int_S \mathbf{E}\cdot\hat{\mathbf{e}}_n\, dS \tag{4.9}$$

If the conductor has a uniform cross-sectional area A,

$$I = \sigma A E \tag{4.10}$$

If the potential difference maintained across the ends of a conductor is 'V' and the length of the conductor is 'l', the electric field within the conductor will be uniform and given by

$$E = \frac{V}{l} \tag{4.11}$$

$$\therefore \qquad I = \frac{\sigma A V}{l} \text{ and, hence,}$$

$$\frac{V}{I} = \frac{l}{\sigma A} = R \tag{4.12}$$

where R is called the **resistance** of the conductor, and is measured in **Ohms.**

The equation (4.12) is known as **Ohm's law**. The materials obeying this law are said to be **Ohmic.**

There are, however, many instances when Ohm's law fails: As an example we may cite the behaviour of a space charge limited dipole discussed in the preceding chapter. We have shown there that the potential at a point in the region between the electrodes, is given by

$$V = \left(\frac{3xm^{1/4}}{2^{5/4}e^{1/4}\epsilon_0^{1/2}}\right)^{4/3} J^{2/3} \quad \text{(see 3.147)}$$

The current density j is

$$J = \frac{I}{A}$$

$$\therefore \qquad V = \left(\frac{3x\,m^{1/4}}{2^{5/4}\,e^{1/4}\epsilon_0^{1/2}}\right)^{4/3} \frac{I^{2/3}}{A^{2/3}}$$

$$\text{or} \qquad I = \frac{4}{9}\epsilon_0 A \left(\frac{2e}{m}\right)^{1/2} \frac{V^{3/2}}{x^2} \tag{4.13}$$

$$\text{i.e.} \qquad I \propto V^{3/2} \tag{4.14}$$

You can see that this is not in conformity with Ohm's law.

Such non-Ohmic behaviour also occurs at semi-conductor junctions.

4.3 The Calculation of Resistance

We will consider in this section a simple example to show how the resistance of a conductor is calculated. We emphasize that the calculation is valid only for Ohmic materials.

Consider a coaxial cable (Fig. 4.1) generally used to transmit voltage along its inner wire.

Let the radius of the inner wire be 'a' and that of the outer conducting sheath be 'b'. Let, further, the outer sheath be earthed and the inner wire be at a potential V. If the space between the conductor is filled

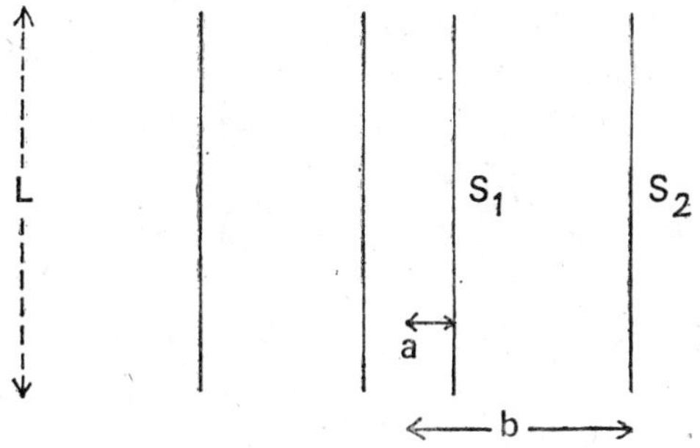

Fig. 4.1

with a material which is not a perfect insulator, there will be a leakage current through the material. The current flowing from the inner wire to the outer sheath is

$$I = \int_{S_1} \mathbf{J} \cdot d\mathbf{S} \tag{4.15}$$

where **J** is the current density and S_1 the surface of the wire

$$\therefore \quad I = \int_{S_1} \sigma \mathbf{E} \cdot d\mathbf{S} = 2\pi a L \sigma E(a) \tag{4.16}$$

where L is the length of the cable and $E(a)$ is the magnitude of the field at the surface S_1, which is perpendicular to the surface.

Now, $\mathbf{E} = -\nabla\Phi$ where Φ is the potential.

$$\therefore \quad \mathbf{J} = -\sigma\nabla\Phi \tag{4.17}$$

Hence,

$$\text{div } \mathbf{J} = -\sigma\nabla^2\Phi \tag{4.18}$$

If the current is steady

$$\text{div } \mathbf{J} = 0 \quad (\text{see } 4.7)$$

$$\therefore \quad \nabla^2\Phi = 0 \tag{4.19}$$

We may remind you that we came across this equation — Laplace's equation—in the last chapter while discussing the problem of electrostatic potential. Its solution is

$$\Phi = A \ln r + C \quad (\text{see } 3.139)$$

The boundary conditions give us the values of the constants A and C.

At $r = a$, $\quad V = A \ln a + C$

and at $r = b$, $\quad 0 = A \ln b + C$

$$\therefore \quad A = V/\ln\left(\frac{a}{b}\right) \quad \text{and } C = -V \ln b/\ln\left(\frac{a}{b}\right)$$

$$\therefore \quad \Phi = \frac{V}{\ln\left(\frac{a}{b}\right)} \ln r - \frac{V}{\ln\left(\frac{a}{b}\right)} \ln b \tag{4.20}$$

From (4.16)

$$I = 2\pi a L\sigma E(a) = -2\pi a L\sigma \nabla\Phi$$

$$= -2\pi a L\sigma \frac{V}{\ln\left(\frac{a}{b}\right)} \frac{1}{a} = -\frac{2\pi L\sigma V}{\ln\left(\frac{a}{b}\right)}$$

Hence, $$R = \frac{V}{I} = \frac{\ln\left(\frac{a}{b}\right)}{2\pi L\sigma}. \quad (4.21)$$

We may draw your attention to the analogy that this example illustrates between the solution of electrostatic problems and steady current problems. As stated above both the cases are based on the solution of Laplace's equation. In the former the two metallic surfaces act as the two plates of a capacitor, while in the later they are the electrodes through which the current enters and leaves the material.

It must be mentioned here that although the technique, as applied to the problem discussed above, did not offer any difficulty in the calculation of the resistance, such a calculation for the material of arbitrary shape is not always easy even if the material is ohmic.

4.4 Magnetic Effects

We are now in a position to discuss the relation between electricity and magnetism. The earliest workers treated electricity and magnetism as two separate subjects. It is true that the interaction of magnets can be explained on the basis of an inverse square law similar to that in electricity viz.

$$F = K\frac{m_1 m_2}{r^2} \quad (4.22)$$

where m_1, m_2 are the pole strengths. But this does not imply that there are magnetic free charges as there are in electricity. The magnetic poles are analogous to the polarization charges in insulators, the smallest entity being a dipole and not a simple pole. Hence, no attempt was made to establish a connection between them. Some events that occurred in the early nineteenth century, however, brought about a rapid development in this direction. Oesterd's discovery that electric current exerts forces on a magnetized needle, similar to those exerted by a permanent magnet, were reported by Arago to the French Academy on 11th September, 1820. This indicated that there are forces, not only between one charge and another, or between one magnet and another, but also between moving charges and magnets. This, for the first time, brought the subjects of electricity and magnetism together and thus "electromagnetism" emerged. Exactly a week after the news of Oesterd's discovery reached Academy, Ampère showed that two parallel wires carrying current attract each other if the currents are in the same direction, and repel each other if the currents are in opposite direction. Ampere continued his researches unabated, and, within three years, produced a brilliant mathematical analysis of the whole situation and developed the theory of the equivalence of magnets with circuits carrying currents. He published his results in a memoir. Half a century later, Maxwell described him as "Newton of electricity" and his work as "one of the most brilliant achievements of science". Ampère gave the name *electrodynamics* to the science which deals with the mutual action of currents.

Assuming that the force between the elements of two different circuits acts along the line joining them, Ampère derived an expression for the force between the circuits which, eventually, proved to be wrong. This was evidently due to the erroneous assumption made by Ampère that force acts along the line joining the elements, for, in the analogous case of the action between two magnetic molecules, we know that the force is not directed along the line joining them. The correct expression is obtained if this restriction is removed.

4.5 The Magnetic Field

We have referred in the preceding section to Ampère's observations that a wire carrying a current exerts a force on another wire also carrying a current and parallel to it, as shown in Fig. 4.2.

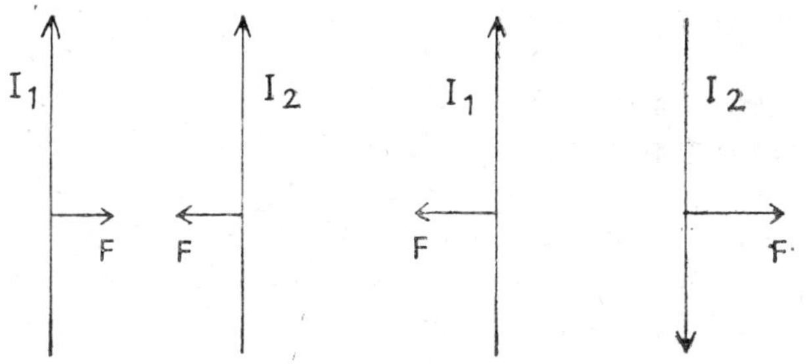

Fig. 4.2

Permanent magnets also exert forces on wires carrying currents. If a wire carrying a current is suspended between the poles of a magnet, it experiences a force perpendicular to the direction of the current and, as a consequence, is deflected. On reversing the direction of the current the deflection is reversed. Thus, a wire carrying a steady current has the same effect on another wire carrying a current as that of a permanent magnet. These effects can be described satisfactorily by introducing the concept of a magnetic field.

The current in one of the wires, say I_1, gives rise to a magnetic field which causes forces to act on the moving charges constituting the current I_2 in the other wire. It has been found experimentally that for a given current I_1 in the first wire, the force on the other wire is proportional to the current I_2 in that wire. Hence, in a given magnetic field, the magnitude of the force on a moving charge is proportional to the charge and to the speed u of the charge. It has also been found to be proportional to the sine of the angle between the vector $\mathbf{u}$ and the magnetic flux density* $\mathbf{B}$, and acts in a direction perpendicular to these vectors. Thus,

$$\mathbf{F} \propto q\mathbf{u} \times \mathbf{B} \quad \text{i.e.} \quad \mathbf{F} = kq\mathbf{u} \times \mathbf{B} \tag{4.23}$$

In Gaussian system, the constant $k = 1/c$; in S.I. units, $k = 1$.

If both electric and magnetic fields are present, the force experienced by a moving charge q is

$$\mathbf{F} = q\mathbf{E} + q\mathbf{u} \times \mathbf{B} \tag{4.24}$$

This force is known as the **Lorentz** force.

The law (4.24) is found to be true even for particles moving at speeds close to the speed of light. This shows that the charge on a body is independent of its speed. In other words, the charge, like rest mass, is relativistically invariant.

We can use (4.24) to define a unit for $\mathbf{B}$. The unit of $\mathbf{E}$ is volt per metre. The term $\mathbf{u} \times \mathbf{B}$ also must have the same unit. Hence, the unit of $\mathbf{B}$ is volt second per metre square. The volt second is known as a Weber (Wb), and, hence, $\mathbf{B}$ is expressed in terms of Webers per square metre. The unit can also be expressed as newton per ampere per metre and is known as the tesla (T).

4.6 Force on a Current

In the preceding section we considered the force on just a single particle. Let us now consider what happens when we have a number of moving charges as, for example, conduction electrons in metals. If the number of electrons per unit volume is N, the force per unit volume will be

$$\mathbf{F} = Nq\mathbf{u} \times \mathbf{B} = \mathbf{j} \times \mathbf{B}$$

(if E is assumed to be zero).

The total force

$$\mathbf{F}_{tot} = \int_V \mathbf{j} \times \mathbf{B}\, d\tau = \int_S \int_l (\mathbf{j} \times \mathbf{B})(d\mathbf{l} \cdot \hat{\mathbf{e}}_n dS) \tag{4.25}$$

where dS is the cross-section of the conductor and dl an element of its length, with its direction the same as that of the current.

Since $\mathbf{j}$ and $d\mathbf{l}$ are parallel

$$\mathbf{F}_{tot} = \int_S \int_l (d\mathbf{l} \times \mathbf{B})(\mathbf{j} \cdot \hat{\mathbf{e}}_n dS) = \int I d\mathbf{l} \times \mathbf{B} \tag{4.26}$$

*We have defined in Chapter 1 electric flux as $\mathbf{E} \cdot \hat{e}_n\, dS$. In the same way, we define magnetic flux as $\mathbf{B} \cdot \hat{e}_n\, dS$.

If the wire carrying the current is in the form of a closed loop

$$\mathbf{F} = \oint I d\mathbf{l} \times \mathbf{B} = I \oint d\mathbf{l} \times \mathbf{B} \tag{4.27}$$

4.7 Biot-Savart Law

A month after Ampère's discovery, Biot and Savart, having analysed the various experiments, formulated appropriate laws relating the magnetic flux density **B** to the current and also the law of force between one current and another:

(i) Let dl_1 be an element of length of a wire with its sense taken in the direction of the current I_1 flowing in the wire (Fig. 4.3). The magnetic

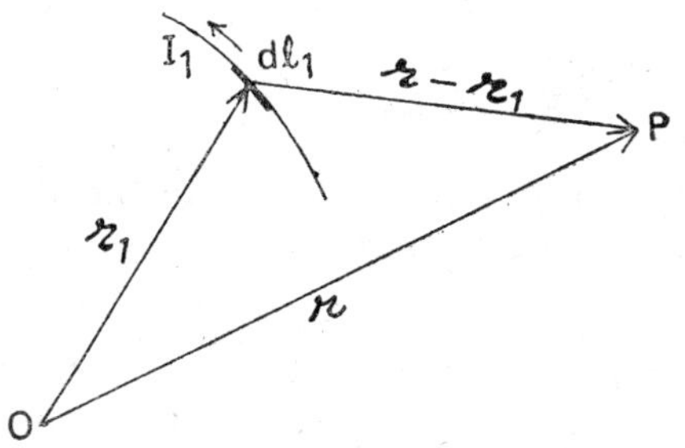

Fig. 4.3

flux density $d\mathbf{B}$ due to this element at a point P specified by the location vector $\mathbf{r}$ is given in magnitude and direction by

$$d\mathbf{B} = k \frac{I_1 d\mathbf{l}_1 \times (\mathbf{r} - \mathbf{r}_1)}{|\mathbf{r} - \mathbf{r}_1|^3} \tag{4.28}$$

where $\mathbf{r}_1$ is the location vector of the element dl. This means, the flux density is directly proportional to the current flowing in the circuit and to the length of the element of the wire and inversely proportional to the square of the distance of the point from the element of the wire. Again, an inverse square law. The constant k depends upon the material as well as on the system of units used. In free space, if the current is measured in e.s.u. and the magnetic flux density in e.m.u., $k = 1/c$. In S.I. units k is taken to be equal to $\frac{\mu_0}{4\pi}$ to ensure compatibility between the experimental law and the S.I. units. The constant μ_0 is termed the permeability of free space and its value is exactly $4\pi \times 10^{-7}$ NA^{-2}

$$\therefore \quad d\mathbf{B} = \frac{\mu_0}{4\pi} \frac{I_1 d\mathbf{l}_1 \times (\mathbf{r} - \mathbf{r}_1)}{|\mathbf{r} - \mathbf{r}_1|^3} \tag{4.29}$$

i.e.,

$$\mathbf{B} = \frac{\mu_0 I_1}{4\pi} \int \frac{d\mathbf{l}_1 \times (\mathbf{r} - \mathbf{r}_1)}{|\mathbf{r} - \mathbf{r}_1|^3} \tag{4.30}$$

Equations (4.29) and (4.30) are the statements of the Biot-Savart law.

(ii) If there are two closed circuits as shown in Fig. 4.4, the force on circuit 2 due to circuit 1 is

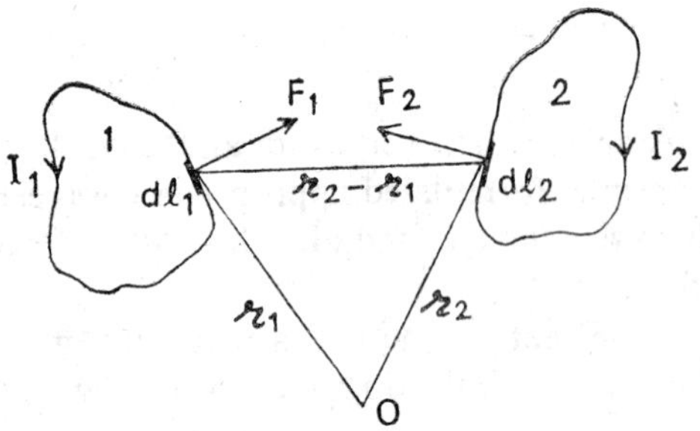

Fig. 4.4

$$\mathbf{F}_2 = \oint I_2 d\mathbf{l}_2 \times \mathbf{B}_2 \tag{4.31}$$

where $\mathbf{B}_2$ is the magnetic flux density due to circuit 1 at the position where circuit 2 is located. We assume $\mathbf{B}_2$ to be constant over the region occupied by the circuit.

$$\therefore \quad \mathbf{F}_2 = \oint_2 I_2\, d\mathbf{l}_2 \times \frac{\mu_0}{4\pi} \oint_1 \frac{I_1 d\mathbf{l}_1 \times (\mathbf{r}_2 - \mathbf{r}_1)}{|\mathbf{r}_2 - \mathbf{r}_1|^3}$$

$$= \frac{\mu_0}{4\pi} I_1 I_2 \oint_2 \oint_1 \frac{d\mathbf{l}_2 \times [d\mathbf{l}_1 \times (\mathbf{r}_2 - \mathbf{r}_1)]}{|\mathbf{r}_2 - \mathbf{r}_1|^3} \tag{4.32}$$

Similarly, the force exerted by the circuit 2 on the circuit 1 is

$$\mathbf{F}_1 = \frac{\mu_0}{4\pi} I_1 I_2 \oint_1 \oint_2 \frac{d\mathbf{l}_1 \times [d\mathbf{l}_2 \times (\mathbf{r}_1 - \mathbf{r}_2)]}{|\mathbf{r}_1 - \mathbf{r}_2|^3} \tag{4.33}$$

But equations (4.32) and (4.33) seem to give rise to an unacceptable situation. One would expect the forces $\mathbf{F}_2$ and $\mathbf{F}_1$ to be equal and opposite. However, since

$$\frac{d\mathbf{l}_2 \times [d\mathbf{l}_1 \times (\mathbf{r}_2 - \mathbf{r}_1)]}{|\mathbf{r}_2 - \mathbf{r}_1|^3} \neq \frac{d\mathbf{l}_1 \times [d\mathbf{l}_2 \times (\mathbf{r}_2 - \mathbf{r}_1)]}{|\mathbf{r}_1 - \mathbf{r}_2|^3}$$

$$\mathbf{F}_2 \neq -\mathbf{F}_1$$

Does this not violate Newton's law? Apparently it does. But let us now expand the integrand in (4.32)

$$\mathbf{F}_2 = \frac{\mu_0}{4\pi} I_1 I_2 \oint_2 \oint_1 \left[\frac{\{d\mathbf{l}_2 \cdot (\mathbf{r}_2 - \mathbf{r}_1)\} d\mathbf{l}_1}{|\mathbf{r}_2 - \mathbf{r}_1|^3} - \frac{(d\mathbf{l}_1 \cdot d\mathbf{l}_2)(\mathbf{r}_2 - \mathbf{r}_1)}{|\mathbf{r}_2 - \mathbf{r}_1|^3}\right]$$

Now, the first integral

$$\oint_2 \frac{d\mathbf{l}_2 \cdot (\mathbf{r}_2 - \mathbf{r}_1)}{|\mathbf{r}_2 - \mathbf{r}_1|^3} = -\oint \nabla_2 \left(\frac{1}{|\mathbf{r}_2 - \mathbf{r}_1|}\right) . d\mathbf{l}_2 = 0$$

(Here ∇_2 operates only on the terms involving r_2)

$$\therefore \quad \mathbf{F}_2 = -\frac{\mu_0}{4\pi} I_1 I_2 \oint_1 \oint_2 \frac{(d\mathbf{l}_1 . d\mathbf{l}_2)(\mathbf{r}_2 - \mathbf{r}_1)}{|\mathbf{r}_2 - \mathbf{r}_1|^3} = -\mathbf{F}_1 \tag{4.34}$$

which, when compared with the expression obtained from (4.33) by similar transformation, shows that it is in conformity with Newton's law.

Let us now examine whether Ampère's observations regarding the forces between parallel wires carrying currents can be accounted for on the basis of Biot and Savart's law.

In Fig. 4.5, we have shown two parallel wires distant 'a' apart and carrying currents I_1, I_2 respectively. Let us assume that in our coordinate system the first wire is along the z-axis and the second, parallel to the first and passing through the point (a, o, o). Let us calculate the magnetic field density at the point $P\ (a, o, z)$, due to the current I_1 in the first wire using (4.29). Let the distance of P from the origin be r.

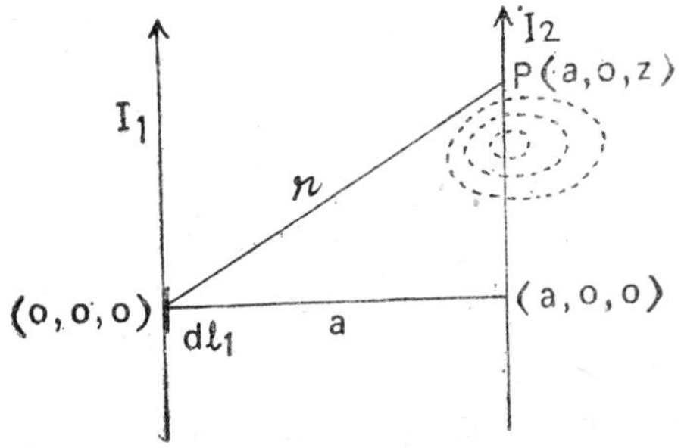

Fig. 4.5

Consider an element dl_1 of the first wire at the origin as shown in the figure. Its components are: (o, o, dz) and those of r are: (a, o, z). Therefore, the components of $d\mathbf{l}_1 \times \mathbf{r}$ are (o, adz, o). This shows that the magnetic field dB at P due to the element dl_1 is parallel to y–axis. Clearly, $\mathbf{B}$ also will be parallel to y–axis with components $B_x = 0$, $B_z = 0$ and

$$B_y = \frac{\mu_0}{4\pi} I_1 \int_{-\infty}^{\infty} \frac{|d\mathbf{l}_1 \times \mathbf{r}|}{|\mathbf{r}|^3} = \frac{\mu_0}{4\pi} I_1 \int_{-\infty}^{\infty} \frac{adz}{(a^2+z^2)^{3/2}}$$

Put $\qquad z = a \tan \theta$

$$\therefore \quad B_y = \frac{\mu_0}{4\pi} I_1 \int_{-\pi/2}^{\pi/2} \frac{a \cdot a \sec^2 \theta}{a^3 \sec^3\theta}\, d\theta = \frac{2\mu_0 I_1}{4\pi a} = \frac{\mu_0 I_1}{2\pi a} \tag{4.35}$$

We see that $\mathbf{B}$ is directly proportional to I_1 and inversely proportional to a. Since $\mathbf{B}$ will always be normal to the plane containing the wire and the radius vector $\mathbf{r}$, the lines of constant $\mathbf{B}$ form closed circles with their centres on the wire in planes perpendicular to the current.

We can use the above result to find the force dF acting on an element dl_2 of the second wire due to the current in the first. The force is given by (4.26). Since B is in positive y-direction and dl_2 in the z-direction, the force acting on dl_2 will be in the negative x-direction, its component in this direction being

$$dF_x = -\frac{\mu_0 I_1 I_2\, dl_2}{2\,\pi a} \tag{4.36}$$

We, thus, see that if the currents are in the same direction, the force is one of attraction. If the currents are opposed, the force is one of repulsion, in conformity with Ampère's observations.

EXAMPLE 4.1. Using Biot and Savart's law find the magnetic flux density **B** due to a circular coil carrying a current I at a point on the axis of the coil.

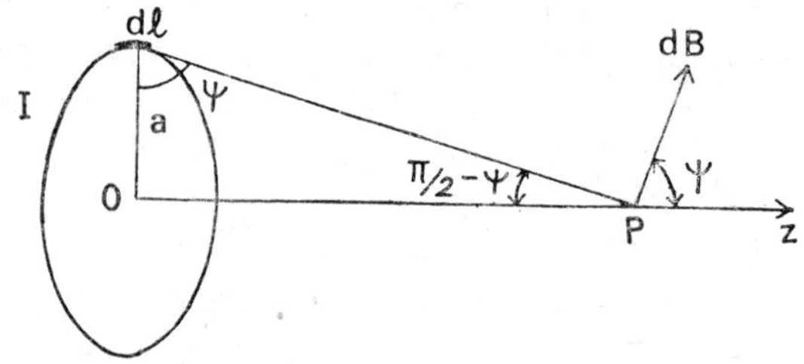

Fig. 4.6

Let 'a' be the radius of the coil and let its axis coincide with the z-axis as in Fig. 4.6. Let us calculate the magnetic flux density at the point P distant r from an element 'dl' of the coil. Since dB, the magnetic field produced by the element dl, is normal to 'dl' as well as 'r', it will act in the direction shown in the figure. On integrating round the coil, the sum of the components of dB, normal to the axis, is zero. The component parallel to the axis is

$$dB = \frac{\mu_0}{4\pi}\frac{Idl}{r^2}\cos\psi$$

where ψ is the angle between the magnetic field dB and the axis.

$$\therefore \qquad B = \frac{\mu_0 I}{4\pi r^2}\int \cos\psi\, dl = \frac{\mu_0 I\cos\psi}{4\pi r^2}\, 2\pi a$$

$$= \frac{\mu_0 I a}{2r^2}\frac{a}{r} = \frac{\mu_0 I a^2}{2(a^2+z^2)^{3/2}} \tag{4.37}$$

EXAMPLE 4.2. Find the field inside a solenoid of length L having N turns uniformly wound round a cylinder of radius 'a'.

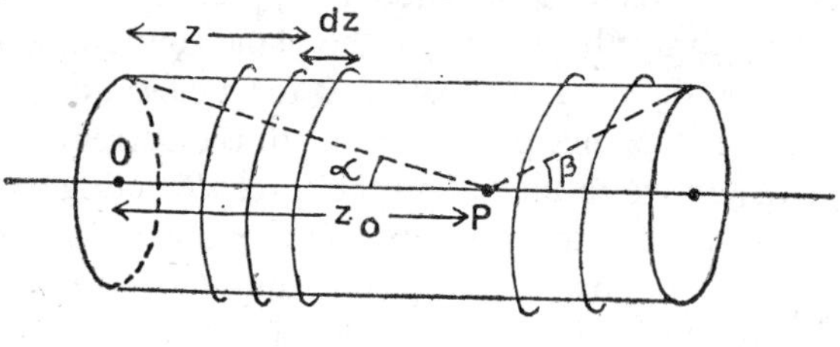

Fig. 4.7

Figure 4.7 shows a solenoid. Let us calculate the field at a point P on the axis of the solenoid. Let P be at a distance z_0 from O. If the turns are closely wound, we may regard the current flowing uniformly round the cylinder. That is, if we consider the length to be divided into elements dz, such as the one shown at a distance z from O, the current in this section of the coil is $\frac{N}{L}\,dzI$, and the field at P due to this element is

$$dB = \frac{1}{2}\frac{\mu_0 NI}{L}\frac{a^2 dz}{[(z_0-z)^2+a^2]^{3/2}} \qquad \text{(see 4.37)}$$

$$\therefore \quad B = \frac{1}{2}\frac{\mu_0 NI}{L}\int_0^L \frac{a^2 dz}{[(z_0-z)^2+a^2]^{3/2}}$$

$$= \frac{1}{2}\frac{\mu_0 NI}{L}\left[-\frac{z_0-z}{\sqrt{(z_0-z)^2+a^2}}\right]_0^L$$

$$= \frac{\mu_0 NI}{2L}\left[\frac{z_0}{\sqrt{z_0^2+a^2}}+\frac{L-z_0}{\sqrt{(L-z_0)^2+a^2}}\right]$$

$$= \frac{\mu_0 NI}{2L}\left[\cos\alpha+\cos\beta\right] \tag{4.38}$$

The angles α, β are shown in Fig. 4.7.

For an infinite solenoid $\alpha = \beta = 0$

and
$$B = \frac{\mu_0 NI}{L} \tag{4.39}$$

We see that B does not depend upon the position of the point P and, hence, we deduce that the flux density has the same value everywhere within the solenoid.

4.8 The Laws of Magnetostatics

The expression for the magnetic flux density (4.30) can be expressed in terms of the current density. Thus

$$\mathbf{B} = \frac{\mu_0 I}{4\pi}\int\frac{d\mathbf{l}_1\times(\mathbf{r}-\mathbf{r}_1)}{|\mathbf{r}-\mathbf{r}_1|^3} = \frac{\mu_0}{4\pi}\int\frac{\mathbf{j}(\mathbf{r}_1)\times(\mathbf{r}-\mathbf{r}_1)}{|\mathbf{r}-\mathbf{r}_1|^3}\,d\tau$$

$$(\because \quad I d\mathbf{l}_1 = \mathbf{j}(\mathbf{r}_1)\,dl_1\cdot\hat{\mathbf{e}}_n\,dS = \mathbf{j}(\mathbf{r}_1)\,d\tau)$$

$$\therefore \quad \mathbf{B} = -\frac{\mu_0}{4\pi}\int_V\left[\mathbf{j}(\mathbf{r}_1)\times\nabla\left(\frac{1}{|\mathbf{r}-\mathbf{r}_1|}\right)\right]d\tau. \tag{4.40}$$

Taking the divergence

$$\nabla\cdot\mathbf{B} = -\frac{\mu_0}{4\pi}\int_V\nabla\cdot\left[\mathbf{j}(\mathbf{r}_1)\times\nabla\left(\frac{1}{|\mathbf{r}-\mathbf{r}_1|}\right)\right]d\tau.$$

Using the identity

$$\nabla\cdot(\mathbf{A}\times\mathbf{C}) = \mathbf{C}\cdot(\nabla\times\mathbf{A}) - \mathbf{A}\cdot(\nabla\times\mathbf{C})$$

we have

$$\nabla \cdot \mathbf{B} = -\frac{\mu_0}{4\pi}\int_V \nabla\left(\frac{1}{|\mathbf{r}-\mathbf{r}_1|}\right)\cdot\{\nabla \times \mathbf{j}(\mathbf{r}_1)\}\, d\tau$$

$$+\frac{\mu_0}{4\pi}\int \mathbf{j}(\mathbf{r}_1)\cdot\left\{\nabla \times \nabla\left(\frac{1}{|\mathbf{r}-\mathbf{r}_1|}\right)\right\} d\tau \tag{4.41}$$

Because ∇ operates only on r, the first integral is zero. The second term contains a factor curl grad $\left(\frac{1}{|\mathbf{r}-\mathbf{r}_1|}\right)$ which is identically zero. Therefore,

$$\nabla \cdot \mathbf{B} = 0 \tag{4.42}$$

This is the first law in magnetostatics corresponding to the relation $\nabla \times \mathbf{E} = 0$ in electrostatics. The relation shows that the magnetic field is solenoidal in contrast to the electric field which is irrotational.

Let us now find the value of $\nabla \times \mathbf{B}$ to complete the analogy with electrostatics.

$$\mathbf{B} = \frac{\mu_0}{4\pi}\int_V \frac{\mathbf{j}(\mathbf{r}_1)\times(\mathbf{r}-\mathbf{r}_1)}{|\mathbf{r}-\mathbf{r}_1|^3}\, d\tau = -\frac{\mu_0}{4\pi}\int_V\left[\mathbf{j}(\mathbf{r}_1)\times\nabla\left(\frac{1}{|\mathbf{r}-\mathbf{r}_1|}\right)\right] d\tau$$

$$= \frac{\mu_0}{4\pi}\nabla \times \int \frac{\mathbf{j}(\mathbf{r}_1)}{|\mathbf{r}-\mathbf{r}_1|}\, d\tau \tag{4.43}$$

$$\therefore \quad \nabla \times \mathbf{B} = \frac{\mu_0}{4\pi}\nabla \times \left(\nabla \times \int \frac{\mathbf{j}(\mathbf{r}_1)}{|\mathbf{r}-\mathbf{r}_1|}\, d\tau\right).$$

Using the identity

$$\nabla \times (\nabla \times \mathbf{A}) = \nabla(\nabla \cdot \mathbf{A}) - \nabla^2 \mathbf{A}$$

we have,

$$\nabla \times \mathbf{B} = \frac{\mu_0}{4\pi}\nabla\left(\nabla \cdot \int \frac{\mathbf{j}(\mathbf{r}_1)}{|\mathbf{r}-\mathbf{r}_1|}\, d\tau\right) - \frac{\mu_0}{4\pi}\nabla^2 \int \frac{\mathbf{j}(\mathbf{r}_1)}{|\mathbf{r}-\mathbf{r}_1|}\, d\tau$$

$$= \frac{\mu_0}{4\pi}\nabla \int \mathbf{j}(\mathbf{r}_1)\cdot\nabla\left(\frac{1}{|\mathbf{r}-\mathbf{r}_1|}\right) d\tau - \frac{\mu_0}{4\pi}\int \mathbf{j}(\mathbf{r}_1)\nabla^2\left(\frac{1}{|\mathbf{r}-\mathbf{r}_1|}\right) d\tau. \tag{4.44}$$

We know that

$$\nabla^2\left(\frac{1}{|\mathbf{r}-\mathbf{r}_1|}\right) = -4\pi\delta(\mathbf{r}-\mathbf{r}_1) \qquad \text{(see 3.162).} \tag{4.45}$$

We have also the relation

$$\nabla\left(\frac{1}{|\mathbf{r}-\mathbf{r}_1|}\right) = -\nabla_1\left(\frac{1}{|\mathbf{r}-\mathbf{r}_1|}\right) \tag{4.46}$$

where ∇_1 operates on r_1 only.

Using (4.45) and (4.46) we can write (4.44) as

$$\nabla \times \mathbf{B} = -\frac{\mu_0}{4\pi}\nabla \int \mathbf{j}(\mathbf{r}_1)\cdot\nabla_1\left(\frac{1}{|\mathbf{r}-\mathbf{r}_1|}\right) d\tau + \frac{\mu_0}{4\pi}\int \mathbf{j}(\mathbf{r}_1)4\pi\delta(\mathbf{r}-\mathbf{r}_1)\, d\tau$$

$$= -\frac{\mu_0}{4\pi} \nabla \int \mathbf{j}(\mathbf{r}_1) \cdot \nabla_1 \left(\frac{1}{|\mathbf{r} - \mathbf{r}_1|} \right) d\tau + \mu_0 \mathbf{j}(\mathbf{r})$$

$$= -\frac{\mu_0}{4\pi} \nabla \int \left[\nabla_1 \cdot \frac{\mathbf{j}(\mathbf{r}_1)}{|\mathbf{r} - \mathbf{r}_1|} - \frac{\nabla_1 \cdot \mathbf{j}(\mathbf{r}_1)}{|\mathbf{r} - \mathbf{r}_1|} \right] d\tau + \mu_0 \mathbf{j}(\mathbf{r})$$

(Note: $\nabla \cdot (\phi \mathbf{A}) = \mathbf{A} \cdot \nabla \phi + \phi \nabla \cdot \mathbf{A}$)

Using Divergence theorem you can show by taking a large enough surface outside the current carrying region that the first term vanishes, while the second is identically zero since $\nabla \cdot \mathbf{j} = 0$ for steady state.

$$\therefore \quad \nabla \times \mathbf{B} = \mu_0 \mathbf{j} \tag{4.47}$$

This is the second law of magnetostatics corresponding to $\nabla \cdot \mathbf{E} = \rho/\epsilon_0$ of electrostatics.

The integral form of this law is readily found. Thus,

$$\int_S \nabla \times \mathbf{B} \cdot \hat{\mathbf{e}}_n \, dS = \mu_0 \int \mathbf{j}(\mathbf{r}) \cdot \hat{\mathbf{e}}_n dS.$$

By Stokes' theorem

$$\int_S \nabla \times \mathbf{B} \cdot \hat{\mathbf{e}}_n \, dS = \oint \mathbf{B} \cdot d\mathbf{l}$$

$$\therefore \quad \oint \mathbf{B} \cdot d\mathbf{l} = \mu_0 \int \mathbf{j}(\mathbf{r}) \cdot \hat{\mathbf{e}}_n dS = \mu_0 I \tag{4.48}$$

This is known as **Ampère's law**: The line integral of flux density round any closed path is equal to μ_0 times the current flowing through the area enclosed by the path.

We summarize the laws of electrostatics and magnetostatics in the following table:

Electrostatics	Magnetostatics
$\nabla \times \mathbf{E} = 0$	$\nabla \cdot \mathbf{B} = 0$
$\nabla \cdot \mathbf{E} = \rho/\epsilon_0$	$\nabla \times \mathbf{B} = \mu_0 \mathbf{J}$

4.9 The Magnetic Potentials

(a) *Magnetic scalar potential*

In electrostatics we have shown that once the electrostatic potential V is found, **E** can easily be calculated. Is there any such potential in magnetostatics, from which magnetic field can be calculated?

Since $\nabla \times \mathbf{E} = 0$, **E** can be expressed as the gradient of a scalar quantity V. The table given in the preceding section shows that if $j = 0$, analogous situation arises in magnetostatics viz. $\nabla \times \mathbf{B} = 0$. We may, therefore, express **B** as a gradient of a scalar quanity Φ_m, i.e.

$$\mathbf{B} = -\nabla \Phi_m \tag{4.49}$$

We call Φ_m **magnetic scalar potential.**

We have the relation $\nabla \cdot \mathbf{B} = 0$. Substituting for $\mathbf{B}$ we get

$$\nabla \cdot \mathbf{B} = -\nabla \cdot \nabla \Phi_m = -\nabla^2 \Phi_m = 0. \tag{4.50}$$

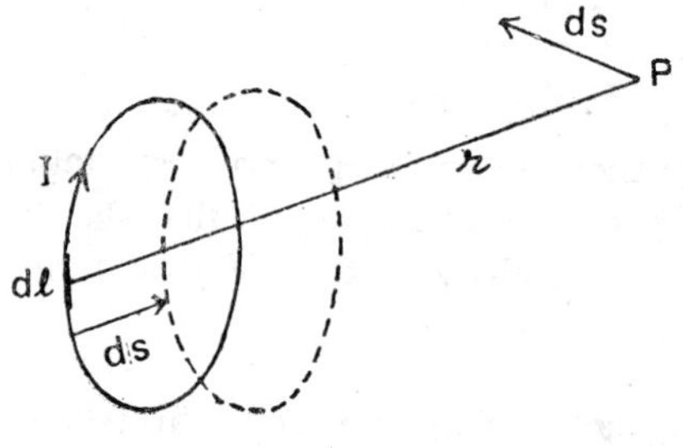

Fig. 4.8

We see that Φ_m satisfies Laplace's equation. In Fig. 4.8, we have shown a circuit carrying a current I. The magnetic flux density at P due to the current in the circuit is given by

$$\mathbf{B} = \frac{\mu_0}{4\pi} I \int \frac{d\mathbf{l} \times \mathbf{r}}{|\mathbf{r}|^3}.$$

If the point P is displaced through a distance $\delta\mathbf{s}$, the change in the potential will be

$$\delta\Phi = -\mathbf{B} \cdot \delta\mathbf{s} = -\delta\mathbf{s} \cdot \frac{\mu_0 I}{4\pi} \int \frac{d\mathbf{l} \times \mathbf{r}}{|\mathbf{r}|^3}$$

$$= -\frac{\mu_0 I}{4\pi} \int \frac{\delta\mathbf{s} \cdot (d\mathbf{l} \times \mathbf{r})}{|\mathbf{r}|^3} \tag{4.51}$$

where δs is taken inside the sign of integration as it is a constant during the integration.

Suppose now that the point P is fixed and the circuit is displaced by an amount $-\delta s$. The change in the potential should be the same as above. In such a displacement an element dl of the circuit sweeps an area $dl\delta s \sin\theta$, if θ is the angle between dl and δs. Vectorially, the area swept out is $-\delta\mathbf{s} \times d\mathbf{l}$. The angle subtended at P by this area is

$$d\Omega = -\frac{\mathbf{r} \cdot (\delta\mathbf{s} \times d\mathbf{l})}{|\mathbf{r}|^3} = -\frac{\delta\mathbf{s} \cdot (d\mathbf{l} \times \mathbf{r})}{|\mathbf{r}|^3}.$$

The relation (4.51), therefore, can be written as

$$\delta\Phi = \frac{\mu_0 I}{4\pi} d\Omega$$

Hence, the potential Φ at $P = \frac{\mu_0 I}{4\pi} \Omega$ (4.52)

Compare this with the potential of an electrostatic double layer (1.72) viz. $\Phi(r) = \frac{D}{4\pi\epsilon_0} d\Omega.$

We conclude that the magnetic scalar potential has mathematically the same properties as the electrostatic double layer potential.

Now $$\int \mathbf{B}\cdot d\mathbf{l} = -\int \nabla\Phi_m \cdot d\mathbf{l} = -\Phi_m \tag{4.53}$$

The quantity $-\Phi_m$ is sometimes known as "magnetomotive force".

(b) *Vector potential*

The magnetic scalar potential is only meaningful in the region of space where there is no current. Magnetic flux density **B** can be found from the scalar function Φ_m only if $j = 0$. Is there any other potential function which is not subject to this limitation and is valid in the region in which j is finite? If there is such a potential, it has to satisfy the fundamental relation

$$\nabla\cdot\mathbf{B} = 0. \tag{4.54}$$

We know that

$$\text{div curl } \mathbf{A} = \nabla\cdot\nabla\times\mathbf{A} = 0.$$

We can, therefore, express **B** as curl of a vector **A**

$$\mathbf{B} = \nabla\times\mathbf{A} \tag{4.55}$$

The vector **A** which satisfies (4.55) is known as "**vector potential**".

One can easily see that **A** is not uniquely defined by Eq. (4.55). For, we could add to **A** any function whose curl is zero, say, gradient of a scalar ψ, and still have the same **B**.

$$\nabla\times(\mathbf{A} + \nabla\psi) = \nabla\times\mathbf{A} + \nabla\times\nabla\psi = \nabla\times\mathbf{A} = \mathbf{B}.$$

We may remind the reader that in electrostatics the scalar potential V was not completely specified by its definition $E = -\nabla V$. If V is the potential for some problem, a different potential $V' = V + C$ where C is a constant, gives the same field

$$-\nabla V' = -\nabla(V + C) = -\nabla V - \nabla C = -\nabla V = \mathbf{E}$$

To make **A** more specific, we have to impose an additional restriction on it. We have to do it without affecting **B**. In magnetostatics a convenient condition that **A** has to satisfy is

$$\nabla\cdot\mathbf{A} = 0. \tag{4.56}$$

The reason for this choice is that it makes calculations easier than with any other choice. In electrodynamics we have to choose it differently.

Thus $\mathbf{B} = \nabla\times\mathbf{A}$ and $\nabla\cdot\mathbf{A} = 0$ together define the vector potential **A** which satisfies the fundamental equation $\nabla\cdot\mathbf{B} = 0$.

We have so far tried to draw a parallel between electrostatics and magnetostatics in respect of the two fundamental equations. We now ask: "is there a relation in magnetostatics corresponding to Poisson's equation $\nabla^2 V = \rho/\epsilon_0$ in electrostatics?"

We have Ampère's law (4.47)

$$\nabla \times \mathbf{B} = \mu_0 \mathbf{j}.$$

Now
$$\nabla \times \mathbf{B} = \nabla \times \nabla \times \mathbf{A} = \nabla (\nabla \cdot \mathbf{A}) - \nabla^2 \mathbf{A}$$
$$= -\nabla^2 \mathbf{A} \quad \text{(Thanks to our choice of the condition } \nabla \cdot \mathbf{A} = 0)$$

$$\therefore \quad \nabla^2 \mathbf{A} = -\mu_0 \mathbf{j}. \tag{4.57}$$

This equation is similar to Poisson's equation except that **A** is a vector. This should not cause any difficulty, since each component of **A** must satisfy the differential equation. Thus,

$$\nabla^2 A_x = -\mu_0 J_x; \; \nabla^2 A_y = -\mu_0 J_y; \; \nabla^2 A_z = -\mu_0 J_z, \tag{4.58}$$

the solution being

$$A_x = \frac{\mu_0}{4\pi} \int \frac{J_x}{|\mathbf{r}|} d\tau \text{ etc.} \tag{4.59}$$

The general solution in the vector form is

$$\mathbf{A} = \frac{\mu_0}{4\pi} \int \frac{\mathbf{j}}{|\mathbf{r}|} d\tau \tag{4.60}$$

Thus, the field produced by a current can be computed by first determining **A** using (4.60) and substituting this in relation (4.55).

EXAMPLE 4.3. Find the vector potential and hence the magnetic flux density **B** due to an infinite wire carrying a current, at a point (i) outside, (ii) inside the wire.

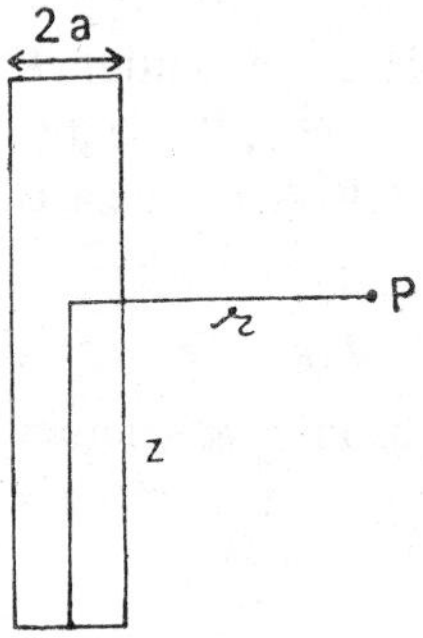

Fig. 4.9

Let the permeability inside the wire be μ_1. Let us further assume that the axis of the wire coincides with the z-axis. The only component of the current density, therefore, is j_z and, hence, the only component of **A** is A_z. It is convenient in this case to use cylindrical coordinates r, ϕ, z.

Inside the wire

$$\nabla^2 A_z = -\mu_1 j_z = -\frac{\mu_1 I}{\pi a^2} \tag{4.61}$$

where 'a' is the radius of cross-section of the wire.

Since A_z is independent of z and ϕ, we can write (4.61) as

$$\frac{1}{r} \frac{\partial}{\partial r}\left(r \frac{\partial A_z}{\partial r} \right) = -\frac{\mu_1 I}{\pi a^2}$$

Integrating

$$r \frac{\partial A_z}{\partial r} = -\frac{\mu_1 I r^2}{2\pi a^2} + \text{a constant} \tag{4.62}$$

Since $r \dfrac{\partial A_z}{\partial r} = 0$ at $r = 0$, the constant vanishes. Integrating again

$$A_z = -\frac{\mu_1 I r^2}{4\pi a^2} + \text{a constant}$$

If we assume A_z to be zero at $r = a$

$$A_z = \frac{\mu_1 I}{4\pi}\left(1 - \frac{r^2}{a^2}\right). \tag{4.63}$$

Since $\mathbf{B} = \nabla \times \mathbf{A}$ and the components of curl $\mathbf{A}$ are

$$(\text{curl } \mathbf{A})_r = \frac{1}{r}\frac{\partial A_z}{\partial \phi} - \frac{\partial A_\phi}{\partial z}$$

$$(\text{curl } \mathbf{A})_\phi = \frac{\partial A_r}{\partial z} - \frac{\partial A_z}{\partial r} \tag{4.64}$$

$$(\text{curl } \mathbf{A})_z = \frac{\partial A_\phi}{\partial r} + \frac{A_\phi}{r} - \frac{1}{r}\frac{\partial A_r}{\partial \phi}$$

we arrive at the result

$$B_\phi = \frac{\mu_1 I r}{2\pi a^2} \tag{4.65}$$

The same result can be arrived at using Ampère's law.

Outside the wire

$$\frac{1}{r}\frac{\partial}{\partial r}\left(r\frac{\partial A_z}{\partial r}\right) = 0$$

$$\therefore \quad \frac{\partial A_z}{\partial r} = \frac{C}{r} \tag{4.66}$$

where C is a constant and $A_z = C \ln r + C'$ (another constant). Since at $r = a$, $A_z = 0$, $C' = -C \ln a$

$$\therefore \quad A_z = C \ln \frac{r}{a} \tag{4.67}$$

The value of C can be found from the boundary condition for $\frac{\partial A_z}{\partial r}$ at $r = a$.

Thus, since $\mathbf{B} = \nabla \times \mathbf{A}$

$$B_\phi = -\frac{\partial A_z}{\partial r}$$

Now B_ϕ must be continuous at $r = a$. From (4.63)

at $r = a$, $-\frac{\partial A_z}{\partial r} = \frac{\mu_1 I}{2\pi a}$ and from (4.66) $\quad -\frac{\partial A_z}{\partial r} = -\frac{C}{a}$

$$\therefore \quad -\frac{C}{a} = \frac{\mu_1 I}{2\pi a} \quad \text{i.e.} \quad C = -\frac{\mu_1 I}{2\pi}$$

$$\text{and } A_z = -\frac{\mu_1 I}{2\pi} \ln \frac{r}{a}. \tag{4.68}$$

From this B can be found.

$$B = \frac{\mu_1 I}{2\pi r} \tag{4.69}$$

4.10 Current Loops in External Fields—Magnetic Dipole

In this section we shall discuss the magnetic properties of small current loops such as those formed by electrons revolving round the nucleus in atoms. The results we arrive at, will be useful in the study of the magnetic behaviour of atoms.

Although for convenience we will be considering rectangular loops, the theory developed is applicable to circular loops also.

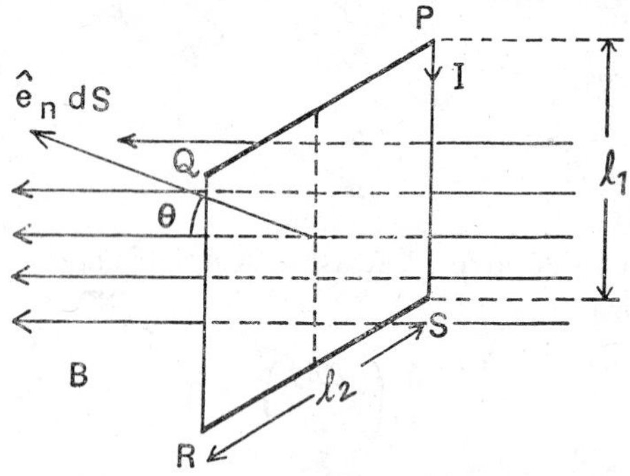

Fig. 4.10

In Fig. 4.10 we have shown a simple loop of wire in rectangular form carrying a current and suspended in a vertical plane in an horizontal magnetic field **B** in such a way that the normal to the plane of the coil makes an angle θ with the direction of the field. Let 'a' be the cross-section of the wire. The current flowing in the wire is

$$I = Neua \tag{4.70}$$

where N is the number of free electrons per unit volume, e the charge on the electrons and **u** its velocity. The force dF_1 acting on a small element dl_1 of the side PS of the loop is

$$d\mathbf{F}_1 = Neau\, d\mathbf{l}_1 \times \mathbf{B} = I d\mathbf{l}_1 \times \mathbf{B} \tag{4.71}$$

where the vector $d\mathbf{l}_1$ is in the direction of the current.

The total force on the side PS is

$$\mathbf{F}_1 = I\mathbf{l}_1 \times \mathbf{B} = I l_1 B \hat{e}_y \quad (\because PS \perp B). \tag{4.72}$$

If we assume that the field B is parallel to x-axis and the current I is parallel to z-axis, the fore F, acts in the positive y-direction as shown in Fig. 4.11. The figure presents a view looking down on the coil. A

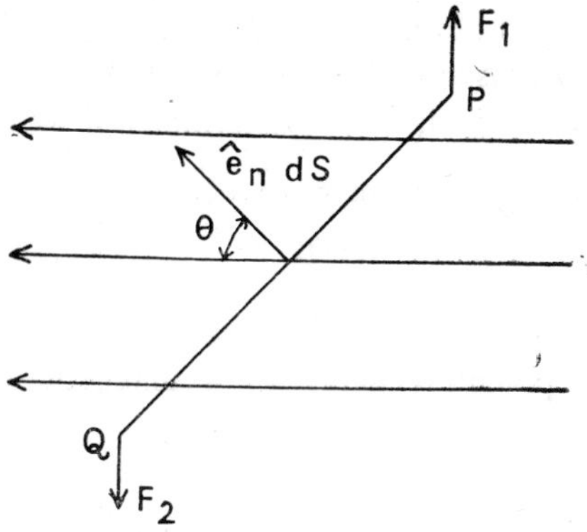

Fig. 4.11

force $\mathbf{F}_2$, equal and opposite to $\mathbf{F}_1$, acts on the side *QR*. In the same way there will be forces acting on the sides *PQ* and *SR* which are equal and opposite. The net force on the loop, therefore, is zero and, hence, there will not be any translational motion of the loop. There will, however, be a torque **T** acting on the coil tending to turn it about the vertical axis. This torque is given by

$$|\mathbf{T}| = Il_1 Bl_2 \sin\theta = IdS\, B \sin\theta \tag{4.73}$$

where $dS = l_1 l_2$ is the area enclosed by the loop. In vector notation

$$\mathbf{T} = I\hat{\mathbf{e}}_n dS \times \mathbf{B} \tag{4.74}$$

The torque tends to decrease the angle θ. If U_P is the potential energy of the loop, the torque tending to increase θ is $-\dfrac{\partial U_P}{\partial \theta}$

$$\therefore\ -\frac{\partial U_P}{\partial \theta} = -IdSB\sin\theta$$

and $$U_P = -IdSB\cos\theta + \text{a constant.}$$

If U_P is zero when $\theta = \pi/2$, we have

$$U_P = -IdSB\cos\theta = -I\hat{\mathbf{e}}_n dS\cdot\mathbf{B} \tag{4.75}$$

The quantity $\Phi = \mathbf{B}\cdot\hat{\mathbf{e}}_n dS$ is the magnetic flux through the circuit.

$$\therefore\ U_P = -I\Phi \tag{4.76}$$

(Note : Φ should not be confused with Φ_m)

We may now compare Eq. (4.75) with Eq. (1.66) in which the potential energy of an electric dipole of dipole moment **p** in an electric field **E** is given by

$$U = -\mathbf{p}\cdot\mathbf{E} \tag{4.77}$$

We see that the magnetic field produced by a current loop is similar in form to the electrostatic field of an electric dipole.

We conclude that the forces and torques acting on a small loop of current, placed in a magnetic field, can be determined if we associate with the loop a magnetic dipole moment $\mathbf{m} = I\hat{e}_n \, dS$

$$\therefore \quad U_P = -\mathbf{m}\cdot\mathbf{B}. \tag{4.78}$$

4.11 Magnetic Dipole in a Non-uniform Magnetic field

We have seen that the potential energy of a magnetic dipole **m** in a magnetic field **B** is

$$U = -(\mathbf{m}\cdot\mathbf{B}).$$

We know that the force is related to the potential energy by the relation

$$\mathbf{F} = -\nabla U$$

$$\therefore \quad \mathbf{F} = \nabla(\mathbf{m}\cdot\mathbf{B}) = (\mathbf{B}\cdot\nabla)\,\mathbf{m} + (\mathbf{m}\cdot\nabla)\,\mathbf{B} + \mathbf{B}\times(\nabla\times\mathbf{m}) + \mathbf{m}\times(\nabla\times\mathbf{B}) \tag{4.79}$$

where we have used an important vector identity (see Appendix A).

Since **m** is not a function of the space coordinates,

$$(\mathbf{B}\cdot\nabla)\,\mathbf{m} = 0, \; \nabla\times\mathbf{m} = 0$$

$$\therefore \quad \mathbf{F} = \mathbf{m}\times(\nabla\times\mathbf{B}) + (\mathbf{m}\cdot\nabla)\,\mathbf{B}$$

But $\quad \nabla\times\mathbf{B} = 0$

$$\therefore \quad \mathbf{F} = (\mathbf{m}\cdot\nabla)\,\mathbf{B}. \tag{4.80}$$

4.12 Magnetic Vector Potential due to a Small Current Loop

Consider a current circuit of radius 'a' and carrying a current I (Fig. 4.12). We will show how the vector potential of such a current loop can be calculated following a simple procedure in which Cartesian coordinates are used. In the next section we will give an alternative method. Let us assume that the origin of the frame of reference coincides with the centre of the coil and its z-axis is perpendicular to the coil. Let 'r' be the

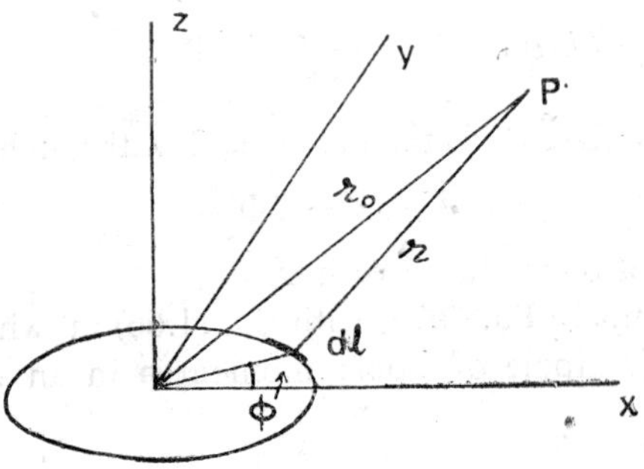

Fig. 4.12

distance of a point P from an element dl of the circuit and ϕ be the azimuthal angle of the element. The components of the element $dl = ad\phi$ are $(-\ a \sin \phi\, d\phi,\ a \cos \phi\ d\phi,\ 0)$

Since
$$\mathbf{A} = \frac{\mu_0}{4\pi} \int \frac{\mathbf{j}}{|\mathbf{r}|}\, d\tau = \frac{\mu_0 I}{4\pi} \int \frac{d\mathbf{l}}{|\mathbf{r}|},$$

its x component is
$$A_x = \frac{\mu_0 I}{4\pi} \int \frac{-\ a \sin \phi\, d\phi}{|\mathbf{r}|}$$

Now
$$r^2 = (x - a \cos \phi)^2 + (y - a \sin \phi)^2 + z^2$$

and because $a << r_0$
$$\frac{1}{r} = \frac{1}{r_0} + \frac{ax \cos \phi + ay \sin \phi}{r_0^3} + \ldots.$$

where r_0 is the distane of P from the origin.

$$\therefore \quad A_x = -\frac{\mu_0 Ia}{4\pi} \int_0^{2\pi} \left(\frac{1}{r_0} + \frac{ax \cos \phi + ay \sin \phi)}{r_0^3} \right) \sin \phi\, d\phi$$

$$= -\frac{\mu_0 I\, a^2 \pi y}{4\pi r_0^3}. \tag{4.81}$$

Similarly, $A_y = \dfrac{\mu_0 I\, a^2 \pi x}{4\pi r_0^3}$ and $A_z = 0.$ (4.82)

The magnitude of the dipole moment equivalent to the circuit is
$$m = \text{area} \times \text{current} = \pi a^2 I$$

and its direction is normal to the plane of the coil i.e. along the z-axis.

$$\therefore \quad m_x = 0,\ m_y = 0,\ m_z = \pi a^2 I$$

Hence, the components of $\mathbf{A}$ (4.81, 4.82) are proportional to those of the vector $\mathbf{m} \times \mathbf{r}$. We can, therefore, write, dropping the subscripts,

$$\mathbf{A} = \frac{\mu_0}{4\pi r^3} \left(\mathbf{m} \times \mathbf{r} \right) = -\frac{\mu_0}{4\pi} \left\{ \mathbf{m} \times \nabla \left(\frac{1}{r} \right) \right\} \tag{4.83}$$

4.13 An Alternative Method for Finding the Vector Potential A and, hence, the Field B due to a Current Loop.

Consider a filament circuit C carrying a direct current I (Fig. 4.13).

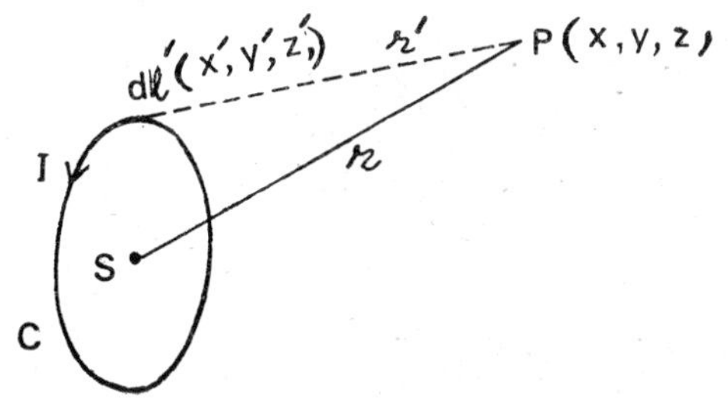

Fig. 4.13

The vector potential **A** at a point $P(x, y, z)$ due to the current loop is

$$\mathbf{A} = \frac{\mu_0 I}{4\pi}\oint \frac{d\mathbf{l}'}{r'} \tag{4.84}$$

where $dl'\,(x', y', z')$ is an element of the circuit.

By Stokes' theorem (see Appendix A)

$$\oint \frac{d\mathbf{l}'}{r'} = \int_S \hat{\mathbf{e}}_n \times \nabla'\left(\frac{1}{r'}\right) dS$$

$$\therefore \quad \mathbf{A} = \frac{\mu_0 I}{4\pi}\int_S \hat{\mathbf{e}}_n \times \nabla'\left(\frac{1}{r'}\right) dS$$

$$= \frac{\mu_0 I}{4\pi}\int_S \left(\hat{\mathbf{e}}_n \times \frac{\hat{\mathbf{e}}_r'}{r'^2}\right) dS \tag{4.85}$$

where $\hat{\mathbf{e}}_n$, $\hat{\mathbf{e}}_r'$ are the unit vectors perpendicular to the surface enclosed by the circuit and along $\mathbf{r}'$ respectively. If the dimensions of the current loop are very small compared with the distance to the point P, the factor $\frac{\hat{\mathbf{e}}_r'}{r'^2}$ is very nearly constant during integration. We write it as $\frac{\hat{\mathbf{e}}_r}{r^2}$

$$\therefore \mathbf{A} = -\frac{\mu_0 I}{4\pi}\frac{\hat{\mathbf{e}}_r}{r^2} \times \int_S \hat{\mathbf{e}}_n\, dS = -\frac{\mu_0 I}{4\pi r^2}\,\hat{\mathbf{e}}_r \times \hat{\mathbf{e}}_n S$$

$$= \left(\frac{\mu_0}{4\pi}\right)\left(\mathbf{m} \times \frac{\hat{\mathbf{e}}_r}{r^2}\right) = \left(\frac{\mu_0}{4\pi}\right)\left(\frac{\mathbf{m} \times \mathbf{r}}{|\mathbf{r}|^3}\right) \tag{4.86}$$

where $\mathbf{m} = \hat{\mathbf{e}}_n IS$, the magnetic dipole moment of the loop.

The magnetic flux density **B** is

$$\mathbf{B} = \nabla \times \mathbf{A} = \frac{\mu_0}{4\pi}\nabla \times \left(\frac{\mathbf{m} \times \mathbf{r}}{|\mathbf{r}|^3}\right)$$

$$= \frac{\mu_0}{4\pi}\left[\mathbf{m}\nabla\cdot\left(\frac{\mathbf{r}}{|\mathbf{r}|^3}\right) - (\mathbf{m}\cdot\nabla)\left(\frac{\mathbf{r}}{|\mathbf{r}|^3}\right)\right]$$

Since $\nabla.\left(\frac{\mathbf{r}}{|\mathbf{r}|^3}\right) = 0$

$$\mathbf{B} = -\frac{\mu_0}{4\pi}(\mathbf{m}\cdot\nabla)\left(\frac{\mathbf{r}}{|\mathbf{r}|^3}\right).$$

Now $\nabla\left(\mathbf{m}\cdot\frac{\mathbf{r}}{|\mathbf{r}|^3}\right) = (\mathbf{m}\cdot\nabla)\frac{\mathbf{r}}{|\mathbf{r}|^3} + \mathbf{m} \times \nabla \times \left(\frac{\mathbf{r}}{|\mathbf{r}|^3}\right).$

The last term vanishes.

$$\therefore \quad (\mathbf{m}\cdot\nabla)\frac{\mathbf{r}}{|\mathbf{r}|^3} = \nabla\left(\mathbf{m}\cdot\frac{\mathbf{r}}{|\mathbf{r}|^3}\right)$$

$$= \left[\frac{\mathbf{m}}{|\mathbf{r}|^3} - \frac{3(\mathbf{m}\cdot\mathbf{r})\mathbf{r}}{|\mathbf{r}|^5}\right]$$

and $$\mathbf{B} = \frac{\mu_0}{4\pi r^3}\left[\frac{3(\mathbf{m}\cdot\mathbf{r})\mathbf{r}}{r^2} - \mathbf{m}\right] \tag{4.87}$$

This relation is similar to the relation (1.65) for the field of an electric dipole.

Ampère believed that all magnetic effects were due to current loops. He proposed the hypothesis that each atom is really a minute current loop and further suggested that the magnetic effects in iron were due to atomic currents. This was indeed a remarkable suggestion, particularly considering that it was made at a time when there was no knowledge of atomic structure.

One can easily verify that expressing **m** in terms of current I, we can write (4.87) in the form

$$\mathbf{B} = -\frac{\mu_0}{4\pi}\,\text{grad}\left(\frac{I\hat{\mathbf{e}}_n S\cdot\mathbf{r}}{|\mathbf{r}|^3}\right)$$

$$= -\,\text{grad}\left\{\frac{\mu_0 I}{4\pi}\frac{\hat{\mathbf{e}}_n S\cdot\mathbf{r}}{|\mathbf{r}|^3}\right\}$$

$$= -\,\text{grad}\left\{\frac{\mu_0 I}{4\pi}\,\Omega\right\} \tag{4.88}$$

where Ω is the solid angle subtended by the loop at **r**.

$$\therefore \qquad \mathbf{B} = -\,\text{grad}\,\Phi m \qquad \text{(see 4.52)} \tag{4.89}$$

in conformity with the relation (4.49).

This expression does not contain any indication of the shape of the loop. It is, therefore, reasonable to assume that it is applicable to a small loop of any shape and the loop does not even have to be plane.

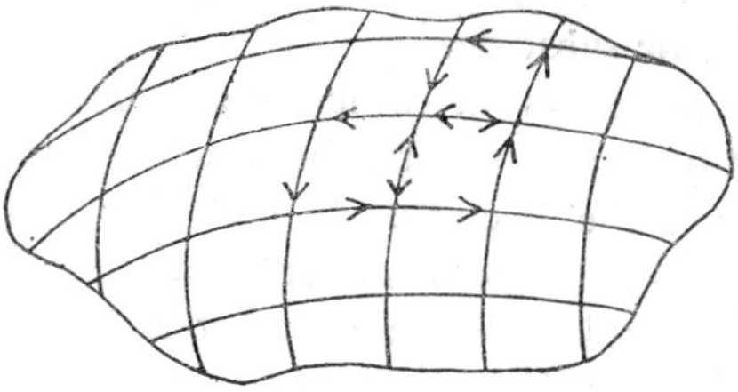

Fig. 4.14

By using the method of sub-division and super-position we can show that the results arrived at above can be extended to the case of a larger loop. Whatever its shape, it is possible to represent a finite current circuit as made up of small meshes (Fig. 4.14). Each mesh may be regarded as having a current I flowing round its edges. Summation of the current in all the meshes leaves only the current in the circuit as the resultant, since the currents in the meshes cancel out everywhere except at the boundary. Each mesh contributes the amount $\frac{\mu_0 I}{4\pi}\,d\Omega$ to the potential. The flux density due to the large loop is the same as the sum of those of the individual small loops.

$$\therefore \qquad \mathbf{B} = -\nabla\left(\frac{\mu_0 I}{4\pi}\Omega\right) \tag{4.90}$$

where Ω is the angle subtended by the loop at P.

We thus have a current loop replaced by a double layer of magnetic poles—a **magnetic shell**.

4.14 Magnetic Media

We have dealt so far with the magnetic effects of currents in vacuum. How is the magnetic field affected when material media are present? We shall discuss this question in the present section.

It has been observed experimentally that a material substance acquires a 'magnetic polarization' when placed in a magnetic field, just as a dielectric medium aquires an electric polarization in an electric field. This is, no doubt, a macroscopic effect. However, it wtll be useful to look first, briefly though, at the physical basis of the effect.

The response of the materials to an applied magnetic field depends on the properties of the individual atoms and molecules, and on their interactions. The orbital motion of the electrons in atoms and molecules provide currents which give rise to the magnetic dipoles. In many materials the small electric currents associated with the orbital motion and spin of the electrons average to zero. When such atoms are placed in a magnetic field, minute electron currents are generated by induction in the clouds of electrons, the direction of which is such that the magnetic field associated with them opposes the inducing field **B** (Chapter 5). These are known as **diamagnetic substances**.

There are other materials in which the atoms do have an intrinsic magnetic moment on account of the fact that the currents from the orbital motions and spins of the electrons do not average to zero. Although electron spins tend to pair and cancel one another, there are atoms in which pairing is incomplete. When such materials are placed in a magnetic field, the magnetic moments of the atoms and the induced magnetism enhances the external field. This effect is more pronounced at low temperatures. Such substances are called **paramagnetic substances.**

There is yet another class of materials which do not shed their magnetic properties even at high temperatures. These are **ferromagnetic substances.** In ferromagnetic materials, the mutual coupling forces between neighbouring molecular dipole moments are sufficiently stronger than the randomizing effect of thermal agitation and the dipoles are all aligned parallel within a small region known as **domain** (Fig. 4.15). The resulting net macroscopic magnetic field in the vicinity of matter can be expected to depend on the relative sizes of these domains and on the orientation of the molecular dipoles within each domain. In a weak external field there is a net alignment of the domains in the field direction giving a very large paramagnetic effect. As the field is increased these domains which are already aligned become enlarged at the expense of the others.

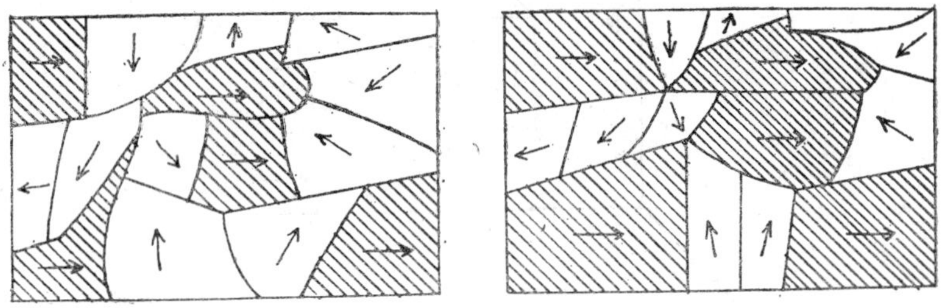

Weak magnetic field   Strong magnetic field

Fig. 4.15

It is no gainsaying the fact that the classical arguments are inadequate to explain the behaviour of magnetic media. A respectable theory of such media must be based on a quantum mechanical formulation. We shall, however, use classical arguments here to explain the behaviour of diamagnetic and paramagnetic substances as they give us some idea of the interactions taking place in such media.

4.15 Magnetization

Every atom or molecule may be regarded as a tiny magnetic dipole with the magnetic moment

$$Q_m\mathbf{l} = \hat{\mathbf{e}}_n I dS \tag{4.91}$$

where Q_m is the magnetic pole strength, 'l' is the pole separation, I is the current and dS is the area of the loop. In material substances these magnetic dipoles are randomly oriented, but if this material is placed in a magnetic field some alignment of these dipoles occurs as in the case of electric dipoles. The material then is said to be *magnetized*. The effect of atomic magnets is described by a quantity called **magnetization M** which is defined as the magnetic dipole moment per unit volume. For the purpose of defining **M**, it is convenient to assume that the medium has a continuous distribution of infinitesimal magnetic dipoles, although the dipoles actually are of discrete, finite size. The assumption of continuous magnetization leads to no appreciable error provided the element of volume is assumed to contain a statistically large number of atoms. However, the element is assumed to be so small on the macroscopic scale that **M** may be regarded as a vector point function. Note also that the magnetization **M** is not necessarily uniform throughout the substance.

Thus,
$$\mathbf{M} = \frac{dm}{d\tau} \tag{4.92}$$

Consider an elementary volume with sides δx, δy and δz at a point

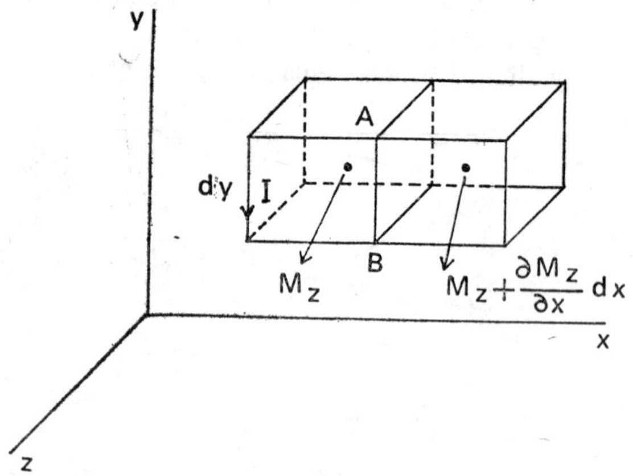

Fig. 4.16

(*x*, *y*, *z*) within a magnetized body as shown in Fig. 4.16. The component of the magnetic moment associated with this volume in the *z*-direction is

$$M_z \, \delta x \, \delta y \, \delta z$$

Now, from (4.91) $M_z \, \delta x \, \delta y \, \delta z = I \, \delta x \, \delta y$ (4.93)

where I is the current flowing in the loop in a plane parallel to XY plane

$$\therefore \quad I = M_z \, \delta z \tag{4.94}$$

The current in the adjacent loop at $(x + \delta x, y, z)$ is

$$I' = I + \frac{\partial I}{\partial x} \delta x = M_z \delta z + \frac{\partial M_z}{\partial x} \delta x \delta z.$$

Hence, the current in the interface AB of the two elements has a component in the *y*-direction given by

$$I - I' = -\frac{\partial M_z}{\delta x} \delta x \delta z$$

In a similar manner, by considering a current loop in the YZ plane, we find another component of current in the *y*-direction, having magnitude $\frac{\partial M_x}{\partial z} \delta x \delta z$. The total current in the *y*-direction is now

$$\left(\frac{\partial M_x}{\partial z} - \frac{\partial M_z}{\partial x} \right) \delta x \delta z.$$

This can be written in terms of the *y*-component of the magnetization current density $\mathbf{j}_M$. Thus

$$j_{My} \delta x \delta z = \left(\frac{\partial M_x}{\partial z} - \frac{\partial M_z}{\partial x} \right) \delta x \delta z$$

$$\therefore \quad i_{My} = \frac{\partial M_x}{\partial z} - \frac{\partial M_z}{\partial x} \tag{4.95}$$

In the same way we can find the other components

$$j_{Mx} = \frac{\partial M_z}{\partial y} - \frac{\partial M_y}{\partial z} \tag{4.96}$$

and

$$j_{Mz} = \frac{\partial M_y}{\partial x} - \frac{\partial M_x}{\partial y}$$

Clearly, the expression on the right-hand side of (4·95) and (4.96) are the components of curl **M**

$$\therefore \quad \mathbf{j}_M = \nabla \times \mathbf{M} \tag{4.97}$$

Notice that magnetizations current flow exists only if **M** is varying. If **M** is uniform $\mathbf{j}_M = 0$.

The same result can be arrived at using vector potential. From Eq. (4.83)

$$\mathbf{A} = \frac{\mu_0}{4\pi}\int\left\{\mathbf{M}(\mathbf{r}) \times \nabla\left(\frac{1}{|\mathbf{r}|}\right)\right\} d\tau \tag{4.98}$$

Using the identity

$$\nabla \times (\phi\mathbf{P}) = \phi\nabla \times \mathbf{P} - \mathbf{P} \times \nabla\phi,$$

we can write (4.98) in the form

$$\mathbf{A} = \frac{\mu_0}{4\pi}\int \frac{\nabla \times \mathbf{M}}{|\mathbf{r}|}\, d\tau - \frac{\mu_0}{4\pi}\int\left\{\nabla \times \left(\frac{\mathbf{M}}{|\mathbf{r}|}\right)\right\} d\tau \tag{4.99}$$

We shall now digress for a while to prove an important relation regarding vectors. This will be useful to us in transforming Eq. (4.99) to a suitable form.

The relation we desire to establish is that for any vector **a**,

$$\int (\nabla \times \mathbf{a})\, d\tau = -\int \mathbf{a} \times \hat{\mathbf{e}}_n dS \tag{4.100}$$

Let '**b**' be a constant vector

$$\therefore \quad \mathbf{b}\cdot\int_S \mathbf{a} \times \hat{\mathbf{e}}_n\, dS = \int_S (\mathbf{b} \times \mathbf{a})\cdot\hat{\mathbf{e}}_n\, dS$$

$$= \int_V \nabla\cdot(\mathbf{b} \times \mathbf{a})\, d\tau \qquad \text{(by Divergence theorem)}$$

$$= -\,\mathbf{b}\cdot\int \nabla \times \mathbf{a}\, d\tau$$

Since **b** is an arbitrary vector

$$\int_S \mathbf{a} \times \hat{\mathbf{e}}_n\, dS = -\int_V \nabla \times \mathbf{a}\, d\tau$$

Using this, we can now write (4.99) as

$$\mathbf{A} = \frac{\mu_0}{4\pi}\int \frac{\nabla \times \mathbf{M}}{|\mathbf{r}|}\, d\tau + \frac{\mu_0}{4\pi}\int_S \frac{\mathbf{M} \times \hat{\mathbf{e}}_n\, dS}{|\mathbf{r}|}$$

Since **M** is localized, the surface integral taken over a surface outside the region in which the current flows, vanishes. Hence,

$$\mathbf{A} = \frac{\mu_0}{4\pi}\int\frac{\nabla \times \mathbf{M}}{|\mathbf{r}|}\,d\tau \tag{4.101}$$

But we have already shown that

$$\mathbf{A} = \frac{\mu_0}{4\pi}\int\frac{\mathbf{j}(\mathbf{r})}{|\mathbf{r}|}\,d\tau$$

$$\therefore\quad \mathbf{j}_M = \nabla \times \mathbf{M}$$

4.16 Magnetic Field Vector

In general, if the medium is electrically conducting and also magnetizable, there will be present both a real current density **j** and the magnetization current density $\mathbf{j}_M$ and both these must be taken into account. That is

$$\nabla \times \mathbf{B} = \mu_0\,(\mathbf{j} + \mathbf{j}_M)$$

$$= \mu_0\,(\mathbf{j} + \nabla \times \mathbf{M}) \tag{4.102}$$

$$\therefore\quad \nabla \times (\mathbf{B} - \mu_0\mathbf{M}) = \mu_0\mathbf{j} \tag{4.103}$$

We now introduce a new vector **H**, defined by

$$\mathbf{B} - \mu_0\,\mathbf{M} = \mu_0\mathbf{H} \tag{4.104}$$

or

$$\mathbf{B} = \mu_0\,(\mathbf{M} + \mathbf{H}) \tag{4.105}$$

Hence, Eq. (4.103) can be written as

$$\nabla \times \mathbf{H} = \mathbf{j} \tag{4.106}$$

This is the new form of Ampère's law which holds both in vacuum and in a medium and, hence, is more general than Eq. (4.47). In vacuum

$$\mathbf{B} = \mu_0\mathbf{H} \qquad \therefore\quad \nabla \times \mathbf{B} = \mu_0\,\mathbf{j}$$

which is a particular case of (4.106).

In integral form it is

$$\int_S \nabla \times \mathbf{H}\cdot\hat{\mathbf{e}}_n\,dS = \oint \mathbf{H}\cdot d\mathbf{l} = \int \mathbf{j}\cdot\hat{\mathbf{e}}_n\,dS$$

or

$$\oint \mathbf{H}\cdot d\mathbf{l} = I \tag{4.107}$$

The quantity H is called the **magnetic field intensity**.

Eq. (4.107) shows that the dimensions of H are ampere metre^{-1}. From Eq. (4.105) we conclude that M must have the same dimensions as H. Dimensions of H, however, are different from those of B.

We have shown earlier in this chapter that **B** is the magnetic analogue of the electric field **E**. We now see that **H** is analogous to the electric displacement **D**. It may be noted that the advantage of using **H** is that curl **H** is determined solely by the real current density, while **B** is related to the sum of real current density and magnetization current density.

4.17 Magnetic Susceptibility and Permeability

In many materials it is found that the magnetization **M** is linearly proportional to **H**, that is

$$\mathbf{M} = \chi_m \mathbf{H} \qquad (4.108)$$

where χ_m is a dimensionless constant and is called the **magnetic susceptibility** of the material. The susceptibility χ_m is a function of temperature. However, for paramagnetic and diamagnetic substances it is quite small. For paramagnetic substances it is positive, while for diamagnetic substances it is negative.

Now
$$\mathbf{B} = \mu_0 (\mathbf{H} + \mathbf{M})$$
$$= \mu_0 (1 + \chi_m) \mathbf{H}$$
$$= \mu_0 \mu_r \mathbf{H} = \mu \mathbf{H} \text{ where } \mu_r = 1 + \chi_m \qquad (4.109)$$

Here μ is known as the **magnetic permeability** of the medium. Since μ_0 is the magnetic permeability of vacuum, $\mu_r = \dfrac{\mu}{\mu_0}$ is called the **relative permeability** of the medium. The permeability μ differs from unity only by few parts for non-ferromagnetic substances. For paramagnetic substances $\mu > 1$ and for diamagnetic $\mu < 1$. For ferromagnetic substances μ has a large value around 1000.

4.18 Boundary Conditions

In electrostatics we have seen that the vectors **E** and **D** satisfy certain boundary conditions. Analogously **B** and **H** also satisfy certain conditions at the boundary between two media in which the permeability is different.

By Divergence theorem

$$\int_V \nabla \cdot \mathbf{B}\, d\tau = \int_S \mathbf{B} \cdot \hat{\mathbf{e}}_n\, dS \qquad (4.110)$$

The condition $\nabla \cdot \mathbf{B} = 0$ is unaltered by the presence of magnetic materials

$$\therefore \qquad \int \mathbf{B} \cdot \hat{\mathbf{e}}_n dS = 0 \qquad (4.111)$$

This is Gauss' theorem. The equation means that the flux of the field **B** out of any closed surface is zero. Consider a small disc of height 'h' that sits astride the boundary between two media (Fig. 4.17). If h is very small ($h \to 0$), the integral $\int \mathbf{B} \cdot \hat{\mathbf{e}}_n\, dS = 0$ has contribution from the top and bottom ends only, i.e.

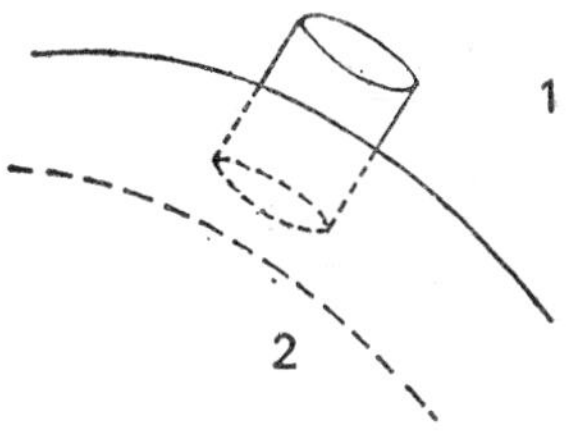

Fig. 4.17

$$\int_1 \mathbf{B} \cdot \hat{\mathbf{e}}_n dS - \int_2 \mathbf{B} \cdot \hat{\mathbf{e}}_n dS = 0 \qquad (4.112)$$

Now $\mathbf{B}\cdot\hat{\mathbf{e}}_n dS = B_\perp\, dS$ where $B_\perp$ is the component of $\mathbf{B}$ normal to dS

$$\therefore \quad \int_1 \mathbf{B}\cdot\hat{\mathbf{e}}_n dS = \int \mathbf{B}_{(1)\perp} dS \text{ and } \int_2 \mathbf{B}\cdot\hat{\mathbf{e}}_n dS = -\int \mathbf{B}_{(2)\perp} dS$$

$$\therefore \quad \int \mathbf{B}_{(1)\perp} dS = \int \mathbf{B}_{(2)\perp} dS$$

This equation is true whatever the size of dS

$$\therefore \quad B_{(1)\perp} = B_{(2)\perp} \quad \text{i.e.} \quad B_\perp \text{ is continuous.} \tag{4.113}$$

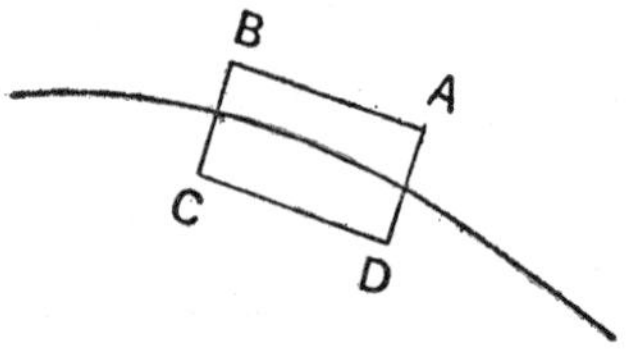

Fig. 4.18

The condition for $\mathbf{H}$ can be found by means of Ampère's law. Consider a small circuit $ABCDA$ (Fig. 4.18). The sides BC, DA are extremely small and $AB = CD = dl$.

Now
$$\oint \mathbf{H}\cdot d\mathbf{l} = I \tag{4.114}$$

where I is the current. Since BC, DA tend to zero

$$H_{(1)\parallel}\, dl - H_{(2)\parallel}\, dl = I \tag{4.115}$$

where $H_{(1)\parallel}$, $H_{(2)\parallel}$ are the tangential components of $\mathbf{H}$ in the two media.

Thus, there is a discontinuity in the component of the magnetic field $\mathbf{H}$ equal to the surface current. If there is no surface current

$$H_{(1)\parallel} = H_{(2)\parallel} \tag{4.116}$$

It would be interesting to compare the boundary conditions in electrostatics and magnetostatics:

1. The normal component of $\mathbf{B}$ is strictly continuous across the boundary; while the normal component of $\mathbf{D}$ is continuous only if there is no surface charge, and

2. The tangential component of $\mathbf{E}$ is strictly continuous, while that of $\mathbf{H}$ is continuous across the boundary only if there is no surface current.

4.19 Uniformly Magnetized Sphere in External Magnetic Field

We have solved the problem of a polarizable sphere in a uniform electric field, using spherical harmonics. In this section we shall discuss the corresponding magnetic problem.

In Fig. 4.19 we have a sphere placed in a uniform field $\mathbf{H}_0$, the direction of which is parallel to z-axis. Let μ_1, μ_2 be the relative permeabilities inside and outside the sphere respectively. Let us assume the potential inside and outside the sphere to be

$$\Phi_1 = -H_1 r \cos\theta \qquad (r < a) \tag{4.117}$$

$$\Phi_2 = -H_0 r \cos\theta + A r^{-2} \cos\theta \qquad (r > a) \tag{4.118}$$

where $\mathbf{H}_1$ is the field inside the sphere and θ the angle that the radius vector makes with the field direction. Notice that the term in $r^{-2}\cos\theta$ is not included in Φ_1, since with the inclusion of such a term, Φ_1 would become infinite at $r = 0$.

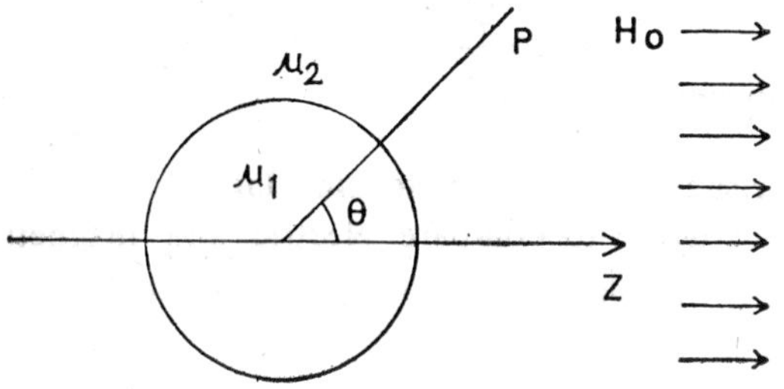

Fig. 4.19

One of the boundary conditions is that the tangential component of **H** be continuous at the boundary. This is equivalent to making $\Phi_1 = \Phi_2$ at $r = a$.

$$\therefore \qquad -H_1 a\cos\theta = -H_0 a\cos\theta + Aa^{-2}\cos\theta$$

i.e.
$$H_1 = H_0 - Aa^{-3} \tag{4.119}$$

We assume that the magnetization $\mathbf{M}_1$ inside the sphere is parallel to the field $\mathbf{H}_0$ and is made up of two components: a permanent component $\mathbf{M}_0$ and a component induced by the field $\mathbf{H}_1$ and given by

$$\mathbf{M}' = \chi_m \mathbf{H}_1 = (\mu_1 - 1)\mathbf{H}_1$$

$$\therefore \qquad \mathbf{M}_1 = (\mu_1 - 1)\,\mathbf{H}_1 + \mathbf{M}_0.$$

Hence,
$$\mathbf{B}_1 = \mu_0(\mathbf{H}_1 + \mathbf{M}_1) = \mu_0\{\mathbf{H}_1 + (\mu_1 - 1)\,\mathbf{H}_1 + \mathbf{M}_0\}$$

$$= \mu_0\mu_1\,\mathbf{H}_1 + \mu_0\mathbf{M}_0 \tag{4.120}$$

and
$$\mathbf{B}_2 = \mu_0\mu_2\mathbf{H}_0 \tag{4.121}$$

The radial components of **B** are

$$-\mu_0\mu_1\left(\frac{\partial\Phi_1}{\partial r}\right) + \mu_0 M_0\cos\theta = \mu_0(\mu_1 H_1 + M_0)\cos\theta \quad \text{(inside)} \tag{4.122}$$

and
$$-\mu_0\mu_2\left(\frac{\partial\Phi_2}{\partial r}\right) = \mu_0\mu_2(H_0 + 2Ar^{-3})\cos\theta \quad \text{(outside).} \tag{4.123}$$

By the boundary conditions, these two components must be equal to one another at $r = a$

$$\mu_0(\mu_1\mathbf{H}_1 + \mathbf{M}_0)\cos\theta = \mu_0\mu_2(\mathbf{H}_0 + 2Aa^{-3})\cos\theta$$

$$\therefore \qquad \mu_1\,\mathbf{H}_1 + \mathbf{M}_0 = \mu_2\mathbf{H}_0 + 2\mu_2 Aa^{-3}.$$

But
$$A = \frac{H_0 - H_1}{a^{-3}} \qquad \text{(see 4.119)}$$

$$\therefore \qquad \mu_1\mathbf{H}_1 + \mathbf{M}_0 = \mu_2\mathbf{H}_0 + 2\mu_2\,(\mathbf{H}_0 - \mathbf{H}_1)$$

i.e.
$$(\mu_1 + 2\mu_2)\,\mathbf{H}_1 = 3\mu_2\mathbf{H}_0 - \mathbf{M}_0$$

i.e.
$$\mathbf{H}_1 = \frac{3\mu_2}{\mu_1 + 2\mu_2}\mathbf{H}_0 - \frac{1}{\mu_1 + 2\mu_2}\mathbf{M}_0. \tag{4.124}$$

The difference

$$\mathbf{H}_d = \mathbf{H}_1 - \mathbf{H}_0 \tag{4.125}$$

is called the *demagnetizing field.*

4.20 A Comparison of Static Electric and Magnetic Fields

It would be instructive to compare electric and magnetic fields. We give in Table 4.1 a partial comparison of these two fields involving the relations, developed so far.

Table 4.1 A comparison of static electric and magnetic field equations

	Electric field	*Magnetic field*
Force	$\mathbf{F} = q\mathbf{E}$	$d\mathbf{F} = (I d\mathbf{l} \times \mathbf{B})$
Basic relation for the field	$\nabla \times \mathbf{E} = 0$	$\nabla \cdot \mathbf{B} = 0$
Derivation of the field from potential	$\mathbf{E} = -\nabla \Phi$ $\Phi = \frac{1}{4\pi\varepsilon_0}\int \frac{\rho d\tau}{r}$	$\mathbf{B} = \nabla \times \mathbf{A}$ $\mathbf{A} = \frac{\mu_0}{4\pi}\int \frac{\mathbf{j}}{\lvert \mathbf{r} \rvert} d\tau$
Constitutive relations	$\mathbf{D} = \varepsilon \mathbf{E}$	$\mathbf{B} = \mu \mathbf{H}$
Sources of the fields	$\nabla \cdot \mathbf{D} = \rho$	$\nabla \times \mathbf{H} = \mathbf{j}$

PROBLEMS

4.1 The current flowing into a conductor varies in time according to the equation $I = I_0 e^{-\alpha t}$ where I_0 and α are constants. Find the charge Q which has accumulated on the conductor after a time t_0.

4.2 Assuming that each copper atom contributes one free electronic charge to the current in a wire, estimate the mean drift velocity of the charges when the wire has a diameter of 1 mm and carries a current of 1 A.
(At. mass of Cu = 63.6 amu; density of Cu = 8.9×10^3 kg/m^3).

4.3 The total transfer of positive charge from the ionosphere to the earth due to the current in fine weather is about 90 C/km^2 per year. What is the approximate fine weather current in A falling on 1 mm^2 of the earth and how many electronic charges per second does this represent?

4.4 Calculate the force between two small plane circular coils, each of one turn of radius 'a' carrying a current I, with common axis and separated by a distance z ($a << z$).

4.5 Two infinite thin wires {one at (0, 0, 0) and the other at (R, 0, 0)} parallel to the z-axis carry current I in opposite directions where R is very large. Show that the magnetic potential at a point (r, ϕ, z) is

$$\Phi_m = -\frac{I}{2\pi}(\pi - \phi)$$

4.6 Calculate the flux density at the centre of a square loop of wire of sides 10 cm carrying a current of 10 A.

4.7 A coaxial cable has core of radius 'a' and sheath of radius 'b'. A current I flows along the core, uniformly distributed across it, and returns along the sheath, uniformly distributed around it. Find the magnetic flux density (i) within the core ($r < a$) (ii) within the core-sheath space ($a \leqslant r \leqslant b$) and (iii) outside the sheath ($r > b$).

4.8 A current is flowing in a long circular conductor of radius 'a'. The current is distributed in the wire in such a way that the current density at a distance 'r' from the axis is given by

$$\mathbf{j} = \mathbf{j}_0 \left(1 + \frac{r^2}{a^2} \right).$$

Find the total current in the wire and the magnetic flux density both inside and outside the wire.

4.9 A particle of mass M is rotating in a circular orbit of radius r with angular velocity ω. If the particle carries a charge 'q', show that a magnetic dipole moment is associated with the motion of the charge given by

$$\mathbf{m} = \left(\frac{q}{2M} \right) G$$

where $G = Mr^2w$ is the angular momentum of the particle.

4.10 A sphere uniformly charged, rotates with a constant angular velocity ω. Show that the magnetic flux density $\mathbf{B}$ at the centre of the sphere is

$$\mathbf{B} = \frac{\mu_0 \rho \omega a^2}{3}$$

where 'a' is the radius and 'ρ' the charge density. Find also the vector potential both inside and outside the sphere.

4.11 Show that the lines of magnetic field strength $\mathbf{H}$ are refracted at a change of media. Show further that

$$\mu_1 \cot \theta_1 = \mu_2 \cot \theta_2$$

where μ_1, μ_2 are the permeabilities of the media and θ_1, θ_2 are the angles made with the normal.

4.12 Find the magnetic field strength inside a cylindrical cavity of radius 'b' in a long infinite cylindrical conductor of radius 'a', with an electric current whose density $\mathbf{j}$ is the same through any section of the conductor. The axes of the cavity and the coducting cylinder are parallel and a distance 'c' apart with $a > b + c$.

Chapter 5

Electromagnetic Induction

In the last chapter we initiated a discussion on the relationship between electricity and magnetism and obtained a good deal of information in this regard. The information, thus obtained, however, is incomplete, because we considered only time-independent magnetic fields. In this chapter we shall investigate the behaviour of time-dependent electric and magnetic fields. We shall show how a time varying magnetic field gives rise to an electric field and vice-versa.

5.1 Electromotive Force

For a steady current flow in a single conductor, a potential difference must exist between its two ends. This potential difference is maintained by a source of **electromotive force** (e.m.f.). There are several types of devices which provide electromotive force such as a battery, a dynamo, etc. They all convert some other kind of energy into electrical energy. For example, a battery uses stored chemical energy to drive a current in the circuit. If the battery maintains a constant potential difference of V volts between its terminals, its e.m.f. is V volts. If the terminals of the battery are connected by a conductor, an electric field is established in the conductor which is static. The line integral of the field over a path between two points (A, B) is equal to the potential difference between the points. If these points are the terminals of the battery, the e.m.f. of the battery is

$$V = \int_A^B \mathbf{E} \cdot d\mathbf{l} \tag{5.1}$$

Now, when the current flows in the conductor, the battery must do work to keep the potential difference between its terminals constant. If a charge q moves from one terminal of the battery to the other through the conductor, the work done by the battery is

$$Vq = \int_A^B q\mathbf{E} \cdot d\mathbf{l}$$

$\therefore$ The e.m.f. of the battery is

$$V = \frac{1}{q}\int_A^B q\mathbf{E} \cdot d\mathbf{l} \tag{5.2}$$

Note that the term **electromotive force**, in the context it has been used, is not entirely appropriate inasmuch as it is not a **force** but a quantity which has the dimensions energy per charge. However, the term is firmly entrenched into common usage.

Consider now a metal bar moving with a constant velocity **u** in a direction perpendicular to a uniform magnetic field **B** (Fig. 5.1).

There will be a magnetic force on each electron in the bar, given by (4.24). The free electrons will move towards the end of the bar and collect there setting a field **E** given by

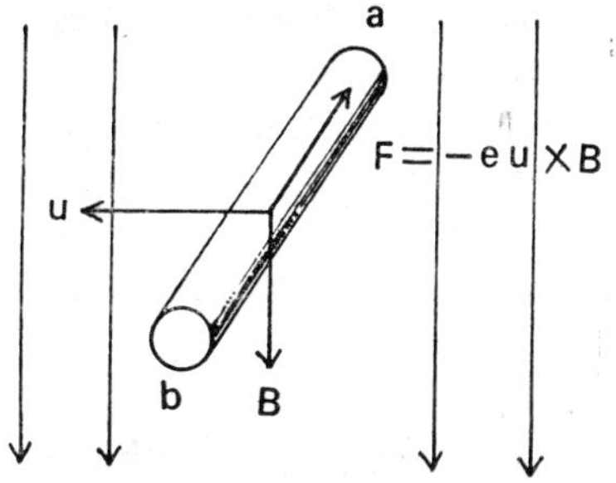

Fig. 5.1

$$\mathbf{E} = -\mathbf{u} \times \mathbf{B} \tag{5.3}$$

The potential difference between the ends of the bar is

$$V_{ba} = \int_b^a \mathbf{E} \cdot d\mathbf{l} = uBL \tag{5.4}$$

where L is the length of the bar. This potential difference will produce no current flow. If, however, the bar forms a part of a circuit as shown in Fig. 5.2, a current will flow.

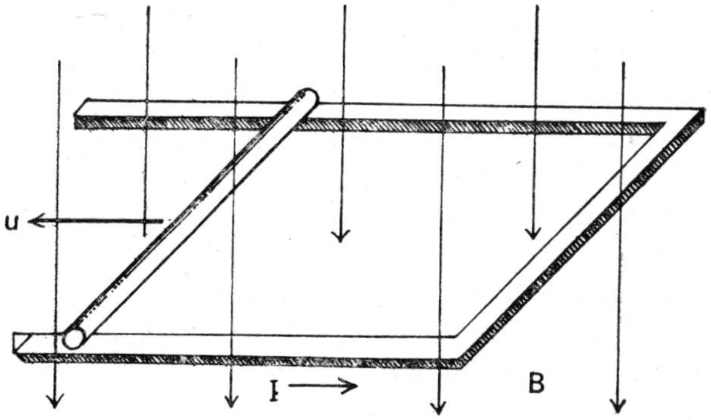

Fig. 5.2

The line integral of the force on a charge q round the circuit is

$$\oint (q\mathbf{E} \cdot d\mathbf{l}) = quBL.$$

The e.m.f. induced in the closed circuit, due to the motion of the conductor, therefore, is

$$V = \frac{1}{q} \oint (q\mathbf{E} \cdot d\mathbf{l}) = uBL. \tag{5.5}$$

This may be called a **motional** e.m.f., since it depends on the velocity of the conductor and not its position. We, thus, have two sorts of voltage: electrostatic potential difference due to stationary charges and electromotive force due to moving charges.

Now uBL is the magnetic flux through the area swept by the bar in unit time. If Φ is the total magnetic flux through the circuit, uBL is the rate of change of flux through the circuit. We can, therefore, write

$$|\text{e.m.f.}| = uBL = \frac{d\Phi}{dt}. \tag{5.6}$$

Experimental observations in this respect were available from the work of Faraday in U.K. and Henry in U.S.A. working independently.

5.2 Faraday's Law of Electromagnetic Induction

Faraday and Henry observed that:

(i) If a magnet be moved about in the neighbourhood of a wire in the form of a closed circuit but with no battery in the circuit, then a current is produced in the wire **so long as the movement lasts**, but disappears on cessation of movement.

(ii) The same effect is observed if the magnet is kept still and the circuit is moved.

This gives an impression that for a current to be produced in the wire, there must be a relative movement. But the current can be produced without any mechanical movement.

(iii) A transient current is induced in a loop of wire if the stationary current in an adjacent circuit is turned on and off.

In other words, the transient current flowed when the flux through the circuit is altered. The changing flux induces an electric field and, hence, an e.m.f., in the circuit which causes the current to flow. Faraday called this phenomenon **electromagnetic induction.** Faraday's results were summed up in what is known as the "flux rule".

"When the magnetic flux through a circuit is changing, an electromotive force is induced in the circuit, the magnitude of which is proportional to the rate of change of flux".

Thus, if $\mathcal{E}$ is the e.m.f. and Φ, the flux

$$|\mathcal{E}| \propto \frac{d\Phi}{dt} \tag{5.7}$$

The direction of the induced e.m.f. is provided by **Lenz's law.** This law states that:

"The direction of the induced e.m.f. is such that the magnetic flux associated with the current generated by it, opposes the original change of flux causing the e.m.f."

Thus, in Fig. 5.3, if the magnet is moved in the direction of the arrow, i.e. towards the loop, thus increasing the magnetic flux through the coil, the induced current will flow in the direction shown, so that its own flux opposes the increase in the flux of the magnet. This law is a particular case of a very general physical principle—**Le Chatelier's Principle**—

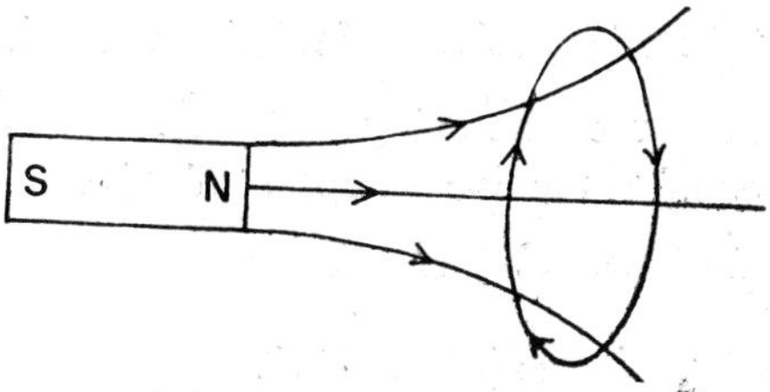

Fig. 5.3

which states that a physical system always reacts to oppose any change that is impressed from outside.

Hence, $$\mathcal{E} = -\frac{d\Phi}{dt} \tag{5.8}$$

(In SI units the constant of proportionality is found to be unity.)
The induced e.m.f. is often called the **back e.m.f.**

5.3 Induction Law for Moving Circuits

As an example of electromagnetic induction, we consider a circuit of any arbitrary shape moving in a time-independent magnetic field **B**, with a velocity **v** which need not be uniform (Fig. 5.4). An element $d\mathbf{l}$ (PQ) will move in time dt a distance $\mathbf{v}\,dt$ to a position $P'Q'$. If the velocity of the conducting electrons relative to the wire is **u**, their velocity relative to the field is $\mathbf{v} + \mathbf{u}$. The force exerted on each electron, therefore, is $e\,(\mathbf{v} + \mathbf{u}) \times \mathbf{B}$. The component of this force along the element $d\mathbf{l}$ is $\{e(\mathbf{v} + \mathbf{u}) \times \mathbf{B}\}\cdot\hat{\mathbf{e}}_l$ where $\hat{\mathbf{e}}_l$ is a unit vector in the direction PQ. Since $\hat{\mathbf{e}}_l$ is parallel to **u**, $\mathbf{u} \times \mathbf{B}\cdot\hat{\mathbf{e}}_l = 0$.

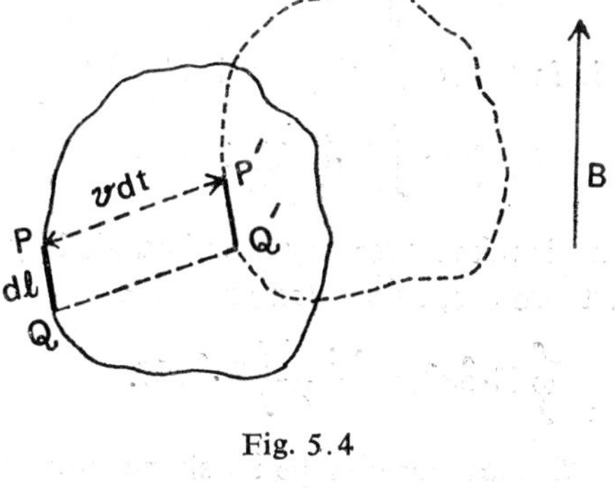

Fig. 5.4

$$\therefore\quad \{e(\mathbf{v} + \mathbf{u}) \times \mathbf{B}\}\cdot\hat{\mathbf{e}}_l = e\mathbf{v} \times \mathbf{B}\cdot\hat{\mathbf{e}}_l + e\mathbf{u} \times \mathbf{B}\cdot\hat{\mathbf{e}}_l$$

$$= e\mathbf{v} \times \mathbf{B}\cdot\hat{\mathbf{e}}_l \tag{5.9}$$

This shows that there is an electric field $\mathbf{E} = \mathbf{v} \times \mathbf{B}$ induced in the wire, its component along the wire being $\mathbf{v} \times \mathbf{B} \cdot \hat{\mathbf{e}}_l$.

The induced e.m.f. is the line integral of this field round the circuit.

$\therefore$ The induced e.m.f. $= \mathcal{E}$

$$= \oint (\mathbf{v} \times \mathbf{B}) \cdot \hat{\mathbf{e}}_l dl. \tag{5.10}$$

Now as the circuit moves, the element of the circuit dl sweeps in time dt an area $PP'Q'Q = \mathbf{v}\, dt \times \hat{\mathbf{e}}_l dl$. The flux across this element is

$$(\mathbf{v}\, dt \times \hat{\mathbf{e}}_l\, dl) \cdot \mathbf{B}.$$

The flux over the entire band

$$d\Phi = \oint \mathbf{B} \cdot (\mathbf{v}\, dt \times \hat{\mathbf{e}}_l\, dl)$$

i.e.
$$\frac{d\Phi}{dt} = \oint \mathbf{B} \cdot (\mathbf{v} \times \hat{\mathbf{e}}_l\, dl)$$

$$= -\oint (\mathbf{v} \times \mathbf{B}) \cdot \hat{\mathbf{e}}_l\, dl \tag{5.11}$$

Comparing (5.10) and (5.11), we have

$$\mathcal{E} = -\frac{d\Phi}{dt}$$

5.4 Integral and Differential Form of Faraday's Law

The induced e.m.f. is equal to the line integral of the induced electric field $\mathbf{E}$ around the coil i.e.

$$\mathcal{E} = \oint \mathbf{E} \cdot \hat{\mathbf{e}}_l dl$$

and the magnetic flux through the coil is equal to

$$\Phi = \int_S \mathbf{B} \cdot \hat{\mathbf{e}}_n dS$$

where the integral is taken over any area bounded by the circuit. Equation (5.8) may, therefore, be written as

$$\oint \mathbf{E} \cdot \hat{\mathbf{e}}_l dl = -\frac{d}{dt}\int \mathbf{B} \cdot \hat{\mathbf{e}}_n dS.$$

Since the surface S does not change its shape nor position with time, we can write the above equation as

$$\oint \mathbf{E} \cdot \hat{\mathbf{e}}_l dl = -\int_S \frac{d\mathbf{B}}{dt} \cdot \hat{\mathbf{e}}_n dS \tag{5.12}$$

This is the integral form of Faraday's law.

Using Stokes' theorem

$$\oint \mathbf{E}\cdot\hat{\mathbf{e}}_l dl = \int_S \text{curl } \mathbf{E}\cdot\hat{\mathbf{e}}_n dS = -\int \frac{\partial \mathbf{B}}{\partial t}\cdot\hat{\mathbf{e}}_n dS$$

$$\therefore \qquad \int\left(\text{curl } \mathbf{E} + \frac{\partial \mathbf{B}}{\partial t}\right)\cdot\hat{\mathbf{e}}_n dS = 0 \qquad (5.13)$$

The total time derivative has been replaced by partial derivative as we are only concerned with the changes in field **B** with time at the fixed position of the elemental area dS.

Since, equation (5.13) must hold for any arbitrary surface

$$\text{curl } \mathbf{E} = \nabla \times \mathbf{E} = -\frac{\partial \mathbf{B}}{\partial t}. \qquad (5.14)$$

This is the differential form of Faraday's law of electromagnetic induction.

The equations (5.14) and (1.30) show that the electric field has a non-conservative part due to changing magnetic flux density as well as a conservative part due to electric charge density.

We may state here, without proof, a theorem useful in the formulation of electromagnetism in terms of vector calculus.

We have seen that the sources of the electromagnetic field are of two kinds. The first kind is associated with a system such as electrostatics in which energy is conserved during a cyclic process within the system. The sources of such a conservative or irrotational system are described by the equation div $\mathbf{E} = \rho/\epsilon_0$, where $\mathbf{E} = -\nabla\Phi$. Such a field has no curl

$$\because \qquad \text{curl grad } \Phi = 0$$

The second kind of sources is associated with a system in which there is energy transferred in a cyclic process (e.g., magnetic field of a solenoid). Such a field is specified by the curl sources and has no divergence. In general, electromagnetic fields have both types of sources and, hence, the complete specification of such a vector field should include both types of sources. Such a specification is not only necessary but sufficient. **Any vector field is uniquely determined if its divergence and curl sources are given.** This is known as **Helmholtz theorem.**

Taking divergence of (5.14) we find

$$\nabla\cdot\nabla \times \mathbf{E} = -\frac{\partial}{\partial t}(\nabla\cdot\mathbf{B}) = 0$$

Hence, $\nabla\cdot\mathbf{B}$ is necessarily independent of time at every point in space and this condition can reasonably be satisfied if we assume

$$\nabla\cdot\mathbf{B} = 0. \qquad (5.15)$$

That is, **B** is always solenoidal.

Faraday's law (5.14), thus, has two important consequences:

(i) The electric field **E** is no longer a conservative field when the magnetic field varies with time. It eventually turns out that energy can flow between electric and magnetic forms through time-varying fields, and

(ii) No free magnetic poles can exist. All magnetic poles occur in pairs, positive and negative.

Faraday's law of induction shows how the electric and magnetic fields are interrelated. Their independent nature disappears when we consider their time-dependence. Hence, it would be proper to look upon these two fields as a single field—**electromagnetic field**.

5.5 Self-inductance and Mutual Inductance

When a current flows in a circuit, there will be a magnetic flux Φ through the circuit arising from its own magnetic field. The flux is proportional to the current I, i.e.

$$\Phi = LI \tag{5.16}$$

where L is a constant.

Consider a situation in which the circuit is stationary but the magnetic field and, hence, the flux changes with time. Suppose, for example, that the current in the circuit is time-dependent. The flux, then, is a function of time $\Phi = \Phi\ (t)$

$$\therefore \quad \frac{d\Phi}{dt} = \frac{d\Phi}{dI}\frac{dI}{dt} = L\frac{dI}{dt} \quad \text{(by 5.16)}$$

$$\therefore \quad L = \frac{d\Phi}{dI} \tag{5.17}$$

The quantity L is called the **self-inductance**.

Equation (5.8) can now be written as

$$\mathcal{E} = -L\frac{dI}{dt} \tag{5.18}$$

The constant L depends upon the geometry of the circuit. A wire wound in the form of a solenoid has much larger self-inductance then when it is unwound. It depends also on the permeability of the medium. Self-inductance can be defined as the total flux through the circuit when a unit current is flowing. Hence, a circuit has unit self-inductance if it is threaded by one **weber** of flux when the current flowing is one ampere; or, a circuit has unit self-inductance when the e.m.f. of 1 volt is generated in it by a current varying at the rate of one ampere per second. The unit of self-inductance is called the **henry**.

Suppose a coil carries a current I_1. If a second coil is brought near this coil there will be a magnetic flux Φ_2 through the second coil due to the current in the first. Since Φ_2 is linearly proportional to I_1,

$$\Phi_2 = L_{21}I_1 \tag{5.19}$$

where L_{21} is a constant. There will also be a flux through the first circuit due to the current I_2 in the second circuit

$$\therefore \qquad \Phi_1 = L_{12}I_2 \tag{5.20}$$

where L_{12} is again a constant. What is the relationship between these two constants? The potential energy of the system as given by equation (4.76) is

$$U_P = -\Phi_2 I_2 = -L_{21}I_1I_2 = -\Phi_1 I_1 = -L_{12}I_2I_1$$

$$\therefore \qquad L_{21} = L_{12} \tag{5.21}$$

The constant $L_{12} = L_{21} = M$ is called the *mutual inductance* between the two circuits.

Now $U_P = -\Phi_1 I_1 = -I_1 \int \mathbf{B}_1 \cdot \hat{\mathbf{e}}_n dS$

$$= -I_1 \int \nabla \times \mathbf{A}_{12} \cdot \hat{\mathbf{e}}_n dS = -I_1 \int \mathbf{A}_{12} \cdot \hat{\mathbf{e}}_{l1} dl_1$$

(By Stokes' theorem)

$$= -I_1 \int \left(\frac{\mu}{4\pi} \int \frac{I_2 \hat{\mathbf{e}}_{l2} dl_2}{|\mathbf{r}|} \right) \cdot \hat{\mathbf{e}}_{l2} dl_1$$

$$= -\frac{\mu}{4\pi} I_1 I_2 \iint \frac{\hat{\mathbf{e}}_{l1} dl_1 \cdot \hat{\mathbf{e}}_{l2} dl_2}{|\mathbf{r}|} = -L_{12}I_1I_2$$

$$\therefore \qquad L_{12} = L_{21} = M = \frac{\mu}{4\pi} \iint \frac{\hat{\mathbf{e}}_{l1} dl_1 \cdot \hat{\mathbf{e}}_{l2} dl_2}{|\mathbf{r}|} \tag{5.22}$$

This is known as **Neumann's formula.** Clearly, the unit of M is also the **henry**.

EXAMPLE 5.1. A car ignition coil consists of two insulated coils, one of 16000 turns and the other of 400 turns, wound over each other. The length of each coil is 10 cm and the turns have the radius of 3 cm. A current of 3A is passed through the primary coil and broken in about 10^{-4} seconds. Calculate the voltage induced in the secondary circuit.

The magnetic flux density inside the solenoid is given by

$$B = \mu_0 N_1 I_1$$

where N_1 is the number of turns per unit length and I_1 is the current flowing through the coil. The magnetic flux through each turn of the top coil in $\mu_0 N_1 I_1 \pi r^2$ and the total flux is

$$\Phi_2 = \mu_0 N_1 N_2 l \pi r^2 I_1$$

where N_2 is the number of turns per unit length of the second coil.

The induced e.m.f. is given by

$$\mathcal{E} = -\frac{d\Phi_2}{dt} = -\mu_0 N_1 N_2 l \pi r^2 \frac{dI_1}{dt}$$

$$= - M \frac{dI_1}{dt}$$

Now $\frac{dI_1}{dt} = 3 \times 10^4 \text{ As}^{-1}$

$\therefore$

$$\mathcal{E} = 4\pi \times 10^{-7} \times 16 \times 10^4 \times 4 \times 10^3 \times 10^{-1}\pi$$
$$\times (0.03)^2 \times 3 \times 10^4$$
$$= 6814 \text{ volts.}$$

5.6 Energy in Magnetic Fields

The energy in a field is by definition the total work done to establish it. To set up a magnetic field from a steady current requires first switching on the current. There is evidently an interval during which the current and the field are brought to their final value. Since the field during this interval is time-dependent, there will be induced e.m.f.s which cause the current source to do work and this contribution must be taken into consideration while calculating the energy.

(*a*) *Magnetic energy stored in an inductor*

Consider a simple circuit shown in Fig. 5.5 in which R is the resistance, L an inductive coil with self-inductance L and $\mathcal{E}$ is the e.m.f. of the

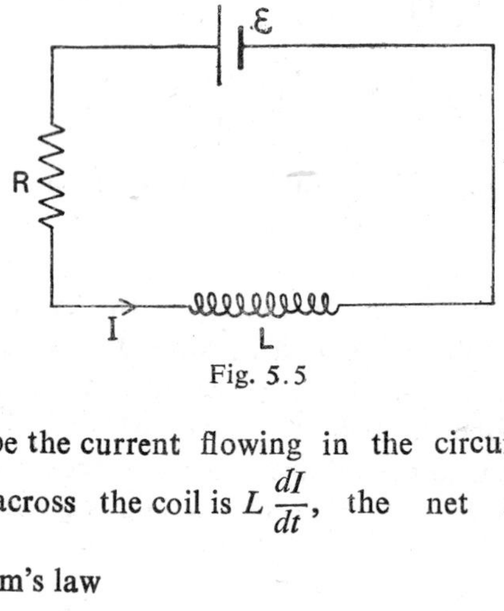

Fig. 5.5

battery. Let $I(t)$ be the current flowing in the circuit at time t. Since the voltage drop across the coil is $L\frac{dI}{dt}$, the net forward e.m.f. is $\mathcal{E} - L\frac{dI}{dt}$. By Ohm's law

$$\mathcal{E} - L\frac{dI}{dt} = RI. \tag{5.23}$$

Let us consider the work done by the e.m.f. $\mathcal{E}$ in moving a small amount of charge dQ through the circuit

$$dW = \mathcal{E}dQ = \mathcal{E}Idt$$

$\therefore$

$$\frac{dW}{dt} = \mathcal{E}I = LI\frac{dI}{dt} + RI^2 \tag{5.24}$$

∴ The total work done by the battery in time interval T in which the current changes from 0 to I_T is given by

$$W = \int_0^T \mathcal{E} I dt = L\int_0^T I \frac{dI}{dt} dt + R\int_0^T I^2 dt$$

$$= \frac{1}{2} L I_T^2 + R\int_0^T I^2 dt. \tag{5.25}$$

The first term on the right-hand side is the energy stored in the inductance in time T and the second term is the energy dissipated as heat in the resistance.

(b) The magnetic energy stored in a series of inductances

Let us now derive an expression for the energy stored in a series of inductances in a more general way.

We assume that currents in all the circuits were initially zero and that they attain their equilibrium value in all the circuits simultaneously at $t = T$, so that at any instant t, in the interval $0 \leqslant t \leqslant T$, the current in each circuit $I_k(t)$ and the flux through it $\Phi_k(t)$, will be some fraction α of their ultimate values

i.e. $$I_k(t) = \alpha I_k,\ \Phi_k(t) = \alpha \Phi_k$$

The induced e.m.f. in a kth circuit is

$$\mathcal{E}_k = \frac{d\Phi_k(t)}{dt}.$$

Therefore, the total work done by the circuit k is

$$W_k = \int_0^T \mathcal{E}_k I_k(t) dt = I_k \Phi_k \int_0^T \alpha \frac{d\alpha}{dt} dt$$

$$= I_k \Phi_k \int_0^1 \alpha d\alpha = \tfrac{1}{2} I_k \Phi_k. \tag{5.26}$$

On summing over all the circuits

$$W = \frac{1}{2} \sum_k I_k \Phi_k. \tag{5.27}$$

Since the magnetic flux depends on the self as well as mutual inductances, we can write

$$\Phi_k = L_k I_k + \sum_{j \neq k} M_{kj} I_j \tag{5.28}$$

$$\therefore \quad W = \frac{1}{2}\sum_k L_k I_k^2 + \frac{1}{2}\sum_k \sum_{\substack{j \\ j\neq k}} M_{kj} I_k I_j \tag{5.29}$$

Thus, for a pair of coils

$$W = \tfrac{1}{2} L_1 I_1^2 + \tfrac{1}{2} L_2 I_2^2 + M I_1 I_2$$

where $M = M_{12} = M_{21}$.

We shall now show how the result (5.29) is expressed in a more general and standard form.

For any circuit

$$\Phi_k = \int_k \mathbf{B}\cdot\hat{\mathbf{e}}_n dS = \oint_k \text{curl } \mathbf{A}\cdot\hat{\mathbf{e}}_n dS = \oint_k \mathbf{A}\cdot\hat{\mathbf{e}}_l dl \qquad (5.30)$$

Equation (5.27), therefore, changes to

$$W = \frac{1}{2}\sum_k I_k\Phi_k = \frac{1}{2}\sum_k I_k \oint_k \mathbf{A}\cdot\hat{\mathbf{e}}_l dl.$$

We may make the transition from the discrete to the continuous case by using the relation $I\hat{\mathbf{e}}_l dl = \mathbf{j}d\tau$ and taking the integral over all space, since contributions arise only from regions where j is finite; i.e. by changing

$$\sum_k \oint_k \rightarrow \int_V$$

$$\therefore \qquad W = \frac{1}{2}\int_V (\mathbf{A}\cdot\mathbf{j})\, d\tau \qquad (5.31)$$

Now

$$\mathbf{j} = \nabla \times \mathbf{H}$$

$$\therefore \qquad W = \frac{1}{2}\int_V \left\{\mathbf{A}\cdot(\nabla \times \mathbf{H})\right\} d\tau. \qquad (5.32)$$

Since,

$$\nabla\cdot(\mathbf{A}\times\mathbf{H}) = \mathbf{H}\cdot(\nabla\times\mathbf{A}) - \mathbf{A}\cdot(\nabla\times\mathbf{H})$$

we can write (5.32) as

$$W = \frac{1}{2}\int_V \left\{\mathbf{H}\cdot(\nabla\times\mathbf{A})\right\} d\tau - \frac{1}{2}\int\left\{\nabla\cdot(\mathbf{A}\times\mathbf{H})\right\} d\tau. \qquad (5.33)$$

Using the Divergence theorem to transform the second integral, (5.33) can be written as

$$W = \frac{1}{2}\int_V \left\{\mathbf{H}\cdot\mathbf{B}\right\} d\tau - \frac{1}{2}\int_V (\mathbf{A}\times\mathbf{H})\cdot\hat{\mathbf{e}}_n dS \qquad (5.34)$$

Since the volume integral is to be taken over all space, the surface integral must be taken over the sphere at infinity. Because **H** and **A** fall off rapidly at large distances ($H \sim r^{-3}$, $A \sim r^{-2}$) the surface integral vanishes as $r \rightarrow \infty$.

Hence

$$W = \frac{1}{2}\int_V (\mathbf{H}\cdot\mathbf{B})\, d\tau. \qquad (5.35)$$

This equation is analogous to the Eq. (2.43) in electrostatics and shows that the magnetic energy may be regarded as distributed throughout the region occupied by the field with density $\frac{1}{2}(\mathbf{H}\cdot\mathbf{B})$. Since $\mathbf{B} = \mu\mathbf{H}$, the density is equal to $\frac{1}{2}\mu H^2$ or $\frac{1}{2}\frac{B^2}{\mu}$.

Note that the relation has been obtained on the assumption that **B** is linearly proportional to **H** i.e. the media has values of μ which are independent of field strength. For non-linear materials the analysis will have to be modified suitably.

In establishing the foregoing relations we have used a scaffolding of circuits. However, the electric forces which are brought into existence by a change of magnetic flux cannot depend on whether there is an actual flow of current. The flow of current is an effect, not a cause. Hence, if the scaffolding is taken away, Faraday's equation stands on its own. We may still imagine an e.m.f. around an arbitrary mathematical curve in space equal to the integral of the tangential component of **E** around the curve. Maxwell, therefore, assumed that the equation holds for any closed curve whatever. Experiments have abundantly justified this assumption.

5.7 Maxwell's Equations

Until Maxwell's work, the known basic laws of electricity and magnetism were:

(1) Gauss' law applied to electrostatics

$$\nabla \cdot \mathbf{D} = \rho, \tag{5.36}$$

(2) Corresponding result for magnetic field

$$\nabla \cdot \mathbf{B} = 0, \tag{5.37}$$

(3) Faraday's law of induction

$$\nabla \times \mathbf{E} = -\frac{\partial \mathbf{B}}{\partial t}, \tag{5.38}$$

(4) Ampère's law for magnetomotive force

$$\nabla \times \mathbf{H} = \mathbf{j}. \tag{5.39}$$

The first three of these are general equations and are valid for static as well as dynamic fields. The fourth equation was derived from steady-state observations and we have to examine its validity for time-varying fields.

Taking the divergence of both sides of (5.39) we have

$$\nabla \cdot (\nabla \times \mathbf{H}) = \nabla \cdot \mathbf{j} = 0 \tag{5.40}$$

This, indeed, is true for steady-state phenomena. However, when the currents are changing with time, the result is incompatible with the principle of conservation of charge, reflected in the equation of continuity

$$\nabla \cdot \mathbf{j} + \frac{\partial \rho}{\partial t} = 0. \tag{5.41}$$

Maxwell appreciated this situation and suggested a way out. He realised that the difficulty arose from an incomplete definition of the total current density in equation (5.39). Using Gauss' law (5.36), we can write (5.41) as

$$\nabla \cdot \mathbf{j} = -\frac{\partial \rho}{\partial t} = -\frac{\partial}{\partial t}(\nabla \cdot \mathbf{D}) = \nabla \cdot \left(-\frac{\partial \mathbf{D}}{\partial t}\right)$$

i.e.

$$\nabla \cdot \left(\mathbf{j} + \frac{\partial \mathbf{D}}{\partial t}\right) = 0 \tag{5.42}$$

Maxwell replaced $\mathbf{j}$ in Ampère's law by $\mathbf{j} + \frac{\partial \mathbf{D}}{\partial t}$. With this modification Ampère's law takes the form

$$\nabla \times \mathbf{H} = \mathbf{j} + \frac{\partial \mathbf{D}}{\partial t}. \tag{5.43}$$

The law in this form is valid for steady-state phenomena and is also compatible with the equation of continuity for time dependent fields. The term $\mathbf{j}$ is generally called "conduction current density". The second term $\frac{\partial \mathbf{D}}{\partial t}$ which arises from the variation of electric displacement with time is called "displacement current density".

What exactly is implied by the displacement current? The displacement current does not have the significance of a current in the sense of being the motion of charges. It can be vividly demonstrated by considering a simple circuit such as in Fig. 5.6 which shows a charged capacitor whose plates are joined by a conducting wire. The current flowing in the conducting wire is equal to the rate of change of charge on the plates, i.e.

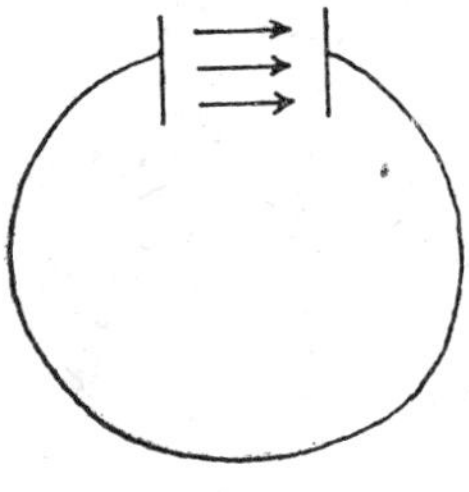

Fig. 5.6

$$I = \frac{dQ}{dt} \tag{5.44}$$

where Q is the charge on the positive plate of the capacitor. The charge on the plates is related to the field in the capacitor by the relation

$$E = \frac{\sigma}{\epsilon_0} = \frac{Q}{\epsilon_0 A} \tag{5.45}$$

where A is the area of the plates

$$\therefore \qquad I = \frac{dQ}{dt} = \epsilon_0 A \frac{\partial E}{\partial t} = A \frac{\partial D}{\partial t}$$

i.e.

$$\frac{I}{A} = \frac{\partial D}{\partial t}. \tag{5.46}$$

Now I/A gives the current density. Hence, the quantity $\frac{\partial D}{\partial t}$ can be interpreted as the density of some current which corresponds, for example, to the "current" which must flow in the space—even in vacuum—between a pair of plates of the capacitor when the charged plates are connected by a wire, thus completing the conduction current.

An idea of the relative size of the two types of current in conductors can be obtained by considering a copper wire in which there is an electric field $\mathbf{E} = \mathbf{E}_0 e^{-i\omega t}$

$$\mathbf{j} = \sigma\mathbf{E} = \sigma\mathbf{E}_0 e^{-i\omega t} \quad \therefore \quad |j|^2 = \sigma^2 |E_0|^2$$

and
$$\mathbf{D} = \epsilon_0 \mathbf{E}_0 e^{-i\omega t}$$

$$\therefore \quad \frac{\partial D}{\partial t} = -i\omega\epsilon_0 \mathbf{E}_0 e^{-i\omega t}. \text{ Hence, } \left|\frac{\partial D}{\partial t}\right|^2 = \omega^2 \epsilon_0^2 |E_0|^2$$

$$\therefore \quad \left|\frac{\mathbf{j}}{\frac{\partial \mathbf{D}}{\partial t}}\right| = \frac{\sigma}{\omega\epsilon_0}.$$

For copper $\quad \sigma = 5.9 \times 10^7 \quad \therefore \quad \dfrac{\sigma}{\omega\epsilon_0} \sim \dfrac{10^{19}}{\omega}.$

Thus, the ratio is very large for all frequencies. Displacement currents, therefore, are not as significant as the currents due to the motion of free charges in the study of the continuous circuits. They have, however, far reaching consequences in other respects. It is mainly due to the additional term in Ampère's law that a rich variety of new electromagnetic phenomena was uncovered.

The four equations which the field vectors **E, D, B, H** satisfy everywhere are:

(i) $\nabla \cdot \mathbf{D} = \rho$,

(ii) $\nabla \cdot \mathbf{B} = 0$,

(iii) $\nabla \times \mathbf{E} + \dfrac{\partial \mathbf{B}}{\partial t} = 0$,

(iv) $\nabla \times \mathbf{H} = \mathbf{j} + \dfrac{\partial \mathbf{D}}{\partial t}$. (5.47)

These equations are the fundamental equations of electromagnetic field and are known as **Maxwell's equations.**

5.8 Decay of Free Charge

One of the important deductions that can be made from Maxwell's equations is regarding the decay of free charges.

The equation (iv) of (5.47) can be written as

$$\nabla \times \mathbf{H} = \mathbf{j} + \frac{\partial \mathbf{D}}{\partial t} = \sigma\mathbf{E} + \epsilon \frac{\partial \mathbf{E}}{\partial t}. \tag{5.48}$$

Taking the divergence of each side and assuming that σ and ϵ are constants we have

$$\nabla \cdot \nabla \times \mathbf{H} = \sigma \nabla \cdot \mathbf{E} + \epsilon \frac{\partial}{\partial t} (\nabla \cdot \mathbf{E})$$

$$= \frac{\sigma \rho}{\epsilon} + \frac{\partial \rho}{\partial t} = 0 \quad \text{(by (i) of 5.47).}$$

Integrating this last equation, we have

$$\rho = \rho_0 e^{-t/\tau} \tag{5.49}$$

where

$$\tau = \epsilon / \sigma \tag{5.50}$$

and is known as the **relaxation time.** The relation shows that any original distribution of charge decays exponentially at a rate which is independent of any other electromagnetic disturbances that may be taking place.

5.9 Potentials of Electromagnetic Field

Complete description of an electromagnetic field can be obtained by solving Maxwell's equations. The process becomes simple if the equations are written in a suitable form. It is often convenient to reduce the number of equations by introducing new quantities called "electromagnetic potentials". We have already adopted this technique in the treatment of static fields: The electrostatic field was expressed in terms of scalar potential ($\mathbf{E} = -\text{grad } \Phi$) and the magnetic fields in terms of a vector potential ($\mathbf{B} = \nabla \times \mathbf{A}$). We shall now consider potentials in electromagnetic fields when electric and magnetic fields are time-varying.

In the time-dependent case the equation (ii) of (5.47)

$$\nabla \cdot \mathbf{B} = 0 \tag{5.51}$$

still holds and hence we can express **B** in terms of a vector potential,

i.e.
$$\mathbf{B} = \nabla \times \mathbf{A}. \tag{5.52}$$

We consider next equation (iii) which does not involve any currents or charges

$$\nabla \times \mathbf{E} = -\frac{\partial \mathbf{B}}{\partial t} = -\frac{\partial}{\partial t} (\nabla \times \mathbf{A}) = -\nabla \times \frac{\partial \mathbf{A}}{\partial t}$$

$$\therefore \quad \nabla \times \left(\mathbf{E} + \frac{\partial \mathbf{A}}{\partial t} \right) = 0 \tag{5.53}$$

Since the curl of the gradient of a scalar function vanishes, the quantity within the brackets can be expressed as a gradient of a scalar function Φ

$$\mathbf{E} + \frac{\partial \mathbf{A}}{\partial t} = -\text{grad } \Phi$$

or
$$\mathbf{E} = -\text{grad } \Phi - \frac{\partial \mathbf{A}}{\partial t} \tag{5.54}$$

We, thus, have solved the two homogeneous equations (ii) and (iii) of

(5.47) in terms of **A** and Φ. Once **A** and Φ are determined **B** and **E** can can be found.

We have seen in the preceding chapter that equation (5.52) does not completely define **A**. If we add the gradient of any arbitrary scalar function to the vector potential, i.e. say if **A** is changed to

$$\mathbf{A}' = \mathbf{A} + \nabla\psi \tag{5·55}$$

the magnetic field remains unchanged. But will **E** remain unchanged? It will certainly change if some special precautions are not taken. In order that the addition of $\nabla\psi$ should not affect the electric field, the scalar potential Φ must be simultaneously transformed to Φ' where

$$\Phi' = \Phi - \frac{\partial\psi}{\partial t} \tag{5.56}$$

You can verify this by substituting $\mathbf{A}'$ and Φ' in (5.54).

Any physical law that can be expressed in terms of the electromagnetic potentials **A** and Φ remains unaffected by the transformations of the type (5.55) and (5.56). These transformations are called **gauge transformations.** Clearly, equations involving potentials must be **gauge invariant.**

In electrostatics we adopted the condition $\nabla\cdot\mathbf{A} = 0$ which together with $\mathbf{B} = \nabla\times\mathbf{A}$ specified **A**. In electromagnetism we have to make a different choice. In order to specify **A** we have to impose an additional condition on **A** in such a way that it does not change the physics. In other words, it must be consistent with the transformation (5.55) and (5.56) so that **E** and **B** remain unaffected. In order to do this, we turn our attention to the remaining two inhomogeneous equations viz (i) and (iv) of (5.47).

Substituting (5.54) in equation (i) we have

$$\nabla\cdot\mathbf{D} = \nabla\cdot\epsilon_0\mathbf{E} = \epsilon_0\nabla\cdot\left(-\nabla\Phi - \frac{\partial\mathbf{A}}{\partial t}\right) = \rho$$

i.e.

$$-\nabla^2\Phi - \frac{\partial}{\partial t}(\nabla\cdot\mathbf{A}) = \rho/\epsilon_0. \tag{5.57}$$

Equation (iv) can be written as

$$\nabla\times\mathbf{H} - \frac{\partial\mathbf{D}}{\partial t} = \mathbf{j}.$$

i.e.

$$\nabla\times\frac{\mathbf{B}}{\mu} - \epsilon_0\frac{\partial\mathbf{E}}{\partial t} = \mathbf{j}$$

i.e.

$$\frac{1}{\mu}\nabla\times(\nabla\times\mathbf{A}) - \epsilon_0\frac{\partial}{\partial t}\left(-\nabla\Phi - \frac{\partial\mathbf{A}}{\partial t}\right) = \mathbf{j}$$

i.e.

$$\nabla\times(\nabla\times\mathbf{A}) - \mu\epsilon_0\frac{\partial}{\partial t}\left(-\nabla\Phi - \frac{\partial\mathbf{A}}{\partial t}\right) = \mu\mathbf{j}$$

i.e.

$$-\nabla^2\mathbf{A} + \nabla(\nabla\cdot\mathbf{A}) + \mu\epsilon_0\frac{\partial}{\partial t}(\nabla\Phi) + \mu\epsilon_0\frac{\partial^2\mathbf{A}}{\partial t^2} = \mu\mathbf{J} \tag{5.58}$$

where we have made use of the indentity.

$$\nabla \times (\nabla \times \mathbf{A}) = \nabla (\nabla \cdot \mathbf{A}) - \nabla^2 \mathbf{A}.$$

We have yet to exercise our freedom of choosing the condition to be imposed upon **A**.

We choose **A** and Φ such that

$$\nabla \cdot \mathbf{A} = -\mu\epsilon_0 \frac{\partial \Phi}{\partial t} = -\frac{1}{c^2}\frac{\partial \Phi}{\partial t}. \tag{5.59}$$

We see at once that with this substitution the two middle terms of (5.58) cancel and the equation reduces to

$$\nabla^2 \mathbf{A} - \mu\epsilon_0 \frac{\partial^2 \mathbf{A}}{\partial t^2} = -\mu \mathbf{j} \tag{5.60}$$

With the condition (5.59) the equation (5.57) becomes

$$-\nabla^2 \Phi - \frac{\partial}{\partial t}(\nabla \cdot \mathbf{A}) = -\nabla^2 \Phi + \mu\epsilon_0 \frac{\partial^2 \Phi}{\partial t^2} = \rho/\epsilon_0,$$

or

$$\nabla^2 \Phi - \mu\epsilon_0 \frac{\partial^2 \Phi}{\partial t^2} = -\rho/\epsilon_0. \tag{5.61}$$

Evidently, we have chosen our condition judiciously. The choice has yielded two independent equations: one for **A** (5.60) and the other for Φ (5.61). **A** is connected with the vector **j** and Φ with the scalar quantity ρ. Furthermore, both the equations have the same form i.e. both potentials satisfy the same equations. The condition, thus, introduces complete symmetry between the scalar and vector potentials. For the steady-state, the time derivatives vanish and we have

$$\nabla^2 \mathbf{A} = -\mu \mathbf{j} \text{ and } \nabla^2 \Phi = -\rho/\epsilon_0$$

The condition (5.59) is known as **Lorentz gauge condition**. The gauge used in magnetostatics viz. $\nabla \cdot \mathbf{A} = 0$ is called **Coulomb gauge.**

We have seen that the electric field **E** and the magnetic field **B** are invariant under transformation (5.55) and (5.56). The potentials thus transformed will have to satisfy the Lorentz condition. Hence, the gauge function ψ which so far remained arbitrary will have to satisfy a certain condition.

Since the original and the transformed potentials have to satisfy Lorentz condition, we have

$$\nabla \cdot \mathbf{A} + \mu\epsilon_0 \frac{\partial \Phi}{\partial t} = 0 \tag{5.62}$$

and

$$\nabla \cdot \mathbf{A}' + \mu\epsilon_0 \frac{\partial \Phi'}{\partial t} = 0 \tag{5.63}$$

i.e. $$\nabla\cdot(\mathbf{A}+\nabla\psi)+\mu\epsilon_0\frac{\partial}{\partial t}\left(\Phi-\frac{\partial\psi}{\partial t}\right)=0$$

i.e. $$\nabla\cdot\mathbf{A}+\nabla^2\psi+\mu\epsilon_0\frac{\partial\Phi}{\partial t}-\mu\epsilon_0\frac{\partial^2\psi}{\partial t^2}=0$$

Hence,

$$\nabla^2\psi-\mu\epsilon_0\frac{\partial^2\psi}{\partial t^2}=0. \tag{5.64}$$

Thus, the restricted gauge transformation

$$\mathbf{A}'\to\mathbf{A}+\nabla\psi$$
$$\Phi'\to\Phi-\frac{\partial\psi}{\partial t} \tag{5.65}$$

where ψ satisfies the equation

$$\nabla^2\psi-\mu\epsilon_0\frac{\partial^2\psi}{\partial t^2}=0$$

preserve the Lorentz condition.

5.10 More about the Lorentz Gauge Condition

The Lorentz gauge condition is not quite as arbitrary as it appears at first sight. We can relate it to the first principles of electromagnetic theory: Coulomb's law, Biot-Savart's law and the principle of conservation of energy.

While deriving the Ampère's circuital law from Biot-Savart's law for stationary current distribution, the Coulomb gauge condition $\nabla\cdot\mathbf{A}=0$ appears in a natural way. We now extend this derivation for quasi-stationary conditions.

The Biot-Savart's law leads us to the expression

$$\mathbf{A}(\mathbf{r}_2)=\frac{\mu_0}{4\pi}\int_{V_1}\frac{\mathbf{j}(\mathbf{r}_1)}{|\mathbf{r}_2-\mathbf{r}_1|}\,d\tau \tag{5.66}$$

where the subscripts 1 and 2 refer to the source and field coordinates. Hence,

$$\nabla_2\cdot\mathbf{A}(\mathbf{r}_2)=\frac{\mu_0}{4\pi}\int_{V_1}\mathbf{j}(\mathbf{r}_1)\cdot\nabla_2\left[\frac{1}{|\mathbf{r}_2-\mathbf{r}_1|}\right]d\tau \tag{5.67}$$

where ∇_2 operates on $\mathbf{r}_2$ only.

Since $$\nabla_2\left[\frac{1}{|\mathbf{r}_2-\mathbf{r}_1|}\right]=-\nabla_1\left[\frac{1}{|\mathbf{r}_2-\mathbf{r}_1|}\right]$$

$$\begin{aligned}\nabla_2\cdot\mathbf{A}(\mathbf{r}_2)&=-\frac{\mu_0}{4\pi}\int_{V_1}\mathbf{j}(\mathbf{r}_1)\cdot\nabla_1\left[\frac{1}{|\mathbf{r}_2-\mathbf{r}_1|}\right]d\tau\\&=\frac{\mu_0}{4\pi}\left[\int_{V_1}\left\{\frac{1}{|\mathbf{r}_2-\mathbf{r}_1|}\right\}\nabla_1\cdot\mathbf{j}(\mathbf{r}_1)d\tau_1\right.\\&\left.\quad-\int_{V_1}\nabla_1\cdot\left\{\frac{\mathbf{j}(\mathbf{r}_1)}{|\mathbf{r}_2-\mathbf{r}_1|}\right\}d\tau_1\right]\end{aligned} \tag{5.68}$$

Using Divergence theorem, the second integral can be writtten for bounded source distribution as

$$\int_{V_1} \nabla_1 \cdot \left\{ \frac{\mathbf{j}(\mathbf{r}_1)}{|\mathbf{r}_2 - \mathbf{r}_1|} \right\} d\tau_1 = \int \frac{\mathbf{j}(\mathbf{r}_1)}{|\mathbf{r}_2 - \mathbf{r}_1|} . \hat{\mathbf{e}}_n dS = 0$$

$$\therefore \quad \nabla_2 \cdot \mathbf{A}(\mathbf{r}_2) = \frac{\mu_0}{4\pi} \int_{V_1} \left\{ \frac{1}{|\mathbf{r}_2 - \mathbf{r}_1|} \right\} \nabla_1 \cdot \mathbf{j}(\mathbf{r}_1)\, d\tau_1. \tag{5.69}$$

The conservation of charge relation gives

$$\nabla \cdot \mathbf{j} = -\frac{\partial \rho}{\partial t}$$

$$\therefore \quad \nabla_2 \cdot \mathbf{A}(\mathbf{r}_2) = -\frac{\mu_0}{4\pi} \int_{V_1} \left\{ \frac{1}{|\mathbf{r}_2 - \mathbf{r}_1|} \right\} \frac{\partial \rho(\mathbf{r}_1)}{\delta t}\, d\tau_1. \tag{5.70}$$

From Coulomb's law we have

$$\Phi(\mathbf{r}_2) = \frac{1}{4\pi\epsilon_0} \int_{V_1} \frac{\rho(\mathbf{r}_1)}{|\mathbf{r}_2 - \mathbf{r}_1|}\, d\tau_1$$

$$\therefore \quad \nabla_2 \cdot \mathbf{A}\,(\mathbf{r}_2) = -\mu_0\epsilon_0 \frac{\partial}{\partial t} \left\{ \frac{1}{4\pi\epsilon_0} \int_{V_1} \frac{\rho(\mathbf{r}_1)}{|\mathbf{r}_2 - \mathbf{r}_1|}\, d\tau_1 \right\}$$

$$= -\mu_0\epsilon_0 \frac{\partial \Phi(\mathbf{r}_2)}{\partial t} \tag{5.71}$$

or more generally,

$$\nabla \cdot \mathbf{A} = -\mu_0\epsilon_0 \frac{\partial \Phi}{\partial t} \tag{5.72}$$

which is the Lorentz Gauge condition.

Now
$$\nabla_2 \times \mathbf{B}(\mathbf{r}_2) = \nabla_2 \times \{\nabla_2 \times \mathbf{A}(\mathbf{r}_2)\}$$
$$= \nabla_2 \{\nabla_2 \cdot \mathbf{A}(\mathbf{r}_2)\} - \nabla_2^2 \mathbf{A}(\mathbf{r}_2).$$

The first term
$$= -\nabla_2 \left[\mu_0\epsilon_0 \frac{\partial \Phi(\mathbf{r}_2)}{\partial t} \right] = \mu_0 \frac{\partial}{\partial t} \left[-\epsilon_0 \nabla_2\, \Phi\,(\mathbf{r}_2) \right]$$

$$= \mu_0 \frac{\partial}{dt} \left[\epsilon_0 \mathbf{E} \right] = \mu_0 \frac{\partial \mathbf{D}}{\partial t} \tag{5.73}$$

The second term
$$= -\nabla_2^2 \mathbf{A}(\mathbf{r}_2) = -\frac{\mu_0}{4\pi} \int_{V_1} \mathbf{j}(\mathbf{r}_1) \nabla_2^2 \left\{ \frac{1}{|\mathbf{r}_2 - \mathbf{r}_1|} \right\} d\tau$$

$= \mu_0 \mathbf{j}(\mathbf{r}_2)$ where we have used (3.162)

$$\therefore \quad \nabla \times \mathbf{B} = \mu_0 \left[\mathbf{j} + \frac{\partial \mathbf{D}}{\partial t} \right] \tag{5.74}$$

We see that this treatment also leads to the displacement current.

5.11 Field Energy and Field Momentum

We recall that the general expressions for the electrostatic and magneto static field energies are:

$$W_E = \frac{1}{2}\int_V (\mathbf{E}\cdot\mathbf{D})\, d\tau \text{ and } W_M = \frac{1}{2}\int_V (\mathbf{B}\cdot\mathbf{H})\, d\tau \tag{5.75}$$

We shall now find the expression for the electromagnetic energy in time-dependent situations.

The force on a moving charge q is given by

$$\mathbf{F} = q(\mathbf{E} + \mathbf{v} \times \mathbf{B}). \tag{5.76}$$

The rate at which the work is done on this charge is

$$\mathbf{F}\cdot\mathbf{v} = q(\mathbf{E} + \mathbf{v} \times \mathbf{B})\cdot\mathbf{v} = q\mathbf{E}\cdot\mathbf{v}. \tag{5.77}$$

The magnetic field does no work as it is perpendicular to the velocity of the charge.

If there exists a continuous distribution of charge, the total rate at which the work is done in a given volume is

$$\int \rho(\mathbf{E}\cdot\mathbf{v})\, d\tau = \int (\mathbf{E}\cdot\mathbf{j}) d\tau. \tag{5.78}$$

Substituting for $\mathbf{j}$ from (iv) of (5.47), we have

$$\int (\mathbf{E}\cdot\mathbf{j}) d\tau = \int \left[\mathbf{E}\cdot\left(\nabla \times \mathbf{H} - \frac{\partial \mathbf{D}}{\partial t}\right)\right] d\tau \tag{5.79}$$

Now $$\mathbf{E}\cdot(\nabla \times \mathbf{H}) = \nabla\cdot(\mathbf{H} \times \mathbf{E}) + \mathbf{H}\cdot(\nabla \times \mathbf{E})$$

$$= \nabla\cdot(\mathbf{H} \times \mathbf{E}) - \mathbf{H}\cdot\frac{\partial \mathbf{B}}{\partial t} \quad \text{[by (iii) of 5.47]}$$

$$\therefore \quad \int (\mathbf{E}\cdot\mathbf{j})\, d\tau = \int \left[\nabla\cdot(\mathbf{H} \times \mathbf{E}) - \mathbf{E}\cdot\frac{\partial \mathbf{D}}{\partial t} - \mathbf{H}\cdot\frac{\partial \mathbf{B}}{\partial t}\right] d\tau$$

$$= \int \left[\nabla\cdot(\mathbf{H} \times \mathbf{E})\right] d\tau - \int \left(\mathbf{E}\cdot\frac{\partial \mathbf{D}}{\partial t} + \mathbf{H}\cdot\frac{\partial \mathbf{B}}{\partial t}\right) d\tau \tag{5.80}$$

Using Divergence theorem to transform the first integral, we have

$$\int (\mathbf{E}\cdot\mathbf{j})\, d\tau = \int_S (\mathbf{H} \times \mathbf{E})\cdot\hat{\mathbf{e}}_n dS - \int_V \left(\mathbf{E}\cdot\frac{\partial \mathbf{D}}{\partial t} + \mathbf{H}\cdot\frac{\partial \mathbf{B}}{\partial t}\right) d\tau$$

where S is the surface boundary of the volume V. Therefore,

$$-\int_V \left((\mathbf{E}\cdot\frac{\partial \mathbf{D}}{\partial t} + \mathbf{H}\cdot\frac{\partial \mathbf{B}}{\partial t}\right) d\tau = \int (\mathbf{E}\cdot\mathbf{j})\, d\tau + \int_S (\mathbf{E} \times \mathbf{H})\cdot\hat{\mathbf{e}}_n dS. \tag{5.81}$$

If ϵ and μ are assumed to be constant, we have for the **linear media**

$$\int_V \left(\mathbf{E}\cdot\frac{\partial \mathbf{D}}{\partial t} + \mathbf{H}\cdot\frac{\partial \mathbf{B}}{\partial t}\right) d\tau = \frac{\partial}{\partial t}\int_V \tfrac{1}{2}[\mathbf{E}\cdot\mathbf{D} + \mathbf{B}\cdot\mathbf{H}]\, d\tau$$

Therefore,

$$-\frac{\partial}{\partial t}\int_V \tfrac{1}{2}[\mathbf{E}\cdot\mathbf{D} + \mathbf{B}\cdot\mathbf{H}]\, d\tau = \int (\mathbf{E}\cdot\mathbf{j})\, d\tau + \int (\mathbf{E} \times \mathbf{H})\cdot\hat{\mathbf{e}}_n dS \tag{5.82}$$

Since an electromagnetic field consists of electric and magnetic fields, it is reasonable to assume that the sum of the energies given in (5.75) represents the total electromagnetic energy. Hence, we may take $\mathcal{E}_M = \frac{1}{2}[\mathbf{E}\cdot\mathbf{D} + \mathbf{B}\cdot\mathbf{H}]$ as the electromagnetic energy density. We may, therefore, interpret the left-hand side of the equation (5.82) as the rate at which the energy stored in the electromagnetic field diminishes. The first term on the right-hand side gives, as stated above, the work done by the field forces on the charges contained in the volume. The vector $[\mathbf{E} \times \mathbf{H}]$ has the dimensions of $\frac{\text{energy}}{\text{area} \times \text{time}}$ and, hence, the last term may be interpreted as representing the energy that flows out of the boundary per unit time. The vector $[\mathbf{E} \times \mathbf{H}]$ which gives the rate at which the energy flows across unit area of the boundary, is called the **Poynting vector** and is represented by the symbol **N**.

The equation (5.82) may also be written in differential form as

$$\frac{d\mathcal{E}_M}{dt} + \operatorname{div} \mathbf{N} = -\mathbf{E}\cdot\mathbf{j}. \tag{5.83}$$

If the medium has zero conductivity $\mathbf{j} = \sigma\mathbf{E} = 0$.

Hence,

$$\frac{d\mathcal{E}_M}{dt} + \operatorname{div} \mathbf{N} = 0. \tag{5.84}$$

This equation has exactly the same form as the equation of continuity (5.41). The volume density of conserved quantity here is $\mathcal{E}_M$ and the "current density" is **N**. This analogy leads us to the same conclusion as above regarding the identification of the flux of electromagnetic energy with the Poynting vector $\mathbf{N} = \mathbf{E} \times \mathbf{H}$.

Thus, equation (5.82) represents the law of conservation of energy. It states that the decrease of electromagnetic energy per unit time in a certain volume V is equal to the work done by the field forces per unit time plus the flux flow outwards per unit time. This is known as **Poynting Theorem.**

Similar computations show that an electromagnetic field possesses momentum.

The force on a region containing both charges and currents is

$$\mathbf{F} = \int_V (\rho\mathbf{E} + \mathbf{j} \times \mathbf{B})d\tau. \tag{5.85}$$

If P_{mech} is the sum of momenta of all the particles

$$\frac{dP_{mech}}{dt} = \int_V (\rho\mathbf{E} + \mathbf{j} \times \mathbf{B})d\tau. \tag{5.86}$$

From Maxwell's equations

$$\rho = \nabla\cdot\mathbf{D};\ \mathbf{j} = \nabla \times \mathbf{H} - \frac{\partial\mathbf{D}}{\partial t}$$

$$\therefore \qquad \frac{dP_{mech}}{dt} = \int_V \left\{ (\nabla \cdot \mathbf{D})\mathbf{E} + \left(\nabla \times \mathbf{H} - \frac{\partial \mathbf{D}}{\partial t} \right) \times \mathbf{B} \right\} d\tau$$

$$= \int_V \left\{ (\nabla \cdot \mathbf{D})\mathbf{E} + \mathbf{B} \times \frac{\partial \mathbf{D}}{\partial t} - \mathbf{B} \times (\nabla \times \mathbf{H}) \right\} d\tau$$

Since $\frac{\partial}{\partial t}(\mathbf{D} \times \mathbf{B}) = \mathbf{D} \times \frac{\partial \mathbf{B}}{\partial t} + \frac{\partial \mathbf{D}}{\partial t} \times \mathbf{B}$

$$\frac{dP_{mech}}{dt} = \int_V \left[(\nabla \cdot \mathbf{D})\mathbf{E} + \left(\mathbf{D} \times \frac{\partial \mathbf{B}}{\partial t} \right) - \frac{\partial}{\partial t}(\mathbf{D} \times \mathbf{B}) - \mathbf{B} \times (\nabla \times \mathbf{H}) \right] d\tau$$

Because $\nabla \cdot \mathbf{B} = 0$, addition of $(\nabla \cdot \mathbf{B})\mathbf{H}$ to the square bracket does not alter the result.

$$\therefore \qquad \frac{dP_{mech}}{dt} + \frac{d}{dt}\int_V (\mathbf{D} \times \mathbf{B}) d\tau$$

$$= \int_V \left[(\nabla \cdot \mathbf{D})\mathbf{E} + (\nabla \cdot \mathbf{B})\mathbf{H} - \{\mathbf{D} \times (\nabla \times \mathbf{E})\} - \{\mathbf{B} \times (\nabla \times \mathbf{H})\} \right] d\tau \qquad \left(\because \ \nabla \times \mathbf{E} = -\frac{\partial \mathbf{B}}{dt} \right) \qquad (5.87)$$

Clearly, the integral in the second term of the left-hand side represents momentum. Since it is not associated with the mass of particles and consists only of fields, we identify it as the electromagnetic momentum P_{field}. The vector $\mathbf{g} = [\mathbf{D} \times \mathbf{B}]$ is called **electromagnetic momentum density.** The right-hand side can be converted into a surface integral and identified as momentum flow. We conclude from equation (5.87) that the total momentum of the closed system consisting of a field and particles is conserved.

It may be noted that the momentum density vector **g** is related to the Poynting vector **N**.

Thus,

$$\mathbf{g} = [\mathbf{D} \times \mathbf{B}] = [\epsilon\mathbf{E} \times \mu\mathbf{H}] = \mu\epsilon[\mathbf{E} \times \mathbf{H}] = \mu\epsilon\mathbf{N} \qquad (5.88)$$

PROBLEMS

5.1 A spatially uniform magnetic field $\mathbf{B} = \mathbf{B}_0 \sin \omega t$ is directed at an angle θ to the normal of the plane of a circular loop. Calculate the e.m.f. induced in the loop.

5.2 A plane circular disc of radius 'a' rotates at a speed of f revolutions per second about an axis though its centre, normal to its plane. A uniform magnetic field **B** exists parallel to the axis. Show that there is an e.m f. between the centre of the disc and its rim of magnitude $\Phi = f B\pi a^2$ (Faraday's disc).

5.3 A solenoid 300 mm in length and 15 mm in diameter has a uniform winding of 2,500 turns. The solenoid is placed in a uniform field of flux density $4T$ and a current of 2 A is passed through the solenoid winding. Find the magnetic moment of the solenoid and the torque acting on it.

5.4 Electrons in a betatron move in a circular orbit in a vacuum chamber in a magnetic field B and are accelerated by the increasing flux linking the orbit.

Show that for a stable circular orbit the magnetic field at the orbit must be exactly one half the average magnetic field over the duration of the acceleration.

5.5 Consider a charge moving with a uniform velocity along the axis of a circle and calculate the displacement current that flows through the circle. Next, calculate the conduction current through the circle and show that the two results are identical.

5.6 A transmission line consists of a pair of nonpermeable parallel wires of radius r_1 and r_2. A current flows down one wire and back the other and is uniformly distributed over the cross-section of the wire. If the wires are separated by a distance t find the vector-potential of the system.

5.7 A small magnet of magnetic moment m is rotating about its centre with an angular velocity $\boldsymbol{\omega}$. Show that $\frac{d\mathbf{m}}{dt} = \boldsymbol{\omega} \times \mathbf{m}$. Show further that the motion of the magnet gives rise to an electric field **E** given by

$$\mathbf{E} = \frac{\mu_0}{4\pi r^3}\left\{(\mathbf{r}.\mathbf{m})\,\boldsymbol{\omega} - (\mathbf{r}.\boldsymbol{\omega})\,\mathbf{m}\right\}$$

5.8 A dielectric sphere of dielectric constant ε and radius 'a', rotates with an angular velocity $\boldsymbol{\omega}$ about the z-axis. The centre of the sphere is at the origin. A uniform electric field $\mathbf{E}_0$. is applied in the x-direction. Show that there is a magnetic field

$$\mathbf{H} = \frac{2}{3}\,\sigma\,\omega\,a \quad \text{for } r < a,$$

$$= \frac{3(\mathbf{m}.\mathbf{r})\,\mathbf{r}}{4\pi r^5} - \frac{\mathbf{m}}{4\pi\, r^3} \quad \text{for } r > a,$$

where $\mathbf{m} = \frac{4\pi}{3}\,a^4\,\sigma\omega$, σ being the charge induced on the surface of the sphere.

Chapter 6

Electromagnetic Waves

6.1 Plane Waves in Non-conducting Media

We have seen that Maxwell's equations provide us with all the information that can be drawn from the classical theory of electric and magnetic fields. We shall show in this chapter that fields generated by the moving charges can leave the source and travel through space in the form of waves. This is one of the important features of Maxwell's equations.

Assume that ϵ, μ and σ are constant. The curl of equation (iii) of (5.47) gives

$$\nabla \times (\nabla \times \mathbf{E}) = -\nabla \times \frac{\partial \mathbf{B}}{\partial t} = -\mu \frac{\partial}{\partial t}(\nabla \times \mathbf{H}).$$

Using a well known vector identity to transform the left-hand side and equation (iv) of (5.47) to transform the right-hand side of the above equation, we have

$$\nabla(\nabla \cdot \mathbf{E}) - \nabla^2 \mathbf{E} = -\mu \frac{\partial \mathbf{j}}{\partial t} - \mu\epsilon \frac{\partial^2 \mathbf{E}}{\partial t^2}$$

i.e.

$$\nabla^2 \mathbf{E} - \mu\epsilon \frac{\partial^2 \mathbf{E}}{\partial t^2} - \mu\sigma \frac{\partial \mathbf{E}}{\partial t} = \nabla(\rho/\epsilon_0). \tag{6.1}$$

In a region in which there is no free charge $\rho = 0$

$$\therefore \quad \nabla^2 \mathbf{E} - \mu\epsilon \frac{\partial^2 \mathbf{E}}{\partial t^2} - \mu\sigma \frac{\partial \mathbf{E}}{\partial t} = 0 \tag{6.2}$$

Assuming $\mathbf{E}(\mathbf{r}, t) = \mathbf{E}(\mathbf{r})e^{-i\omega t}$ we can compare the relative magnitudes of the second and the third term in (6.2). Thus,

$$\frac{\left|\mu\sigma \dfrac{\partial \mathbf{E}}{\partial t}\right|}{\left|\mu\epsilon \dfrac{\partial^2 \mathbf{E}}{\partial t^2}\right|} \simeq \frac{\sigma}{\epsilon\omega} = \frac{1}{\omega\tau} = \frac{T}{2\pi\tau} \tag{6.3}$$

where we have replaced ϵ/σ by the relaxation time τ (5.50) and $\omega = \dfrac{2\pi}{T}$,

T being the period of oscillation. If $\tau \ll T$, which is the case for a conducting medium, the third term in (6.2) is dominant and we can write (6.2) as

$$\nabla^2 \mathbf{E} - \mu\sigma \frac{\partial \mathbf{E}}{\partial t} = 0 \tag{6.4}$$

a diffusion equation.

If $\tau \gg T$, the term involving σ can be neglected and we have for the non-conducting medium the equation

$$\nabla^2 \mathbf{E} - \mu\epsilon \frac{\partial^2 \mathbf{E}}{\partial t^2} = 0. \tag{6.5}$$

Exactly similar equations can be obtained for $\mathbf{H}$, viz.,

$$\nabla^2 \mathbf{H} - \mu\epsilon \frac{\partial^2 \mathbf{H}}{\partial t^2} - \mu\sigma \frac{\partial \mathbf{H}}{\partial t} = 0 \tag{6.6}$$

and

$$\nabla^2 \mathbf{H} - \mu\epsilon \frac{\partial^2 \mathbf{H}}{\partial t^2} = 0 \tag{6.7}$$

These equations are of the type of wave equations we are familiar with. The general wave-equation is

$$\nabla^2 \psi - \frac{1}{v^2} \frac{\partial^2 \psi}{\partial t^2} = 0 \tag{6.8}$$

where v is the velocity of propagation.

Comparison of (6.5) and (6.7) with (6.8) shows that the velocity in our case is

$$v = (\epsilon\mu)^{-1/2} \tag{6.9}$$

In vacuum

$$v = (\epsilon_0\mu_0)^{-1/2} \tag{6.10}$$

We conclude that any time variations in electric or magnetic field are propagated with the same velocity

$$v = (\epsilon\mu)^{-1/2}.$$

It should be noted that the equation (6.8) is a scalar equation; while equations (6.5) and (6.7) are vector equations, which means that the later equations are valid for each component of $\mathbf{E}$ and $\mathbf{H}$.

The simplest type of wave that is a solution of (6.5) or (6.7) is a plane wave. The relative simplicity of plane wave solutions of Maxwell's equations enables some of the important elementary physical characteristics of the electromagnetic field to be elucidated, without appeal to the other than quite straightforward mathematics. Since plane waves are good approximations to actual waves in many situations, we shall discuss the plane wave solutions of the above equations.

A plane wave is one in which the wave amplitude—the field vector component—is constant over all points of a plane normal to the direction of propagation. This plane constitutes a wave-front which advances with a velocity v in a direction normal to itself. The field vector components that lie in given plane are functions of perpendicular distance of the plane from the origin and also of time. We may choose our coordinate system in such a way that the direction of propagation coincides with, say, x–axis. Since the wave-function in this case does not depend on y and z, the wave-equation takes one dimensional form such as

$$\frac{\partial^2\psi}{\partial x^2} - \frac{1}{v^2}\frac{\partial^2\psi}{\partial t^2} = 0 \tag{6.11}$$

This has a general solution

$$\begin{aligned}\psi(x, t) &= Ae^{i(kx-\omega t)} + Be^{-i(kx+\omega t)}\\ &= f(x - vt) + g(x + vt)\end{aligned} \tag{6.12}$$

where A and B are constants (generally complex) and $k = \omega/v$ where v is the phase velocity of the wave. Equation (6.12) represents waves travelling to the right and to the left with velocity v. We may assume, that the plane wave fields are of the form

$$\mathbf{E}(x, t) = \mathbf{E}_0 e^{i(kx-\omega t)} \tag{6 13}$$

$$\mathbf{H}(x, t) = \mathbf{H}_0 e^{i(kx-\omega t)}. \tag{6.14}$$

One may ask: "We find from equations (6.13) and (6.14) that you have assumed **E** and **H** to be in phase. Is this correct?" The justification for the assumption made will be found in the following argument.

If possible let α be the phase difference between **E** and **H**, i.e.

$$\mathbf{E} = \mathbf{E}_0 e^{i(kx-\omega t)}$$

$$\mathbf{H} = \mathbf{H}_0 e^{i(kx-\omega t+\alpha)}.$$

The fields **E** and **H** have to satisfy Maxwell's equation

$$\nabla \times \mathbf{E} = -\frac{\partial \mathbf{B}}{\partial t}$$

$$\therefore \qquad \frac{\partial E_z}{\partial x} = \frac{\partial B_y}{\partial t};\ \frac{\partial E_y}{\partial x} = -\frac{\partial B_z}{\partial t}$$

($\because$ $\frac{\partial}{\partial y}, \frac{\partial}{\partial z}$ of **E** is zero since E travels in x-direction.)

From the second equation

$$ikE_{0y}e^{i(kx-\omega t)} = i\omega\mu H_{0z}e^{i(kx-\omega t+\alpha)}.$$

Taking the real parts

$$kE_{0y}\cos(kx - \omega t) = \omega\mu H_{0z}\cos(kx - \omega t + \alpha.)$$

This is true for all x and t. The only value of α which satisfies this condition is zero. Therefore, **E** and **H** are in phase. The ratio of amplitude of **E** and **H** is given by

$$\frac{E_0}{H_0} = \frac{\mu\omega}{k} = Z_0 \tag{6.15}$$

The ratio Z_0 has dimensions of impedance and is called the *intrinsic impedance* of the medium.

Thus, a monochromatic plane wave always has **E** and **H** in phase and with a ratio of amplitudes at any instant at any point given by

$$\frac{E}{H} = \frac{\mu\omega}{k} = \mu v. \tag{6.16}$$

For a plane wave in free space

$$Z_0 = \mu_0 v_0 = 376.7\Omega. \tag{6.17}$$

As the wave propagates, the two vectors may both change their direction in the y–z plane without violating any of the conditions established so far.

For the propagation of the wave in any arbitrary direction, we have

$$\mathbf{E}(\mathbf{r}, t) = \mathbf{E}_0 e^{i(\mathbf{k}.\mathbf{r} - \omega t)} \tag{6.18}$$

$$\mathbf{H}(\mathbf{r}, t) = \mathbf{H}_0 e^{i(\mathbf{k}.\mathbf{r} - \omega t)}. \tag{6.19}$$

where $\mathbf{E}_0$, $\mathbf{H}_0$ are vectors constant in time and $\mathbf{k} = \hat{\mathbf{e}}_k |k|$ is the propagation vector, $\hat{\mathbf{e}}_k$ being a unit vector in the direction of propagation.

Since, **E** and **H** are real, we are interested only in the real part of (6.18) and (6.19).

Now **E** and **H** obtained as above must also satisfy Maxwell's equations. It must be emphasized that this is not automatic. Maxwell's equations do not completely determine the electromagnetic field.

Substituting (6.18) in (i) of (5.47), we have,

$$\nabla \cdot \{\mathbf{E}_0 e^{i(\mathbf{k}.\mathbf{r} - \omega t)}\} = 0$$

which leads to the relation

$$\mathbf{k} \cdot \mathbf{E} = 0 \tag{6.20}$$

Similarly, substituting (6.19) in (ii) of (5.47), we get

$$\mathbf{k} \cdot \mathbf{H} = 0. \tag{6.21}$$

This shows that both **E** and **H** are perpendicular to the propagation vector **k**. Such a wave is called **transverse wave.** Electromagnetic plane waves are wholly transverse in character.

Let us now substitute (6.18) in (iii) of (5.47)

$$\nabla \times \mathbf{E} = \nabla \times \mathbf{E}_0 e^{i(\mathbf{k}.\mathbf{r} - \omega t)} = \nabla e^{i(\mathbf{k}.\mathbf{r} - \omega t)} \times \mathbf{E}_0$$

$$= i\mathbf{k} \times \mathbf{E} = -\frac{\partial \mathbf{B}}{\partial t} = -\mu \frac{\partial \mathbf{H}}{\partial t} = i\omega\mu\mathbf{H}$$

$$\therefore \quad \mathbf{k} \times \mathbf{E} = \omega\mu\mathbf{H} \tag{6.22}$$

Hence, **H** is perpendicular both to **k** and **E**. In other words, **E** and **H**, which relate to the propagation of the wave, in addition to being perpendicular to the direction of propagation are also perpendicular to one another. The vector **E** × **H** points along the direction of propagation. The vectors **F**, **H**, **k** constitute a right hand orthogonal set (Fig. 6.1).

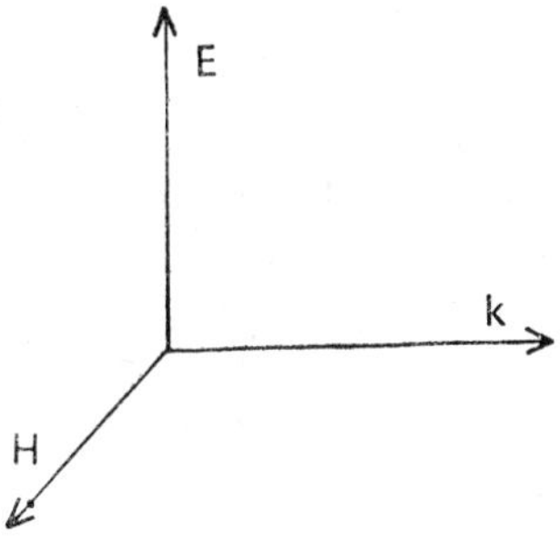

Fig. 6.1

The velocity of the electromagnetic wave in free space is given by

$$v_0 = \frac{1}{(\epsilon_0\mu_0)^{1/2}}.$$

The best value of ϵ_0 is that of Rosa and Porsey viz. $\epsilon_0 = 8.8547 \times 10^{-12}$ and $\mu = 4\pi \times 10^{-7}$. Substituting these in (6.22) we get

$$v_0 = 2.99784 \times 10^8 \text{ ms}^{-1} \tag{6.23}$$

This velocity is the same as the velocity of light in free space. We are, thus, led to conclude that light is simply a form of electromagnetic radiation. X-rays, ultraviolet, infrared, radio and microwave radiations are all electromagnetic radiations, differing only in the order of magnitude of their wave-lengths. Figure 6.2 gives the diagram of the electromagnetic spectrum.

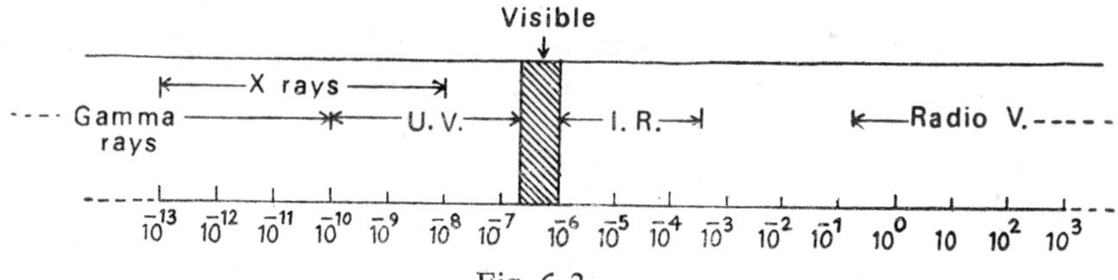

Fig. 6.2

The direct verification of electromagnetic waves predicted by Maxwell was obtained in the experiments of Heinritch Hertz who was among the first to demonstrate the existence of waves generated by the oscillatory spark discharge and to show that they possessed many of the familiar properties of light such as reflection, refraction, interference and polarization.

We identified the phase velocity of electromagnetic waves in a material medium with

$$v = \frac{1}{(\epsilon\mu)^{1/2}} = \frac{1}{(\epsilon_0\epsilon_r\mu_0\mu_r)^{1/2}} = \frac{v_0}{(\epsilon_r\mu_r)^{1/2}} \tag{6.24}$$

which is less than v_0.

For a light wave in a non-dispersive medium

$$v = \frac{c}{n}$$

where n is the refractive index of the medium. Comparison of this relation with (6.24) gives

$$n = (\epsilon_r \mu_r)^{1/2} \tag{6.25}$$

Note that this applies only if n, μ_r, ϵ_r are determined at the same frequency.

6.2 Polarization

In equation (6.18) the direction of **E** is constant in time. Such a wave is said to be *linearly polarized.* The plane of the electric vector **E** (in the case of (6.13) it is, y-z plane) is taken as the plane of polarization*. A linearly polarized monochromatic plane wave can be expressed as a superposition of two linearly independent solutions of the wave equation, e.g.

$$\mathbf{E} = (\hat{\mathbf{e}}_y E_{0y} + \hat{\mathbf{e}}_z E_{0z}) e^{i(kx-\omega t)} \tag{6.26}$$

where $\hat{\mathbf{e}}_y$, $\hat{\mathbf{e}}_z$ are the unit vectors along y and z and are called **polarization vectors.** The amplitudes E_{0y}, E_{0z} are complex amplitudes. Any complex quantity can be expressed as the product of a real quantity and a complex phase factor. Thus,

$$E_{0y} = E_0^y e^{i\alpha}$$

$$E_{0z} = E_0^z e^{i\beta} \tag{6.27}$$

where E_0^y, E_0^z are the real amplitudes. The two independent solutions, therefore, are:

$$\mathbf{E}_y = \hat{\mathbf{e}}_y E_0^y e^{i(kx-\omega t+\alpha)}$$

$$\mathbf{E}_z = \hat{\mathbf{e}}_z E_0^z e^{i(kx-\omega t+\beta)}. \tag{6.28}$$

Therefore,

$$\mathbf{E} = \mathbf{E}_y + \mathbf{E}_z = (\hat{\mathbf{e}}_y E_0^y e^{i\alpha} + \hat{\mathbf{e}}_z E_0^z e^{i\beta}) e^{i(kx-\omega t)} \tag{6 29}$$

or

$$\mathbf{E} = \{\hat{\mathbf{e}}_y E_0^y + \hat{\mathbf{e}}_z E_0^z e^{i(\beta-\alpha)}\} e^{i(kx-\omega t+\alpha)}. \tag{6.30}$$

Let us now consider the following particular cases:

(i) $\mathbf{E}_y$ and $\mathbf{E}_z$ have the same phase, i.e. $\alpha = \beta$ or their phases differ by an integral multiple of π, i.e. $\beta = \alpha \pm m\pi$, where $m = 0, 1, 2 \ldots$, then

$$\mathbf{E} = (\hat{\mathbf{e}}_y E_0^y \pm \hat{\mathbf{e}}_z E_0^z)\ e^{i(kx-\omega t+\alpha)} \tag{6.31}$$

*Before the electromagnetic nature of the radiation was known, the convention in optics was to take the plane of the magnetic vector as the plane of polarization.

This equation represents a linearly polarized wave. The resultant polarization vector oscillates along a line making an angle $\theta = \tan^{-1} \dfrac{E_{0z}}{E_{0y}}$ as shown in Fig. 6.3.

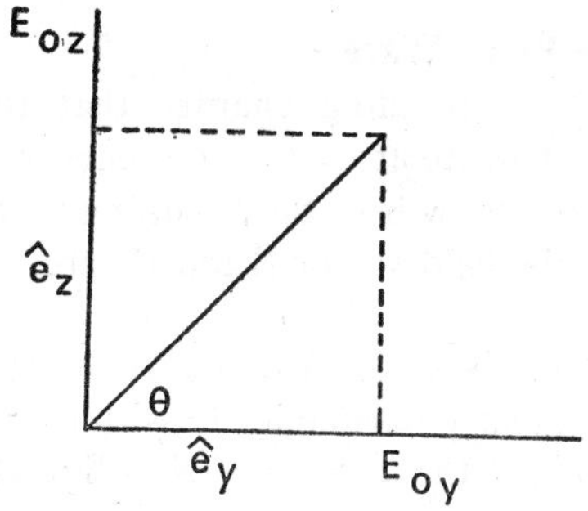

Fig. 6.3

(ii) The amplitudes of the two vectors are equal, i.e. $E_0^y = E_0^z = E_0$ but their phases differ by $\pi/2$.

$$\therefore \qquad \mathbf{E} = E_0(\hat{e}_y \pm i\hat{e}_z)e^{i(kx-\omega t+\alpha)}. \qquad (6.32)$$

The components of the field **E** are (taking only the real part)

$$E_y = E_0 \cos (kx - \omega t + \alpha)$$

$$E_z = \mp E_0 \sin (kx - \omega t + \alpha). \qquad (6.33)$$

This is a wave progressing along the x-axis in which the vector **E** has a constant length $\mathbf{E}_0$ and is a function of time only changing its direction continuously. From (6.33), we have,

$$\frac{E_y^2}{E_0^2} + \frac{E_z^2}{E_0^2} = 1. \qquad (6.34)$$

The vector, therefore, traces out a circle at a frequency ω. Such a wave is said to be **circularly polarized.** The direction of rotation is determined by the sign of E_z in (6.33). If the sign is plus, the rotation is counter clockwise assuming that the observer is facing the oncoming wave. The wave is said to have **left circular polarization** or to have **positive helicity.** If the sign is minus, the rotation is clockwise in which case the wave is **right circularly polarized** or has **negative helicity**.

(iii) $E_0^y \neq E_0^z$ and $\beta = \alpha \pm \pi/2$. The components of **E** are

$$E_y = E_0{}^y e^{i(kx-\omega t+\alpha)}$$

and

$$E_z = E_0{}^z e^{i(kx-\omega t+\alpha\pm\pi/2)}.$$

Taking the real parts

$$E_y = E_0^y \cos (kx - \omega t + \alpha)$$

$$E_z = \mp E_{0z} \sin (kx - \omega t + \alpha)$$

$$\therefore \qquad \frac{E_y^2}{(E_0^y)^2} + \frac{E_z^2}{(E_0^z)^2} = 1 \qquad (6.35)$$

The resultant vector in this case traces an ellipse and the wave is said to be **elliptically polarized.**

6.3 Energy Flux in a Plane Wave

We have shown in the preceding chapter that the Poynting vector $\mathbf{N} = \mathbf{E} \times \mathbf{H}$ represents the rate at which the energy is radiated across unit area. We shall now show how the Poynting vector is calculated for a plane wave in which the field vector **E** and **H** are expressed in terms of complex amplitudes.

The vector $\mathbf{N} = \mathbf{E} \times \mathbf{H}$ gives the instantaneous rate of energy flow. Since the vector **E** and **H** vary harmonically with time, the average energy flow can be found by taking the average of $\mathbf{N} = \mathbf{E} \times \mathbf{H}$ over a complete period, i.e.

$$\langle \mathbf{N} \rangle = \langle R_e\mathbf{E} \times R_e\mathbf{H} \rangle \qquad (6.36)$$

where R_e stands for the real part.

Since **E** and **H** are assumed to be complex quantities, we can express them in the form

$$\mathbf{E} = (\mathbf{E}_1 + i\mathbf{E}_2)e^{-i\omega t} \qquad (6.37)$$

and
$$\mathbf{H} = (\mathbf{H}_1 + i\mathbf{H}_2)e^{-i\omega t}$$

where $\mathbf{E}_1, \mathbf{E}_2, \mathbf{H}_1, \mathbf{H}_2$ are all real.

$$\therefore \qquad R_e\mathbf{E} = \mathbf{E}_1 \cos \omega t + \mathbf{E}_2 \sin \omega t$$

$$R_e\mathbf{H} = \mathbf{H}_1 \cos \omega t + \mathbf{H}_2 \sin \omega t$$

$$\therefore \qquad R_e\mathbf{E} \times R_e\mathbf{H} = (\mathbf{E}_1 \times \mathbf{H}_1) \cos^2 \omega t + (\mathbf{E}_2 \times \mathbf{H}_2) \sin^2 \omega t + \{(\mathbf{E}_1 \times \mathbf{H}_2) + (\mathbf{E}_2 \times \mathbf{H}_1)\} \sin \omega t \cos \omega t$$

Now, over a complete period of oscillation

$$\langle \cos^2 \omega t \rangle = \langle \sin^2 \omega t \rangle = \frac{1}{2}$$

and
$$\langle \sin \omega t \cos \omega t \rangle = 0$$

$$\therefore \qquad \langle R_e\mathbf{E} \times R_e\mathbf{H} \rangle = \frac{1}{2} \{(\mathbf{E}_1 \times \mathbf{H}_1) + (\mathbf{E}_2 \times \mathbf{H}_2)\} \qquad (6.38)$$

Let us now compute $R_e\,(\mathbf{E} \times \mathbf{H}^*)$

$$\mathbf{E} = (\mathbf{E}_1 + i\mathbf{E}_2)e^{-i\omega t} = (\mathbf{E}_1 + i\mathbf{E}_2)(\cos \omega t - i \sin \omega t)$$

$$\mathbf{H}^* = (\mathbf{H}_1 - i\mathbf{H}_2)e^{i\omega t} = (\mathbf{H}_1 - i\mathbf{H}_2)(\cos \omega t + i \sin \omega t)$$

$$\begin{aligned}\therefore \qquad R_e(\mathbf{E} \times \mathbf{H}^*) &= (\mathbf{E}_1 \times \mathbf{H}_1) \cos^2 \omega t + (\mathbf{E}_1 \times \mathbf{H}_2) \cos \omega t \sin \omega t \\ &\quad + (\mathbf{E}_2 \times \mathbf{H}_2) \cos^2 \omega t - (\mathbf{E}_2 \times \mathbf{H}_1) \cos \omega t \sin \omega t \\ &\quad - (\mathbf{E}_1 \times \mathbf{H}_2) \cos \omega t \sin \omega t + (\mathbf{E}_1 \times \mathbf{H}_1) \sin^2 \omega t \\ &\quad + (\mathbf{E}_2 \times \mathbf{H}_1) \cos \omega t \sin \omega t + (\mathbf{E}_2 \times \mathbf{H}_2) \sin^2 \omega t \\ &= (\mathbf{E}_1 \times \mathbf{H}_1) + (\mathbf{E}_2 \times \mathbf{H}_2)\end{aligned} \qquad (6.39)$$

Hence

$$\langle R_e\mathbf{E} \times R_e\mathbf{H}\rangle = \frac{1}{2} R_e(\mathbf{E} \times \mathbf{H}^*) \qquad (6.40)$$

Therefore,

$$\langle \mathbf{N}\rangle = \frac{1}{2} R_e(\mathbf{E} \times \mathbf{H}^*).$$

Since the time factors are automatically cancelled in the product, the time averaged energy can be obtained from the field vectors.

Now $\mathbf{k} \times \mathbf{E} = \omega\mu\mathbf{H}$ (see 6.22)

$$\therefore \quad \mathbf{H}^* = \frac{\mathbf{k} \times \mathbf{E}^*}{\omega\mu} \qquad (6.41)$$

$$\therefore \quad \mathbf{E} \times \mathbf{H}^* = \frac{1}{\omega\mu}\{\mathbf{E} \times (\mathbf{k} \times \mathbf{E}^*)\}$$

$$= \frac{1}{\omega\mu}\{(\mathbf{E}\cdot\mathbf{E}^*)\mathbf{k} - (\mathbf{E}\cdot\mathbf{k})\mathbf{E}^*\}$$

$$= \frac{1}{\omega\mu}\left| E_0 \right|^2 \mathbf{k}.$$

Hence,

$$\langle \mathbf{N}\rangle = \frac{|E_0|^2}{2\omega\mu}\mathbf{k} = \frac{|E_0|^2}{2}\left(\frac{\epsilon}{\mu}\right)^{1/2} \hat{\mathbf{e}}_k \qquad (6.42)$$

[$\because$ $\mathbf{k} = k\hat{\mathbf{e}}_k$ where $\hat{\mathbf{e}}_k$ is the unit vector in the direction of propagation and $k = \frac{\omega}{v} = \omega(\epsilon\mu)^{1/2}$.]

If the wave is moving in the x-direction, the energy flow is in the x-direction and is the same at all points for which z-coordinate is the same. Figure 6.4 shows a rectangular box with length 'c' along the x-axis and 'a' and 'b' being its other two sides. The energy density in an electromagnetic field is $\frac{1}{2}$ $(\mathbf{E}\cdot\mathbf{D}+\mathbf{B}\cdot\mathbf{H})$. Hence, the total energy is given by

$$U = \frac{1}{2}\int_V (\mathbf{E}\cdot\mathbf{D} + \mathbf{B}\cdot\mathbf{H})\, d\tau \qquad (6.43)$$

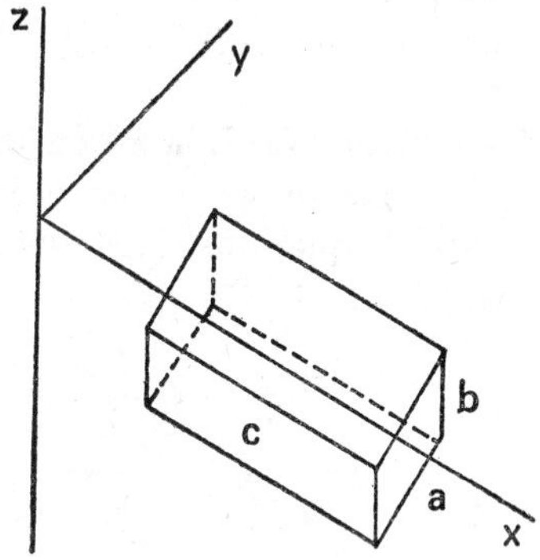

Fig. 6.4

Assuming that the fields **E** and **H** act along y and z-axes respectively, we can write

$$\mathbf{E} = \hat{\mathbf{e}}_y E_0 e^{i(kx-\omega t)} = \hat{\mathbf{e}}_y E_0 \cos(kx - \omega t)$$

$$\mathbf{H} = \hat{\mathbf{e}}_z H_0 \cos(kx - \omega t)$$

$$\therefore \quad U = \frac{1}{2}\int_0^c\int_0^a\int_0^b [\epsilon_0 E_0^2 \cos^2(kx - \omega t) + \mu_0 H_0^2 \cos^2(kx - \omega t)]\, dx\, dy\, dz$$

$$= \frac{ab}{2}\int_0^c [\epsilon_0 E_0^2 \cos^2(kx - \omega t) + \mu_0 H_0^2 \cos^2(kx - \omega t)]\, dx.$$

If T is the period of oscillation, the average energy in the box is

$$\langle U\rangle = \frac{ab}{2T}\int_0^T\int_0^c [\epsilon_0 E_0^2 \cos^2(kx-\omega t)+\mu H_0^2\cos^2(kx-\omega t)]\,dx\,dt$$

$$= \frac{ab}{2}\int_0^c\left[\frac{1}{2}\,\epsilon_0 E_0^2 + \frac{1}{2}\,\mu_0 H_0^2\right]dx$$

$$= \frac{abc}{4}\,[\epsilon_0 E_0^2 + \mu_0 H_0^2] \tag{6.44}$$

Therefore, the time averaged energy density associated with a wave is

$$\langle U_d\rangle = \frac{1}{4}\,(\mathbf{E}\cdot\mathbf{D}^* + \mathbf{B}\cdot\mathbf{H}^*) = \frac{1}{4}\,(\epsilon E^2 + \mu H^2)$$

$$= \frac{1}{4}\left(\epsilon\,|E_0|^2 + \mu\,\frac{\mathbf{k}\times\mathbf{E}}{\omega\mu}\cdot\frac{\mathbf{k}\times\mathbf{E}^*}{\omega\mu}\right)$$

$$= \frac{1}{4}\left\{\epsilon\,|E_0|^2 + \epsilon\,|E_0|^2\right\} = \frac{\epsilon}{2}\,|E_0|^2 \tag{6.45}$$

Hence,

$$\langle \mathbf{N}\rangle = \frac{1}{2}\left(\frac{\epsilon}{\mu}\right)^{1/2}\hat{\mathbf{e}}_k\,|E_0|^2 = \langle U_d\rangle(\epsilon\mu)^{-1/2}\,\hat{\mathbf{e}}_k$$

$$= v\,\langle U_d\rangle\,\hat{\mathbf{e}}_k. \tag{6.46}$$

This shows that the time averaged energy flow is in the direction of propagation of the wave and is equal to the phase velocity of the wave multiplied by the average energy density. The energy, therefore, flows with the same velocity as the wave itself.

6.4 Plane Waves in a Conducting Medium

We have shown in Sec. 6.1 that electric and magnetic fields satisfy identical equations (6.2) and (6.6) in a medium in which there is no free charge

$$\nabla^2\mathbf{E} - \epsilon\mu\,\frac{\partial^2\mathbf{E}}{\partial t^2} - \sigma\mu\,\frac{\partial\mathbf{E}}{\partial t} = 0 \tag{6.2}$$

and

$$\nabla^2\mathbf{H} - \epsilon\mu\,\frac{\partial^2\mathbf{H}}{\partial t^2} - \sigma\mu\,\frac{\partial\mathbf{H}}{\partial t} = 0. \tag{6.6}$$

Both these equations contain damping terms proportional to the conductivity of the medium.

Let us now find the plane wave solutions of Maxwell's equations for a conducting medium. We assume that the field vector **E** and **H** vary harmonically with time, i.e.

$$\mathbf{E}(\mathbf{r}, t) = \mathbf{E}_0 e^{i(\mathbf{k}.\mathbf{r}\,-\,\omega t)} \tag{6.47}$$

$$\mathbf{H}(r, t) = \mathbf{H}_0 e^{i(\mathbf{k}.\mathbf{r}\,-\,\omega t)}. \tag{6.48}$$

Substituting (6.47) in (6.2)

$$-\,k^2\mathbf{E}(\mathbf{r}, t) + \epsilon\mu\omega^2\mathbf{E}(\mathbf{r}, t) + i\sigma\mu\omega\mathbf{E}(\mathbf{r}, t) = 0$$

i.e. $$[k^2 - \epsilon\mu\omega^2 - i\sigma\mu\omega]\mathbf{E}(\mathbf{r}, t) = 0 \tag{6.49}$$

$\therefore$ $$k^2 = \epsilon\mu\omega^2\left[1 + \frac{i\sigma}{\epsilon\omega}\right] \tag{6.50}$$

The first term corresponds to the displacement current and the second to the conduction current.

For any wave there is a functional relationship between the wave-number k and wave frequency ω. This relation is known as **dispersion relation**. For electromagnetic waves in vacuum, the dispersion relation is

$$k = \frac{\omega}{v}.$$

For a conducting medium it is (6.50). If $\sigma = 0$ (i.e. free space) $k^2 = \epsilon\mu\omega^2 = \frac{\omega^2}{v^2}$ which is the relation already established. This relation will provide us with information regarding the nature of the propagation of electro-magnetic waves inside a medium.

We see that in a conducting medium, the propagation vector k is complex. For convenience, we may express it as

$$k = \alpha + i\beta \tag{6.51}$$

Squaring this and comparing with (6.50) we find

$$\begin{aligned} \alpha^2 - \beta^2 &= \epsilon\mu\omega^2 \\ 2\alpha\beta &= \sigma\mu\omega \end{aligned} \tag{6.52}$$

Solving these equations

$$\alpha = \omega\sqrt{\frac{\epsilon\mu}{2}}\left[1 + \left\{1 + \left(\frac{\sigma}{\epsilon\omega}\right)^2\right\}^{1/2}\right]^{1/2} \tag{6.53}$$

$$\beta = \omega\sqrt{\frac{\epsilon\mu}{2}}\left[-1 + \left\{1 + \left(\frac{\sigma}{\epsilon\omega}\right)^2\right\}^{1/2}\right]^{1/2}. \tag{6.54}$$

We have to take positive square root in order to make the solutions yield the proper form for k in free space.

Since $k = \alpha + i\beta$, equation (6.47) and (6.48) can be written as

$$\mathbf{E} = \mathbf{E}_0 e^{-\beta r} e^{i(\alpha r - \omega t)}$$

$$\mathbf{H} = \mathbf{H}_0 e^{-\beta r} e^{i(\alpha r - \omega t)}$$

These equations indicate that a plane wave cannot propagate in a conduction medium without attenuation. When a plane wave is propagating in a conducting medium, the oscillating electric field in the wave sets up currents. Work must be done to drive the currents and some of the energy is dissipated as heat in the medium. This results into the attenuation of the wave. The quantity β is called the **absorption coefficient** and is a measure of this attenuation.

6.5 The Skine Effect

In equation (6.49) the term $i\sigma\mu\omega$ arises from the term involving $\frac{\partial \mathbf{E}}{\partial t}$ in equation (6.2), i.e. from the conduction current; while the term $\epsilon\mu\omega^2$ arises from the term involving $\frac{\partial^2 \mathbf{E}}{\partial t^2}$ in the same equation, i.e. from the displacement current. In almost all conducting media the conduction current dominates the displacement current, and, hence, it is a good approximation to neglect the middle term in equation (6.2). Thus, for a good conducting medium

$$\nabla^2 \mathbf{E} = \sigma\mu \frac{\partial \mathbf{E}}{\partial t} \tag{6.55}$$

and the attenuated solution of this equation is

$$\mathbf{E} = \mathbf{E}_0 e^{-\beta r} e^{i(\alpha r - \omega t)}. \tag{6.56}$$

For a good conductor, if the frequency is not too high $\frac{\sigma}{\epsilon\omega} \gg 1$.

From equation (6.53) and (6.54)

$$\alpha = \beta = \sqrt{\frac{\omega\sigma\mu}{2}} = \frac{1}{\delta}, \text{ where } \delta = \sqrt{\frac{2}{\omega\sigma\mu}} \tag{6·57}$$

$$\therefore \quad \mathbf{E} = \mathbf{E}_0 e^{-r/\delta} e^{i\left(\frac{r}{\delta} - \omega t\right)} \tag{6.58}$$

We find from equation (6.58) that when $r = \delta$ the amplitude decreases in magnitude to $\frac{1}{e}$ times its value at the surface. The quantity δ, therefore, is a measure of the distance of penetration of an electromagnetic wave into a good conductor, before its magnitude drops to $\frac{1}{e}$ times its value at the surface. The distance 'δ' is called the **skin depth.**

The relation (6.57) indicates that the skin depth goes to zero as the conductivity approaches infinity and is small for good conductors at high frequency currents. Thus, for copper $\sigma = 58 \times 10^6$ mhos/m (one mho is reciprocal ohm). The skin depth at various frequencies is

Frequency	Skin depth
60 Hz	8.5×10^{-3}m
1 MHz	6.6×10^{-5}m
30 GHz	3.8×10^{-7}m

The following table will give an idea of the magnitudes of skin depth in different types of materials.

Material	Frequency	δ
Silver	100 MHz	10^{-7}m
Aluminium	50 Hz	1.25×10^{-2}m
Sea water	30 KHz	10^{-1}m

Thus, even a poor conductor can be made a good conductor with a thin coating of silver or copper. The rapid attenuation of waves means that in high frequency circuits current flows only on the surface of the conductor. The relatively higher skin depth in the case of sea water explains why the radio communication with submarines becomes difficult at depth of several metres.

PROBLEMS

6.1 The components of an electric field at a time t in vacuum are given by $E_x = 0$, $E_y = 0$ and $E_z = a \cos kx \cos \omega t$. The magnetic field is $\mathbf{H} = 0$ at $t = 0$. Show that at time t the components of the magnetic field are $H_x = 0, H_y = 0$, $H_z = -a\left(\frac{\varepsilon_0}{\mu_0}\right)^{1/2} \sin kx \sin \omega t$. Show further that there is no mean flux of energy in this problem.

6.2 Two plane monochromatic waves of the same frequency and linearly polarized in perpendicular directions, propagate in the same direction. The plane of one wave leads that of the other by ϕ. Determine the polarization of the resultant wave.

6.3 Show that:
(i) If a monochromatic linearly polarized plane wave is moving in an isotropic non-conducting medium, the time average of its energy density is distributed equally between the magnetic and electric fields.
(ii) If the wave is moving in a conducting medium, the average energy in the magnetic field is greater than that in the electric field.

6.4 A laser beam has a diameter of 10^{-3} m. Calculate the amplitude of the magnetic field in the beam in free space, if the power of the laser is 1 watt.

6.5 Find the skin depth for low frequency radio waves of wave-length 3×10^3 m in sea water, the electrical conductivity of which is $\sim 4\ (\Omega\ \text{m})^{-1}$.

6.6 The gas molecules in the ionosphere are ionized by ultra-violet rays from the sun. Neglecting the magnetic field produced by the motion of the charges and assuming the density of free electrons to be n_0 per unit volume, find the relative permittivity of the ionosphere for waves of frequency ν.

6.7 Assuming that the conductivity, permeability and permittivity of a medium do not change with frequency, does the medium with $\sigma = 0.1$ mho, $\mu_r = 1$, $\varepsilon_r = 40$, behave like a conductor or a dielectric at (a) 50 KHz and (b) 10 GHz?

Chapter 7

Electromagnetic Waves in Bounded Media

We shall discuss in this chapter the behaviour of electromagnetic waves at the boundaries between different media. In a given incident uniform plane wave system, the boundary conditions that prove to be the simplest to treat, are those that are capable of reflecting back another uniform plane wave. This occurs when the boundary is a plane of infinite extent. Curved boundaries tend to scatter incident plane waves into many different directions simultaneously. Although such problems are, in principle, solvable by superposing an infinite series of uniform plane waves, we shall limit our discussion to plane boundaries only.

When plane waves are incident on a boundary between two media, some of its incident energy crosses the boundary and some is reflected. The reflection and refraction of light waves at the surface separating two media of different refractive indices is a familiar phenomenon. We shall now show how electromagnetic theory offers a simple explanation for it. This will also show how optics is contained within the framework of Maxwell's electrodynamics.

7.1 Reflection and Refraction of Plane Waves at a Plane Interface

Consider two non-conducting ($\sigma = 0$) dielectric media referred to as '1' and '2', characterized by constants μ_1, ϵ_1, and μ_2, ϵ_2 and separated by a plane boundary—the plane $x = 0$ (Fig. 7.1).

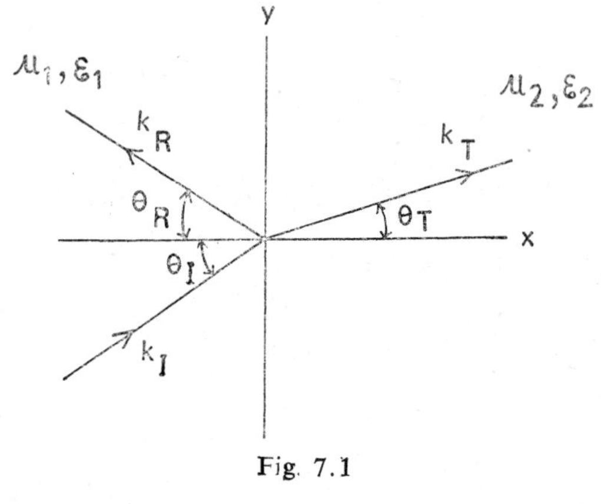

Fig. 7.1

Suppose a plane electromagnetic wave is incident obliquely on the plane boundary. There will in general be both reflected wave and a transmitted wave. We shall now inquire as to what fraction of the energy of the incident wave is reflected and what fraction is transmitted.

We can express the fields for the incident, reflected and transmitted waves as:

$$\mathbf{E}_I = \mathbf{E}_{0I} \exp\{i(\mathbf{k}_I \cdot \mathbf{r} - \omega_I t)\}, \mathbf{H}_I = \frac{\mathbf{k}_I \times \mathbf{E}_I}{\omega_I \mu_1} \tag{7.1}$$

$$\mathbf{E}_R = \mathbf{E}_{0R} \exp\{i(\mathbf{k}_R \cdot \mathbf{r} - \omega_R t)\}, \mathbf{H}_R = \frac{\mathbf{k}_R \times \mathbf{E}_R}{\omega_R \mu_1} \tag{7.2}$$

and

$$\mathbf{E}_T = \mathbf{E}_{0T} \exp\{i(\mathbf{k}_T \cdot \mathbf{r} - \omega_T t)\}, \mathbf{H}_T = \frac{\mathbf{k}_T \times \mathbf{E}_T}{\omega_T \mu_2} \tag{7.3}$$

where the subscripts I, R, T represent incident, reflected and transmitted waves respectively. The quantities E_{I0}, E_{0R}, E_{0T} are time-independent scalar amplitudes which may be complex. The relationship between these can be found by making the total fields obey the boundary conditions in the plane $x = 0$. The tangential components of **E** and **H** can be continuous across the boundary at all points and at all times only if the exponentials are the same at the boundary for all three fields. This is possible if

$$\omega_I = \omega_R = \omega_T$$

i.e. the frequency is unchanged in the reflected and transmitted waves, and

$$\mathbf{k}_I \cdot \mathbf{r} = \mathbf{k}_R \cdot \mathbf{r} = \mathbf{k}_T \cdot \mathbf{r} \tag{7.4}$$

This shows that all the propagation vectors are coplanar. If we choose **r** to lie in the boundary plane (i.e. $\hat{\mathbf{e}}_n \cdot \mathbf{r} = 0$, where $\hat{\mathbf{e}}_n$ is a unit vector normal to the plane) and in the plane of the propagation vector, it follows that

$$k_I \sin\theta_I = k_R \sin\theta_R = k_T \sin\theta_T. \tag{7.5}$$

Now the propagation vectors $\mathbf{k}_I$ and $\mathbf{k}_R$ are in the same medium and, hence, are equal in magnitude

$$\therefore \quad \theta_I = \theta_R \tag{7.6}$$

Since $\quad k_I \sin\theta_I = k_T \sin\theta_T$

$$\frac{\sin\theta_I}{\sin\theta_T} = \frac{k_T}{k_I} = \sqrt{\frac{\epsilon_2 \mu_2}{\epsilon_1 \mu_1}} \qquad \left(\because \quad k = \omega\sqrt{\epsilon\mu}\right).$$

For non-magnetic materials we way assume $\mu_2 = \mu_1$

$$\therefore \quad \frac{\sin\theta_I}{\sin\theta_T} = \sqrt{\frac{\epsilon_2}{\epsilon_1}} = \frac{n_2}{n_1} \quad \text{Snell's law} \tag{7.7}$$

where n_1 n_2 are the refractive indices of the media '1' and '2' respectively. Equations (7.6) and (7.7) are the simpler laws of geometrical optics we are familiar with.

Let us now obtain the relationship between the various field vectors.

Since the divergence equations $\nabla \cdot \mathbf{D} = \rho$ and $\nabla \cdot \mathbf{B} = 0$ can be obtained by applying the divergence operator to the remaining Maxwell's equations involving **E** and **H**, it follows that the boundary conditions on D_n and B_n are automatically satisfied provided the conditions on E_t and H_t are met. The condition are:

$$(\mathbf{E}_I + \mathbf{E}_R) \times \hat{\mathbf{e}}_n = \mathbf{E}_T \times \hat{\mathbf{e}}_n \tag{7.8}$$

and

$$(\mathbf{H}_I + \mathbf{H}_R) \times \hat{\mathbf{e}}_n = \mathbf{H}_T \times \hat{\mathbf{e}}_n. \tag{7.9}$$

This latter equation can be written as

$$(\mathbf{k}_I \times \mathbf{E}_I + \mathbf{k}_I \times \mathbf{E}_R) \times \hat{\mathbf{e}}_n = (\mathbf{k}_T \times E_T) \times \hat{\mathbf{e}}_n \quad (\because \quad \mu_1 = \mu_2) \tag{7.10}$$

Let us now consider two separate situations: (i) E is polarized perpendicular to the plane of incidence (i.e. the plane defined by k and $\hat{\mathbf{e}}_n$); and (ii) E polarized parallel to the plane of incidence. The general case of arbitrary polarization can be obtained by appropriate linear combination of these two.

(*i*) *E polarized perpendicular to the plane of incidence*

The field vectors corresponding to this situation are shown in Fig. 7.2. The electric field vectors are directed away from the viewer. The condition (7.8) and (7.10) give

$$E_{0I} + E_{0R} = E_{0T} \tag{7.11}$$

and

$$k_I E_{0I} \cos \theta_I - k_I E_{0R} \cos \theta_R = k_T E_{0T} \cos \theta_T$$

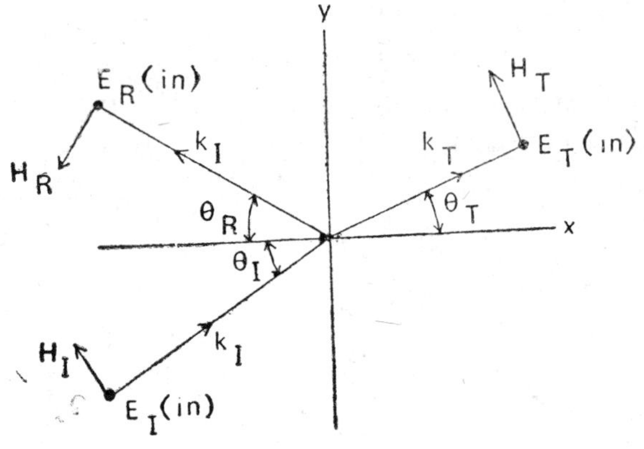

Fig. 7.2

i.e.

$$\left(E_{0I} - E_{0R}\right) \cos \theta_I = \frac{k_T}{k_I} E_{0T} \cos \theta_T \tag{7.12}$$

where E_{0I}, E_{0R} and E_{0T} are the scalar amplitudes of the incident, reflected and transmitted waves respectively.

Solving these two equations, we have

$$\frac{E_{0R}}{E_{0I}} = \frac{\cos \theta_I - \frac{k_T}{k_I} \cos \theta_T}{\cos \theta_I + \frac{k_T}{k_I} \cos \theta_T} = \frac{\cos \theta_I - \frac{n_2}{n_1} \cos \theta_T}{\cos \theta_I + \frac{n_2}{n_1} \cos \theta_T}$$

$$= \frac{\cos \theta_I - \frac{\sin \theta_I}{\sin \theta_T} \cos \theta_T}{\cos \theta_I + \frac{\sin \theta_I}{\sin \theta_T} \cos \theta_T} \quad \text{by Snell's law (7.7)}$$

$$= \frac{\sin (\theta_T - \theta_I)}{\sin (\theta_T + \theta_I)} \tag{7.13}$$

and

$$\frac{E_{0T}}{E_{0I}} = \frac{2 \cos \theta_I}{\cos \theta_I + \frac{k_T}{k_I} \cos \theta_T} = \frac{2 \cos \theta_I \sin \theta_T}{\sin (\theta_I + \theta_T)} \tag{7.14}$$

The equation (7.13) gives the ratio of the amplitudes of the reflected and incident waves. If $n_2 > n_1$, the ratio is negative, indicating that the reflection of the wave results in a phase change of π, i.e. the electric vector of the reflected wave oscillates 180° out of phase with that in the incident wave. The ratio $\frac{E_{0T}}{E_{0I}}$ is always positive.

The reflection coefficient R is defined as the energy flux reflected from the interface divided by the flux incident on it.

$$\therefore \quad R_\perp = \frac{\hat{\mathbf{e}}_n \cdot \langle \mathbf{N}_R \rangle}{\hat{\mathbf{e}}_n \cdot \langle \mathbf{N}_I \rangle} = \frac{|\mathbf{E}_R \times \mathbf{H}_R^*|}{|\mathbf{E}_I \times \mathbf{H}_I^*|} = \frac{E_{0R}^2}{E_{0I}^2} \tag{7.15}$$

where the subscript $\perp$ suggests that E is polarized perpendicular to the plane of incidence, and $\mathbf{N}_R$, $\mathbf{N}_I$ are the Poynting vectors.

$$\therefore \quad R_\perp = \frac{\sin^2 (\theta_T - \theta_I)}{\sin^2 (\theta_T + \theta_I)}. \tag{7.16}$$

Similarly, the transmission coefficient

$$T_\perp = \frac{\hat{\mathbf{e}}_n \cdot \langle \mathbf{N}_T \rangle}{\hat{\mathbf{e}}_u \cdot \langle \mathbf{N}_I \rangle} = \frac{|E_{0T}|^2}{|E_{0I}|^2} \frac{n_2 \cos \theta_T}{n_1 \cos \theta_I}$$

$$= \frac{4 \cos^2 \theta_I \sin^2 \theta_T}{\sin^2 (\theta_I + \theta_T)} \frac{\sin \theta_I \cos \theta_T}{\sin \theta_T \cos \theta_I}$$

$$= \frac{4 \cos \theta_I \cos \theta_T \sin \theta_I \sin \theta_T}{\sin^2 (\theta_I + \theta_T)} = \frac{\sin 2\theta_I \sin 2\theta_T}{\sin^2 (\theta_I + \theta_T)} \quad (7.17)$$

$$\therefore \quad R_\perp + T_\perp = 1. \quad (7.18)$$

For normal incidence, from (7.13) and (7.14) we have

$$R_\perp = \left(\frac{n_1 - n_2}{n_1 + n_2}\right)^2; \; T_\perp = \frac{n_2}{n_1}\left(\frac{2n_1}{n_1 + n_2}\right)^2. \quad (7.19)$$

(*ii*) *E in the plane of incidence*

The boundary conditions give

$$E_{0I} \cos \theta_I - E_{0R} \cos \theta_R = E_{0T} \cos \theta_T$$

i.e.

$$(E_{0I} - E_{0R}) \cos \theta_I = E_{0T} \cos \theta_T \quad (7.20)$$

and

$$k_I E_{0I} + k_I E_{0R} = k_T E_{0T}$$

i.e.

$$E_{0I} + E_{0R} = \frac{n_2}{n_1} E_{0T}. \quad (7.21)$$

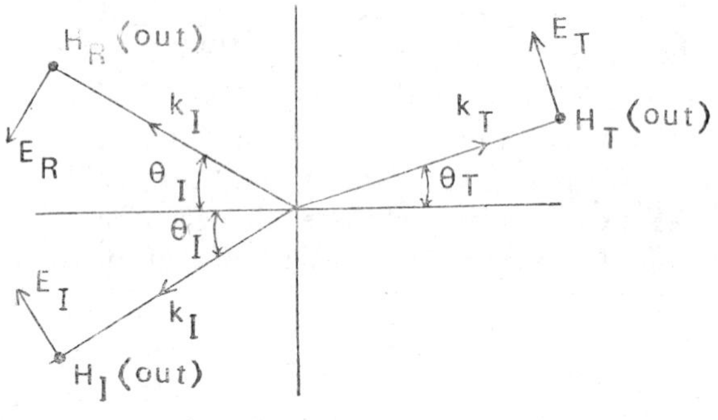

Fig. 7.3

Solving

$$\frac{E_{0R}}{E_{0I}} = \frac{\cos \theta_I - \left(\frac{n_1}{n_2}\right) \cos \theta_T}{\cos \theta_I + \left(\frac{n_1}{n_2}\right) \cos \theta_T} = \frac{\tan (\theta_I - \theta_T)}{\tan (\theta_I + \theta_T)} \quad (7.22)$$

and

$$\frac{E_{0T}}{E_{0I}} = \frac{2 \sin \theta_T \cos \theta_I}{\sin (\theta_I + \theta_T) \cos (\theta_I - \theta_T)}. \quad (7.23)$$

The relations (7.13), (7.14), (7.22) and (7.23) are known as **Fresnel's relations.**

Comparison of (7.13) and (7.22) brings out an important distinction between the two states of polarization of E. Leaving aside the trivial case, when $n_2 = n_1$, i.e. $\theta_2 = \theta_1$, we see from (7.13) that $\frac{E_{0R}}{E_{0I}} \neq 0$ for any angle of incidence θ_I lying between 0 and $\pi/2$. However, when E is polarized in the plane of incidence, equation (7.22) shows that

$$\frac{E_{0R}}{E_{0I}}=0 \tag{7.24}$$

and hence, $R_{\parallel}=0$
when $\theta_I+\theta_T=\pi/2$.
This means, if the wave is incident at an angle $\theta_I=\pi/2-\theta_T$, it crosses the interface without suffering reflection. This angle is known as **Brewster's angle** and is represented by the symbol θ_B. Its magnitude can be found by applying Snell' law, i.e.

$$\frac{\sin\theta_B}{\sin\theta_T}=\frac{\sin\theta_B}{\sin(\pi/2-\theta_B)}=\tan\theta_B=\frac{n_2}{n_1}$$

$$\therefore \qquad \theta_B=\tan^{-1}\left(\frac{n_2}{n_1}\right) \tag{7.25}$$

If the light incident on the interface at Brewster's angle is unpolarized (i.e. it contains a superposition of many components with their electric vectors in random orientations), the only component of **E** polarized perpendicular to the plane of incidence will be reflected and, hence, the reflected wave is plane polarized perpendicular to the plane of incidence and for this reason the angle θ_B is sometimes called **polarizing angle.**

We shall now described an important application of Brewster's angle. In a gas laser we usually have mirrors outside the glass windows as shown in Fig. 7.4. At normal incidence about 92% of the incident intensity is transmitted through a glass window, i.e. about 8% is lost in each traverse. In a laser there are a large number of traverses and, hence very, little of light will be left after a few traverses. To overcome this difficulty the windows are arranged at Brewster angle.

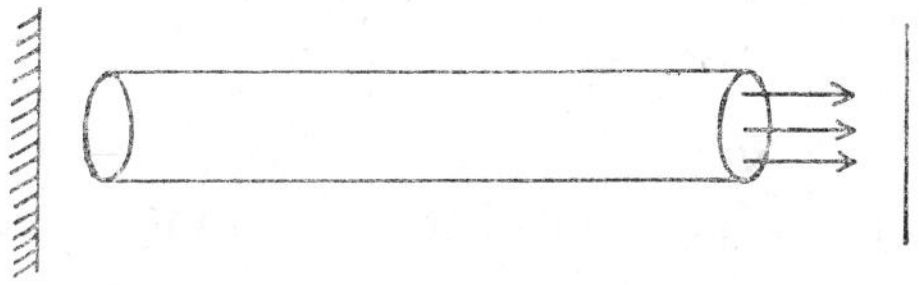

Fig. 7.4

The electric field component polarized parallel to the plane of incidence is transmitted perfectly and suffers negligible loss even after many traverses. The component polarized perpendicular to the plane of incidence is partly reflected and partly transmitted each time it strikes the surface; after a number of traverses it is completely eliminated and, hence, the emerging beam is hundred percent linearly polarized.

7.2 Total Internal Reflection

Let us consider a situation in which radiation is incident from a medium of higher refractive index on the surface of a medium of lower refractive index, i.e. $n_1 > n_2$. We have from Snell's law

$$\sin\theta_T = \frac{n_1}{n_2}\sin\theta_I. \tag{7.26}$$

What happens if θ_I is gradually increased starting from zero? Obviously θ_T will also increase until it attains the value $\pi/2$. Let us represent the value of θ_I, when, this happens, by the symbol θ_C. From (7.26)

$$\sin\theta_C = \frac{n_2}{n_1} \tag{7.27}$$

The angle θ_C is called the **critical angle,** since when $\theta_I = \theta_C$ there is only reflected wave, there being no transmitted wave.

What will happen if θ_C is increased further, i.e. when $\theta_I > \theta_C$?

Let us express θ_T in terms of θ_I and θ_C

$$\cos\theta_T = \sqrt{1-\sin^2\theta_T}$$

But $$\sin\theta_T = \frac{n_1}{n_2}\sin\theta_I = \frac{\sin\theta_I}{\sin\theta_C} \quad \text{(from 7.26)}$$

$$\therefore \quad \cos\theta_T = \sqrt{1-\frac{\sin^2\theta_I}{\sin^2\theta_C}}. \tag{7.28}$$

We see that the value of $\cos\theta_T$ decreases as θ_I is increased; it becomes 0 at $\theta_I = \theta_C$ and for values of θ_I, greater than θ_C, $\cos\theta_T$ becomes an imaginary number.

Let us now calculate the amplitude of the reflected electric vector when $\theta_I > \theta_C$

For $\theta_I > \theta_C$ we write

$$\cos\theta_T = \sqrt{1-\frac{\sin^2\theta_I}{\sin^2\theta_C}} = iQ$$

i.e. $$Q = \sqrt{\frac{\sin^2\theta_I}{\sin^2\theta_C}-1}. \tag{7.29}$$

In the case when **E** is polarized perpendicular to the plane of incidence

$$\frac{E_{0R}}{E_{0I}} = \frac{\cos\theta_I - \frac{n_2}{n_1}\cos\theta_T}{\cos\theta_I + \frac{n_2}{n_1}\cos\theta_T} = \frac{\cos\theta_I - \frac{n_2}{n_1}iQ}{\cos\theta_I + \frac{n_2}{n_1}iQ}.$$

$$\therefore \quad \left|\frac{E_{0R}}{E_{0I}}\right|^2 = 1 \quad \text{i.e.} \quad |E_{0R}| = |E_{0I}|.$$

Similarly when **E** is polarized parallel to the plane of incidence

$$\left|\frac{E_{0R}}{E_{0I}}\right|^2 = 1$$

i.e.

$$|E_{0R}| = |E_{0I}|$$

Thus the wave is totally reflected. This phenomenon is known as **total internal reflection.**

It must be noted that there is a change of phase on reflection; hence, if the incident wave is polarized in a plane intermediate between the plane of incidence and the plane normal to it, the two components will not be in phase after reflection and the wave will be elliptically polarized.

We conclude: when the angle of incidence θ_I is greater than critical angle, there is no refracted wave and all the energy is reflected. That this is so can be shown by computing the average rate of energy flow across the boundary.

The rate of energy flow $= \langle \mathbf{N} \rangle \cdot \hat{\mathbf{e}}_n$

$$= \tfrac{1}{2} Re\left(\mathbf{E}_T \times \mathbf{H}_T^* \right)\cdot \hat{e}_n = \tfrac{1}{2} Re\left(\mathbf{E}_T \times \frac{\mathbf{k}_T \times \mathbf{E}_T^*}{\omega\mu} \right).\hat{\mathbf{e}}_n$$

$$= \frac{1}{2\omega\mu} Re\left[\left(\mathbf{E}_T\cdot\mathbf{E}_T^* \right) \mathbf{k}_T - (\mathbf{E}_T\cdot\mathbf{k}_T)\,\mathbf{E}_I^* \right].\hat{\mathbf{e}}_n$$

$$= \frac{1}{2\omega\mu} Re\left(\mathbf{E}_T\cdot\mathbf{E}_T^* \right)\mathbf{k}_T\cdot\hat{\mathbf{e}}_n \qquad (\because \ \mathbf{E}_T \perp \mathbf{k}_T)$$

$$= \frac{1}{2\mu\omega} Re\,|E_{0T}|^2\,\mathbf{k}_T \cos\theta_T = \frac{1}{2\omega\mu} Re\, iQ\, k_T\,|E_{0T}|^2. \qquad (7.30)$$

Since this expression is purely imaginary, $\langle \mathbf{N} \rangle\cdot\hat{\mathbf{e}}_n = 0$ which justifies our conclusion.

We must not hasten to conclude at this stage that if $n_1 > n_2$ and the wave is incident at an angle greater than the critical angle, there will be no field on the other side of the boundary. Even though there is no energy flow across the surface, the field does exist on the other side of the surface as is shown below.

Consider the equation for the refracted wave

$$\mathbf{E}_T = \mathbf{E}_{0T} \exp\{i(\mathbf{k}_T\cdot\mathbf{r} - \omega t)\} \qquad (7.31)$$

For the coordinate axes shown in Fig. 7.1, this becomes

$$\begin{aligned}\mathbf{E}_T &= \mathbf{E}_{0T} \exp\{i(k_T x \cos\theta_T + k_T y \sin\theta_T - \omega t)\} \\ &= \mathbf{E}_{0T} \exp\{(- k_T x Q \exp\{i(k_T y \sin\theta_T - \omega t)\}. \end{aligned} \qquad (7.32)$$

This shows that the field does exist on the far side of the surface in the medium 2; but is rapidly attenuated as shown in Fig. 7.5.

How far does the wave penetrate the medium? The **penetrating distance** or **skin depth** is given by

$$x = \frac{1}{k_T Q} = \frac{1}{k_T}\left(\frac{\sin^2\theta_I}{\sin^2\theta_C} - 1\right)^{-1/2} \qquad (7.33)$$

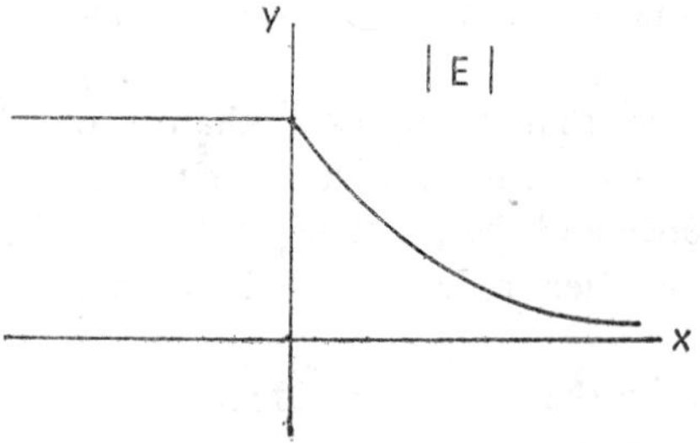

Fig. 7.5

Consider for example a wave going from glass to air. The refractive index of glass is 1.5; hence, the critical angle for glass is $\theta_C = \sin^{-1} 2/3$ i.e. $\sim 42°$. If the light is incident internally from the glass at an angle greater than 42°, say 45°, there will be total internal reflection, and

$$x = \frac{1}{k_T}\left(\frac{1}{2} \times \frac{9}{4} - 1\right)^{-1/2} = \frac{\lambda}{2\pi} 2\sqrt{2} = 0.45\lambda.$$

Thus, the field becomes negligible beyond distances of the order of a few wavelengths. What would be the microscopic explanation of this attenuation? The molecular charges in the medium oscillate due to interaction with the incident wave and give rise to a radiation field (this will be shown later in Chapter 9). The forward wave in this field interferes destructively with the original wave and gives rise to a small transmission.

We have seen above that there is no transport of energy across the boundary. How do you account then for the existence of decaying field on the far side of the boundary? Energy does flow in the second medium since the component of the field in the medium is finite; but during the later part of the cycle the flow is in the opposite direction and the energy is returned to the first medium.

7.3 Reflection from the Surface of a Metal

We shall now show how the arguments of the preceding section can be extended to a boundary surface of a conducting medium. Since the case of oblique incidence is more involved, we shall confine to the simple case of normal incidence only.

We have, for incident, reflected and transmitted waves:

$$\mathbf{E}_I = \mathbf{E}_{0I} \exp\{i(\mathbf{k}_I \cdot \mathbf{r} - \omega t)\}; \mathbf{H}_I = \frac{\mathbf{k}_I \times \mathbf{E}_I}{\omega\mu_1} \tag{7.34}$$

$$\mathbf{E}_R = \mathbf{E}_{0R} \exp\{i(\mathbf{k}_I \cdot \mathbf{r} - \omega t)\}; \mathbf{H}_R = \frac{\mathbf{k}_I \times \mathbf{E}_R}{\omega\mu_1} \tag{7.35}$$

$$\mathbf{E}_T = \mathbf{E}_{0T} \exp\{i(\mathbf{k}_T \cdot \mathbf{r} - \omega t)\}; \mathbf{H}_T = \frac{\mathbf{k}_T \times E_T}{\omega\mu_2} \tag{7.36}$$

Since the medium 2 is a conducting medium, the propagation vector $\mathbf{k}_T$ is given by

$$k_T^2 = \epsilon_2\mu_2\omega^2\left[1 + \frac{i\sigma}{\epsilon_2\omega}\right] \qquad \text{by (6.50)}$$

The boundary conditions require that

$$E_{0I} - E_{0R} = E_{0T} \tag{7.37}$$

$$k_I(E_{0I} + E_{0R}) = k_T E_{0T} \tag{7.38}$$

Since k_T is complex, E_{0R} and E_{0T} cannot both be real and, hence, one should expect phase shifts other than 0 or π in the reflected and transmitted waves.

Solving (7.37) and (7.38) we have

$$E_{0R} = \frac{k_T - k_I}{k_T + k_I}\, E_{0I} \tag{7.39}$$

$$E_{0T} = \frac{2k_I}{k_T + k_I}\, E_{0I}.$$

Substituting $k_I = \omega(\epsilon_1\mu_1)^{1/2}$ and for k_T from (6.50), we have

$$E_{0R} = \frac{\sqrt{\epsilon_2\mu_2\omega^2}\left(1 + \dfrac{i\sigma}{\epsilon_2\omega}\right)^{1/2} - \omega\,(\epsilon_1\mu_1)^{1/2}}{\sqrt{\epsilon_2\mu_2\omega^2}\left(1 + \dfrac{i\sigma}{\epsilon_2\omega}\right)^{1/2} + \omega\,(\epsilon_1\mu_1)^{1/2}}\, E_{0I} \tag{7.40}$$

and

$$E_{0T} = \frac{2\,\omega\,(\epsilon_1\mu_1)^{1/2}}{\sqrt{\epsilon_2\mu_2\omega^2}\left(1 + \dfrac{i\sigma}{\epsilon_2\omega}\right)^{1/2} + \omega(\epsilon_1\mu_1)^{1/2}}\, E_{0I}. \tag{7.41}$$

Let us now examine the case of a perfect conductor for which $\sigma = \infty$. Then

$$E_{0R} = E_{0I} \quad \text{and} \quad E_{0T} = 0$$

Hence, the reflection is complete.

If the conductor is not a perfect conductor but a very good conductor, $\dfrac{\sigma}{\epsilon_2\omega} \gg 1$.

The approximation adopted in the preceding chapter gives

$$k_T = \alpha + i\beta = (1 + i)\sqrt{\frac{\omega\sigma\mu_2}{2}} = \frac{1+i}{\delta}$$

where we have made use of the definition of skin depth δ given in (6.57).

Substituting this in (7.39) we have

$$E_{0R} = \frac{\dfrac{1+i}{\delta} - \omega(\epsilon_1\mu_1)^{1/2}}{\dfrac{1+i}{\delta} + \omega\,(\epsilon_1\mu_1)^{1/2}}\, E_{0I}$$

$$= \frac{\left\{\frac{1}{\delta} - \omega\,(\epsilon_1\mu_1)^{1/2}\right\} + \frac{i}{\delta}}{\left\{\frac{1}{\delta} + \omega(\epsilon_1\mu_1)^{1/2}\right\} + \frac{i}{\delta}}\, E_{0I}.$$

Therefore, the reflection coefficient R is given by

$$R = \frac{|E_{0R}|^2}{|E_{0I}|^2} = \frac{\{1 - \omega\,(\epsilon_1\mu_1)^{1/2}\delta\}^2 + 1}{\{1 + \omega\,(\epsilon_1\mu_1)^{1/2}\delta\}^2 + 1}$$

$$\because \quad \frac{\sigma}{\epsilon_2\omega} \gg 1,\ \omega\,(\epsilon_1\mu_1)^{1/2}\,\delta << 1$$

and, hence,

$$R \simeq 1 - 2\,\omega\,(\epsilon_1\mu_1)^{1/2}\delta \tag{7.42}$$

$$= 1 - 2\sqrt{\frac{2\,\omega\,\epsilon_1\mu_1}{\sigma\mu_2}}\,. \tag{7.43}$$

If the magnetic permeabilities are assumed to be equal

$$R = 1 - 2\sqrt{\frac{2\,\omega\epsilon_1}{\sigma}}. \tag{7.44}$$

The measure of energy transmitted into the conducting medium is obtained by calculating the transmission coefficient T which gives the ratio of the energy transmitted to the energy incident

$$T = 1 - R = 2\sqrt{\frac{2\,\omega\epsilon_1}{\sigma}}\,. \tag{7.45}$$

In order to have an idea of the magnitude of the fraction of the energy transmitted in a good conductor, let us calculate T for copper with the following data:

σ for copper $= 6 \times 10^7$ (Ohm m)$^{-1}$, $\nu = 10^{10}$ sec^{-1}

$$T = 2\sqrt{\frac{2 \times 2\pi \times 10^{10} \times 8.85 \times 10^{-12}}{6 \times 10^7}} \simeq 3 \times 10^{-4}$$

which is extremely small for direct measurements to be made.

The total energy in the medium 1 is

$$\mathbf{E} = \mathbf{E}_I + \mathbf{E}_R$$

Now $\qquad \mathbf{E}_I = \hat{\mathbf{e}}_n\, E_{0I}\, e^{i(k_I x - \omega t)}$

and $\qquad \mathbf{E}_R = -\,\hat{\mathbf{e}}_n\, E_{0I}\, e^{i(k_I x - \omega t)}.$

Note that the amplitudes in both the waves are taken to be equal. Justification for this is found in the fact that for metallic surfaces the reflection coefficient is almost unity.

$$\therefore \qquad \mathbf{E} = \hat{\mathbf{e}}_n \mathbf{E}_{0I}\, e^{-i\omega T}\left\{e^{ik_I x} - e^{-ik_I x}\right\}$$

or taking the real parts

$$\mathbf{E} = 2\,\hat{\mathbf{e}}_n\, E_{0I} \sin \omega t \sin k_I x \tag{7.46}$$

Therefore, the field in medium 1 is represented by a standing wave. the field will behave as shown in Fig. 7.6.

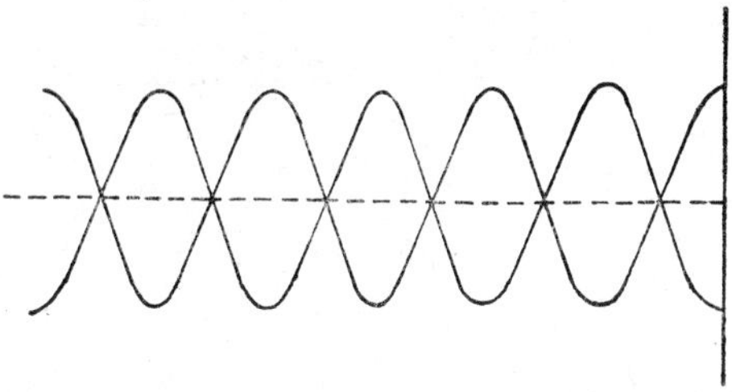

Fig. 7.6

PROBLEMS

7.1 Show that the reflection coefficient for the light passing from glass to air at normal incidence is the same as for the light passing from air to glass. Show further that the phase changes in the two cases are different.

7.2 A plane electromagnetic wave is incident normally on the plane boundary separating two dielectrics of refractive indices n_1 and n_2. If the refractive indices satisfy the relation $(n_2/n_1) = 3 + 2\sqrt{2})$, show that the flux of the reflected energy is the same as that of the transmitted energy.

7.3 Calculate the critical angle and Brewster angle for an electromagnetic wave passing through the following dielectrics

Material	Relative permittivity
Quartz	5
Glass	9
Water	81

7.4 The amplitude of the electric field in a monochromatic plane wave in free space incident normally on a plane surface of medium of refractive index 2 is 10 Vm^{-1}. Find the value of the amplitude of the electric field inside the medium.

7.5 An electromagnetic wave is incident on a dielectric slab bounded by two parallel faces. If the wave is incident on the front surface at Brewster angle show that the refracted wave is incident on the back surface at Brewster angle also.

7.6 A plane electromagnetic wave in free space is incident upon the plane boundary of a thick dielectric at an angle of 56.6°. to the normal to the boundary. The wave is polarized in the plane of incidence. What is the value of the relative permittivity of the dielectric if there is no reflected wave?

Chapter 8

Wave Guides

In the preceding chapter we examined the boundary effects on the propagation of an electromagnetic wave. We shall now discuss the behaviour of such waves in the vicinity of the boundaries so configured as to guide that wave along a certain path. Such systems are called **wave guides**. The most efficient way of transmitting energy over short distances is by using wave guides. Wave guides were first evolved and used in practical communication electronics and now assumed great importance in optical communication—**optoelectronics.**

8.1 Propagation of Waves Between Conducting Planes

Let us first consider the problem of the propagation of waves between two parallel conducting planes which will lead us to a solution of the wave guide problem.

Figure 8.1 shows two parallel plates. We choose our coordinate axes as shown in the figure. One plate is in the plane $y = 0$ and the other in the plane $y = b$. We suppose that the plates are perfectly conducting and are of infinite extent. Let us further suppose that the space between the plates is a vacuum, $\epsilon = \epsilon_0$, $\mu = \mu_0$, $\sigma = 0$, $\rho = 0$. If a wave is introduced in the region between the plates, the walls will reflect the wave to and fro and as a result it will propagate in the direction parallel to the

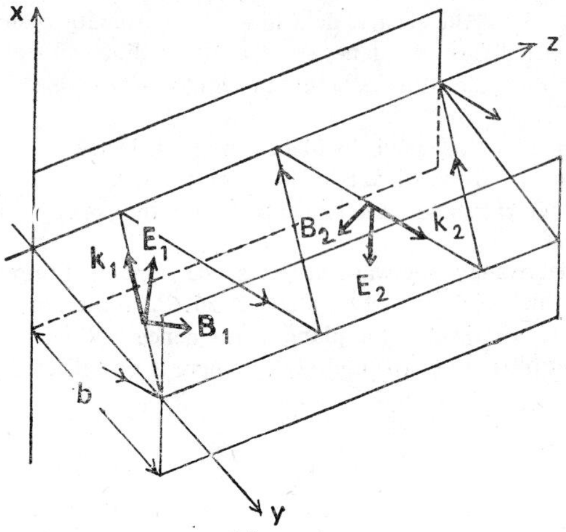

Fig. 8.1

plates. The fields must obey Maxwell's equations in the free space between the plates and must satisfy certain boundary conditions at all points on the wall. The boundary conditions to be satisfied are:

(i) The tangential component of the field **E** must be zero at all points on the wall.

(ii) The normal component of the magnetic field must be zero at all points on the wall.

Note that the normal component of **E** may not be zero, since there may be charges on the conducting surfaces and the tangential component of the magnetic field may not be zero, since there can be surface currents in the conducting walls. Since the electromagnetic waves are transverse, with their electric (**E**) and magnetic (**H**) field vectors being transverse to the direction of propagation (**k**), they are labelled as TEM waves.

Suppose now a linearly polarized plane wave is moving between the plates in a direction specified by the propagation vector **k** in the plane perpendicular to the x-axis and making an angle θ with y-axis. If **E** is in the x-direction, **B** must be in the plane Oyz ($\because$ $\mathbf{B} \perp \mathbf{E}$). In other words $\mathbf{B} = \hat{\mathbf{e}}_y B_y + \hat{\mathbf{e}}_z B_z$, has a component in the direction of the wave i.e. z-direction; whereas **E** does not. Such a wave is called **transverse electric** (TE) wave. If we choose **B** to be along the x-axis, then **E** will have a longitudinal component in the z-direction. Such a wave is known as **transverse magnetic** (TM) wave. Waves propagating within closed conducting regions are either TE or TM modes.

The electric field in the plane wave is

$$\mathbf{E}_I = \hat{\mathbf{e}}_x E_{0I} e^{i(\mathbf{k}.\mathbf{r} - \omega t)} \tag{8.1}$$

where

$$\mathbf{k}\cdot\mathbf{r} = ky \cos \theta + kz \sin \theta \tag{8.2}$$

$$\therefore \quad \mathbf{E}_I = \hat{\mathbf{e}}_x E_{0I} e^{i(ky \cos \theta + kz \sin \theta - \omega t)}. \tag{8.3}$$

The wave will be reflected at the wall. For the reflected wave we have

$$\mathbf{E}_R = \hat{\mathbf{e}}_x E_{0R} e^{i} \; - ky \cos \theta + kz \sin \theta - \omega t) \tag{8.4}$$

The total electric field between the conducting planes is

$$\mathbf{E} = \mathbf{E}_I + \mathbf{E}_R = \hat{\mathbf{e}}_x E_{0I} e^{i(ky \cos \theta + kz \sin \theta - \omega t)}$$

$$+ \hat{\mathbf{e}}_x E_{0R} e^{i(-ky \cos \theta + kz \sin \theta - \omega t)}$$

$$= \hat{\mathbf{e}}_x e^{i(kz \sin \theta - \omega t)} \left(E_{0I} e^{iky \cos \theta} + E_{0R} e^{-iky \cos \theta} \right). \tag{8.5}$$

At the boundary (i.e. at $y = 0$, $E_t = 0$. This condition will be satisfied if $E_{0I} + E_{0R} = 0$, i.e. $E_{0I} = - E_{0R}$

$$\therefore \quad \mathbf{E} = \hat{\mathbf{e}}_x E_{0I} e^{i(kz \sin \theta - \omega t)} \left(e^{iky \cos \theta} - e^{-iky \cos \theta} \right) \tag{8.6}$$

$$= 2i\,\hat{\mathbf{e}}_x\,E_{0I}\sin(ky\cos\theta)\,e^{i\,(kz\sin\theta\,-\,\omega t)}. \tag{8.7}$$

The field must vanish also at $y = b$. This is possible if

$$kb\cos\theta = n\pi \tag{8.8}$$

where n takes only integral values. There will, thus, be several acceptable fields corresponding to the different values of n. The different waves that correspond to different values of n are called **modes.** Now, since $\cos\theta \leqslant 1$

$$\frac{n\pi}{kb} \leqslant 1. \tag{8.9}$$

This condition restricts the maximum value of n for a given frequency of radiation.

If $\frac{\pi}{kb}$ i.e. $\frac{\lambda}{2b} \geqslant 1$, there cannot be any wave of the type (8.7) in the region under consideration. There will thus be a **cut-off wave length** above which such waves cannot be propagated. Each mode has its own cut-off wave length λ_n given by

$$\lambda_n = \frac{2b}{n}. \tag{8.10}$$

The cut-off wave length, thus, depends upon the mode number and the separation of plates.

From (8.8), we get,

$$\sin\theta = \sqrt{1 - \frac{n^2\pi^2}{k^2b^2}}. \tag{8.11}$$

Substituting (8.8) and (8.11) in (8.7), we have

$$\mathbf{E} = 2i\hat{\mathbf{e}}_x\,E_{0I}\,\sin\frac{n\pi y}{b}\,e^{i\left\{kz\sqrt{1-\frac{n^2\pi^2}{k^2b^2}}\,-\,\omega t\right\}}$$

$$= 2i\hat{\mathbf{e}}_x\,E_{0I}\sin\frac{n\pi y}{b}\,e^{i\,(k_g z\,-\,\omega t)} \tag{8.12}$$

where

$$k_g = \left(k^2 - \frac{n^2\pi^2}{b^2}\right)^{1/2}. \tag{8.13}$$

The wave number k_g is known as the **guide wave number** and can be expressed as

$$k_g = \frac{2\pi}{\lambda_g}$$

where λ_g is the **guide wave length.**

As can be seen from Eq. (8.12), **E** does not have a component in the z-direction. The waves, therefore, are transverse electric waves and are represented by TE_n, where n is the mode number.

The magnetic field in TE_n waves can be found using the relation

$$\nabla \times \mathbf{E} = -\frac{\partial \mathbf{B}}{\partial t}.$$

The phase velocity of the wave (8.12) is

$$v_p = \frac{\omega}{k_g} = c\frac{k}{k_g}$$

$$= c\left(1 - \frac{n^2\pi^2}{k^2b^2}\right)^{-1/2}. \tag{8.14}$$

quation (8.14) shows that the phase velocity of the wave exceeds the locity of light in free space. (Note that phase velocity greater than locity of light is possible because it is the velocity of nodes only and ot of energy). The group velocity is given by

$$v_g = \frac{d\omega}{dk_g}.$$

ne energy of the group of waves is moving with this velocity. From e relation (8.13) we have

$$k_g^2 = k^2 - \frac{n^2\pi^2}{b^2}.$$

fferentiating with respect to k_g

$$k_g = k\frac{dk}{dk_g} = \frac{k}{c}\frac{d\omega}{dk_g}$$

$$\therefore \quad v_g = c\frac{k_g}{k} \tag{8.15}$$

ich is less than the speed of light.

The product of v_p and v_g is

$$v_p v_g = c^2. \tag{8.16}$$

In the same way we can obtain another set of solutions in which the gnetic field acts along the x-axes and has not component in the direcn of propagation of wave. These are **transverse magnetic waves** or M_n waves. These different TM_n modes have cut-off wavelengths at the ne value as the TE_n modes given by (8.10).

Waves in Guides of Arbitrary Cross-section

We shall consider in this section the propagation of waves inside a llow conductor of any uniform cross-section, the walls of which are umed to be perfect conductors. We shall obtain the general solution the wave-equation for the electric and magnetic fields within the guide, oject to the requirement that they satisfy Maxwell's equations and also propriate boundary conditions.

The wave equation for the electric field is

$$\left(\nabla^2 - \frac{1}{c^2}\frac{\partial^2}{\partial t^2}\right)\mathbf{E} = 0 \tag{8.17}$$

The desired plane wave solution is of the form

$$\mathbf{E} = \mathbf{E}_0\,(x, y)\, e^{i(k_g z - \omega t)}. \tag{8.18}$$

Similar equation may be written for the magnetic field.

Substituting (8.18) in (8.17) we have

$$\left(\frac{\partial^2}{\partial x^2} + \frac{\partial^2}{\partial y^2} - k_g^{\,2} + \frac{\omega^2}{c^2}\right) E_0\,(x, y) = 0 \tag{8.19}$$

We define a transverse Laplacian operator ∇_T by

$$\nabla_T^{\,2} = \frac{\partial^2}{\partial x^2} + \frac{\partial^2}{\partial y^2} = \nabla^2 - \frac{\partial^2}{\partial z^2}. \tag{8.20}$$

Hence, (8.19) can be written as

$$(\nabla_T^{\,2} + k_c^{\,2})\, E_0\,(x, y) = 0 \tag{8.21}$$

where

$$k_c^{\,2} = -\,k_g^{\,2} + \frac{\omega^2}{c^2}. \tag{8.22}$$

It is convenient to express the fields **E** and **B** in terms of the components parallel (E_z, B_z) and transverse (E_T, B_T) to the axis of the conductor i.e. z-axes, i.e.

$$\mathbf{E} = \mathbf{E}_z + \mathbf{E}_T, \quad \mathbf{B} = \mathbf{B}_z + \mathbf{B}_T$$

where

$$\mathbf{E}_z = \hat{\mathbf{e}}_z\, E_{0z}\,(x, y)\, e^{i\,(k_g z - \omega t)} \tag{8.23}$$

$$\mathbf{E}_T = \mathbf{E}_{0T}\, e^{i\,(k_g z - \omega t)}$$

$$= \{\hat{\mathbf{e}}_x\, E_{0x}\,(x, y) + \hat{\mathbf{e}}_y\, E_{0y}\,(x, y)\}\, e^{i\,(k_g z - \omega t)} \tag{8.24}$$

and similar equations for $\mathbf{B}_z$ and $\mathbf{B}_T$.

The fields must satisfy Maxwell's equations

(i) $\nabla\cdot\mathbf{E} = 0$ i.e. $\dfrac{\partial E_{0x}}{\partial x} + \dfrac{\partial E_{0y}}{\partial y} + ik_g\, E_{0z} = 0$ (8.25)

(ii) $\nabla\cdot\mathbf{B} = 0$ i.e. $\dfrac{\partial B_{0x}}{\partial x} + \dfrac{\partial B_{0y}}{\partial y} + ik_g\, B_{0z} = 0$ (8.26)

(iii) $\nabla \times \mathbf{E} = -\dfrac{\partial \mathbf{B}}{\partial t} = i\omega\mathbf{B}$

i.e.

$$\frac{\partial E_{0z}}{\partial y} - ik_g\, E_{0y} = i\omega\, B_{0x} \tag{8.27}$$

$$ik_g E_{0x} - \frac{\partial E_{0z}}{\partial x} = i\omega B_{0y} \tag{8.28}$$

$$\frac{\partial E_{0y}}{\partial x} - \frac{\partial E_{0x}}{\partial y} = i\omega B_{0z} \tag{8.29}$$

(iv) $\nabla \times \mathbf{B} = \mu \frac{\partial \mathbf{D}}{\partial t} = -\frac{i\omega}{c^2} \mathbf{E}$

:.

$$\frac{\partial B_{0z}}{\partial y} - ik_g B_{0y} = -\frac{i\omega}{c^2} E_{0x} \tag{8.30}$$

$$ik_g B_{0x} - \frac{\partial B_{0z}}{\partial x} = -\frac{i\omega}{c^2} E_{0y} \tag{8.31}$$

$$\frac{\partial B_{0y}}{\partial x} - \frac{\partial B_{0x}}{\partial y} = -\frac{i\omega}{c^2} E_{0z}. \tag{8.32}$$

olving (8.28) and (8.30) for E_{0x} we get

$$E_{0x} = \frac{i}{k_c^2}\left(k_g \frac{\partial E_{0z}}{\partial x} + \omega \frac{\partial B_{0z}}{\partial y}\right) \tag{8.33}$$

imilarly

$$E_{0y} = \frac{i}{k_c^2}\left(k_g \frac{\partial E_{0z}}{\partial y} - \omega \frac{\partial B_{0z}}{\partial x}\right) \tag{8.34}$$

. $\mathbf{E}_{0T} = \hat{\mathbf{e}}_x E_{0x} + \hat{\mathbf{e}}_y E_{0y}$

$$= \frac{i}{k_c^2} k_g \left(\hat{\mathbf{e}}_x \frac{\partial E_{0z}}{\partial x} + \hat{\mathbf{e}}_y \frac{\partial E_{0z}}{\partial y}\right) + \frac{i\omega}{k_c^2}\left(\hat{\mathbf{e}}_x \frac{\partial B_{0z}}{\partial y} - \hat{\mathbf{e}}_y \frac{\partial B_{0z}}{\partial x}\right)$$

$$= \frac{i}{k_c^2}\left[k_g \nabla_T E_{0z} - \omega \hat{\mathbf{e}}_z \times \nabla_T B_{0z}\right] \tag{8.35}$$

here $\nabla_T = \hat{\mathbf{e}}_x \frac{\partial}{\partial x} + \hat{\mathbf{e}}_y \frac{\partial}{\partial y}$.

Similar structural equations can be obtained for B_{0x} and B_{0y}. These uations show that all the transverse components may be completely ecified in terms of the longitudinal components.

Let us now examine whether TEM waves can propagate inside a ollow conductor.

For TEM waves $E_{0z} = B_{0z} = 0$. Therefore, from equations (8.25) d (8.26) we have

$$\frac{\partial E_{0x}}{\partial x} + \frac{\partial E_{0y}}{\partial y} = 0; \quad \frac{\partial B_{0x}}{\partial x} + \frac{\partial B_{0y}}{\partial y} = 0 \tag{8.36}$$

d from equations (8.29) and (8.32)

$$\frac{\partial E_{0y}}{\partial x} - \frac{\partial E_{0x}}{\partial y} = 0; \quad \frac{\partial B_{0y}}{\partial x} - \frac{\partial B_{0x}}{\partial y} = 0 \tag{8.37}$$

Taking the partial derivatives of (8.36) and (8.37) with respect to x and y respectively and combining them suitably we arrive at the result

$$\nabla_T^2 E_{0x} = 0 \; ; \; \nabla_T^2 B_{0x} = 0. \tag{8.38}$$

Since the components of **E** satisfy Laplace's equation, it follows that the surface of the guide is an equi-potential surface and, hence, **E** inside the conductor is zero. There cannot exist, therefore a TEM mode within a hollow guide with perfectly conducting walls. This result is valid for a singly connected surface. For a TEM mode to exist, it is necessary to have two or more unconnected surfaces. In a coaxial cable, for example, a TEM mode is a dominant mode.

We see from (8.21) and (8.38) that for TEM waves $k_c = 0$, i.e. $k_g = k$. This means that the wave-number is real for all frequencies. In other words there is no cut-off frequency for TEM waves.

The fields **E** and **B** have to satisfy boundary conditions

$$E_{\text{tangential}} = \hat{\mathbf{e}}_n \times \mathbf{E} = 0; \; \mathbf{B}_{\text{normal}} = \hat{\mathbf{e}}_n \cdot \mathbf{B} = 0. \tag{8.39}$$

These boundary conditions together with the wave-equations for the field vectors **E** and **B**, lead to the eigen-value problems. That is, for any given frequency ω, only certain values of k will be consonant with the boundary conditions and the wave-equation.

8.3 Wave Guides of Rectangular Cross-section

As an example of the theory given in the preceding section, let us consider a guide of rectangular cross-section used for the propagation of electromagnetic energy at microwave frequencies. Let the dimensions of the cross-section be *a, b* as shown in Fig. 8.2.

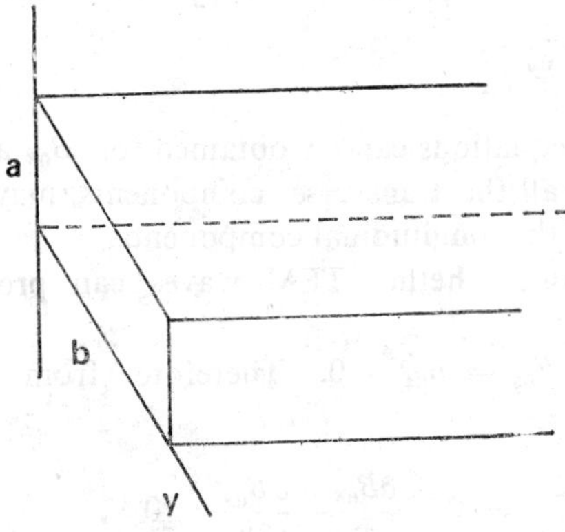

Fig. 8.2

Let us first consider TE waves. For such waves $E_{0z} = 0$. Therefore the field is determined by the solution of the equation

$$(\nabla_T^2 + k_c^2) B_{0z} = 0 \tag{8.40}$$

From (8.35) we find that B_{0z} has to satisfy besides equation (8.40) the condition

$$\left.\frac{\partial B_{0z}}{\partial x}\right|_{x=0,\,a} = 0 \text{ and } \left.\frac{\partial B_{0z}}{\partial y}\right|_{y=0,\,b} = 0 \tag{8.41}$$

so as to make $E_{\text{tangential}} = 0$.

It is easily verified that the solution

$$B_{0z} = B_0 \cos\left(\frac{m\pi x}{a}\right) \cos\left(\frac{n\pi y}{b}\right) \tag{8.42}$$

where m, n, are integers, satisfies both (8.40) and (8.41) provided

$$k_c^2 = \pi^2\left(\frac{m^2}{a^2} + \frac{n^2}{b^2}\right) \tag{8.43}$$

i.e.

$$\omega_{mn} = \pi c \left(\frac{m^2}{a^2} + \frac{n^2}{b^2}\right)^{1/2} \tag{8.44}$$

where ω_{mn} is the cut-off frequency corresponding to the mode numbers m, n. The corresponding mode is designated as TE_{mn} mode. If $m = 0$, $n = 0$, the case corresponds to TEM mode. Hence, TE_{00} mode does not exist. If $a < b$, the lowest cut-off frequency is obtained by putting $m = 0$, $n = 1$

$$\omega_{01} = \frac{\pi c}{b}.$$

This is the **principal** or **dominant mode.**

In general TE_{mn} waves with frequencies greater than the cut-off frequency ($\omega > \omega_{mn}$) will propagate without attenuation. If one wants only the dominant mode to be propagated, the dimensions of the guide must be judiciously chosen.

The field components of the dominant mode TE_{01}, are

$$E_{0x} = \frac{i\omega}{k_c^2}\frac{\partial B_{0z}}{\partial y} = -\frac{i\omega}{k_c^2} B_0 \sin\left(\frac{\pi y}{b}\right)\frac{\pi}{b}$$

$$= -\frac{i\omega b}{\pi} B_0 \sin\left(\frac{\pi y}{b}\right) \quad \text{(see 8.43)}$$

$$E_{0y} = 0, \qquad E_{0z} = 0$$

$$B_{0x} = 0, \qquad B_{0y} = \frac{ik_g}{k_c^2}\frac{\partial B_{0z}}{\partial y} = -\frac{ik_g b}{\pi} B_0 \sin\left(\frac{\pi y}{b}\right)$$

$$B_{0z} = B_0 \cos\left(\frac{\pi y}{b}\right) \tag{8.45}$$

These fields correspond to the waves propagated in the z-direction which is the direction of the Poynting vector. The energy flow in the wave guide is given by

$$\langle \mathbf{N} \rangle_{01} = \frac{1}{2} \text{Real} (\mathbf{E} \times \mathbf{H}^*)$$

$$= \frac{1}{2\mu} R_e (\mathbf{E} \times \mathbf{B}^*)$$

$$= \frac{1}{2\mu} R_e \left(- \hat{\mathbf{e}}_y E_{0x} B^*_{0z} + \hat{\mathbf{e}}_z E_{0x} B^*_{0y} \right).$$

We find from equation (8.45) that $E_{0x} B^*_{0z}$ is purely imaginary.

$$\therefore \langle \mathbf{N} \rangle_{01} = \frac{\hat{\mathbf{e}}_z}{2\mu} \frac{\omega b^2}{\pi^2} B_0^2 k_g \sin^2 \left(\frac{\pi y}{b} \right). \tag{8.46}$$

The total power may be obtained by integrating this expression from $y = 0$ to $y = b$.

The magnetic field described by these equations is shown in Fig. 8.3. The field lines are drawn for the plane parallel to the y-z plane at time $t = 0$. Since the field is independent of the x-coordinate the field pattern will be the same in any plane parallel to the y-z plane.

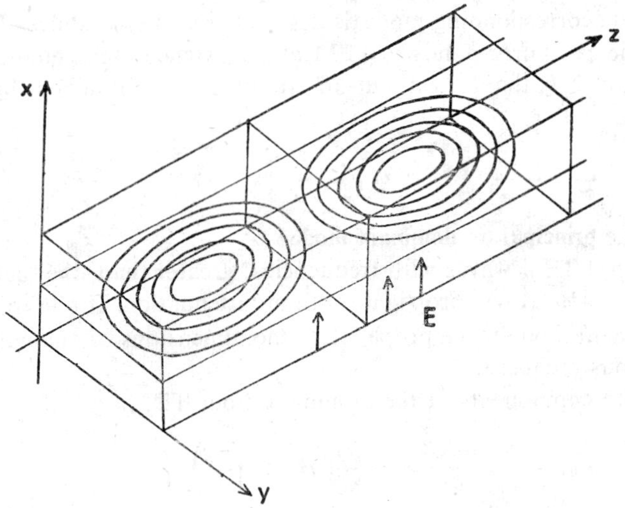

Fig. 8.3

EXAMPLE 8.1. Find the modes of 3 cm wave-length radar waves that would be propagated in a wave guide of rectangular cross section with $a = 1$ cm, $b = 2$ cm. Find also the group velocity of the waves.

The cut-off frequency is given by

$$\omega_{mn} = \pi c \left(\frac{m^2}{a^2} + \frac{n^2}{b^2} \right)^{1/2}, \quad \omega_{mn} = 2\pi \nu_{mn}.$$

The frequency of the given radiation is 10^{10} Hz. The cut-off frequencies for different modes are

$$\nu_{01} = \frac{c}{2} \frac{1}{2 \times 10^{-2}} = 7.5 \times 10^9 \text{ Hz}$$

$$\nu_{10} = \frac{c}{2}\frac{1}{10^{-2}} = 1.5 \times 10^{10}\text{ Hz}$$

$$\nu_{11} = 1.68 \times 10^{10}\text{Hz}.$$

The lowest cut-off frequency for TM mode, i.e. TM_{11} mode is

$$\nu_{11} = 1.68 \times 10^{10}\text{ Hz}.$$

Since only TE_{01} mode has a frequency lower than the frequency of radiation, only TE_{01} mode can be propagated.

The phase velocity of the wave $v_p = \dfrac{\omega}{k_g}$

$$k_g = \sqrt{k_c^2 - \omega^2/c^2} = 1.4\text{ cm}^{-1} \quad \text{(by 8.22)}$$

$$\therefore \quad v_p = 4.5 \times 10^8\text{ ms}^{-1}$$

Now $\quad v_p v_g = c^2 \quad \therefore \quad v_g = 2 \times 10^8\text{ ms}^{-1}.$

8.4 Resonant Cavities

The purpose of wave-guides is to transmit electromagnetic energy efficiently from one point to another. A **resonator,** on the other hand, is an energy storage device and is equivalent to a resonant circuit element. We shall discuss in this section the simplest cavity resonator—a rectangular cavity—whose resonant frequencies can be easily calculated.

Consider a closed box formed by placing end faces on a rectangular wave guide. We will assume that the end surfaces are plane and perpendicular. Because of the reflection at the end faces, the waves in the cavity are standing waves and not progressive. It can be readily verified that the electric field components are of the form

$$\begin{aligned} E_x &= E_1 \cos(k_1 x) \sin(k_2 y) \sin(k_3 z) e^{-i\omega t} \\ E_y &= E_2 \sin(k_1 x) \cos(k_2 y) \sin(k_3 z) e^{-\omega t} \\ E_z &= E_3 \sin(k_1 x) \sin(k_2 y) \cos(k_3 z) e^{-i\omega t}. \end{aligned} \tag{8.47}$$

In order that the boundary conditions be satisfied, it is necessary that k_1, k_2, k_3 have the values given by

$$k_1 = \frac{l\pi}{a},\ k_2 = \frac{m\pi}{b},\ k_3 = \frac{n\pi}{c} \tag{8.48}$$

where a, b, c are the dimensions of the box and l, m, n are integers.

Substitution of any component in the appropriate wave-equation shows that the fields given by (8.47) to be acceptable, the free space wave number has to satisfy the condition

$$k^2 = \frac{\omega^2}{c^2} = \pi^2\left(\frac{l^2}{a^2} + \frac{m^2}{b^2} + \frac{n^2}{c^2}\right) \tag{8.49}$$

Thus, there is an infinite number of resonant frequencies and, hence, the infinite number of modes of the cavity corresponding to the different values of l, m, n.

The magnetic field components can be found from Maxwell's relation

$$\nabla \times \mathbf{E} = -\frac{\partial \mathbf{B}}{\partial t}$$

If $l = 0$, $m = 1$ and $n = 1$, the electric field is transverse to the direction of propagation as can be found from (8.47) and (8.48). This mode is designated as TE_{011} mode. Besides TE_{lmn} modes there are also other modes possible in the cavity. There are TM_{lmn} modes in which the magnetic field is transverse to the direction of propagation. For the same cavity, TE_{lmn} waves and TM_{lmn} waves occur at the same frequency. Hence, at a particular resonant frequency the standing wave in the cavity is the sum of two resonating waves, the TE mode and TM mode. When you do the theory of black-body radiation you will not have any difficulty now in understanding why the factor 2 appears in the density of states function.

We have seen above that for a definite field configuration the cavities have some discrete resonant frequencies. This means that if we try to excite a particular mode of oscillation in a cavity, the right sort of fields will not be built up unless the exciting frequency is equal to the resonance frequency. In practice, however, appreciable excitation occurs over a narrow band of frequencies around the resonant frequency. This smearing out of the sharp frequency of oscillation occurs partly, because of the dissipation of energy in the cavity walls. The measure of these losses are often expressed by giving what is called the "Q" of the cavity defined by

$$Q = \frac{\text{energy stored in the cavity}}{\text{energy lost per cycle to the walls}}. \tag{8.50}$$

The power loss in a cavity can be readily estimated by computing the time average of the Poynting vector into the wall at the surface

$$\langle \mathbf{N} \rangle = \tfrac{1}{2} R_e (E_{||} \times H_{||})$$

where $E_{||}$, $H_{\_|}$ are the tangential components of the electric and magnetic fields.

Cavities are often used as frequency meters.. Since cylindrical cavities can be made more accurately than rectangular cavities, they are often used in accurate frequency measurements. Cavities are also used in experiments where high microwave fields are required as, for example, in electric spin resonance experiments.

8.5 Dielectric Wave-guides

We have discussed so far wave-guides, the walls of which are perfect conductors with fields within the guide. Other guiding structures are also possible. For example, parallel wire transmission line serves to direct electromagnetic waves. A dielectric slab is yet another system which can serve as a wave-guide, with some properties similar to those of the wave-guides discussed so far, but having some differences arising because of the different boundary conditions to be satisfied at the surface. With the advent of lasers, dielectric wave-guides have assumed great importance particularly in the field of optical communication. A laser beam transmitted through the atmosphere suffers severe distortions and, hence, it is expedient to guide the light in some way so as to make optical communication effective. For this purpose dielectric guides are found to be very useful.

The principle involved is that of total internal reflection in which the wave propagating in a dielectric is incident at an angle greater than the critical angle on an interface with another dielectric of lower refractive index.

One of the earlier methods was to use what is known as a "gas lens", in which a pipe was filled with a gas and a suitable temperature gradient was established. This, in turn, produced a radial gradient in the refractive index which helped, to some extent, the guiding of waves along the axis. The method was cumbersome and not very effective. This was soon replaced by light "pipes" made of glass fibres of desired refractive index surrounded by a cladding layer with slightly lower refractive index (Fig 8.4). The light waves reflected internally at the core-cladding interface and thus helped the propagation of the modes along the axis. If the diameter of the core is much larger than the wave-length of light, then there is a possibility of large number of modes being propagated. For a single mode to be transmitted the dimensions of the core must be judiciously chosen.

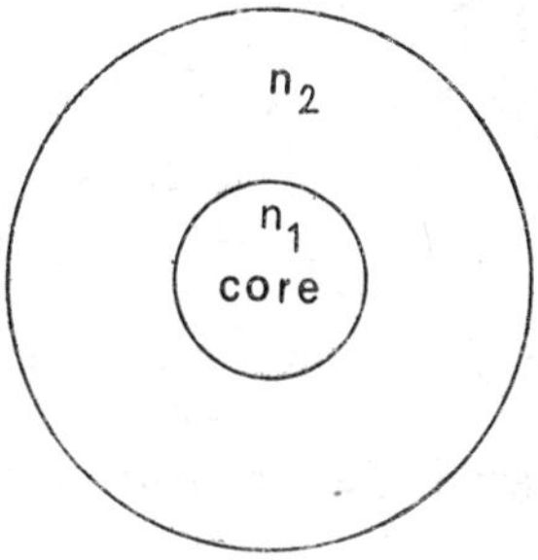

Fig. 8.4

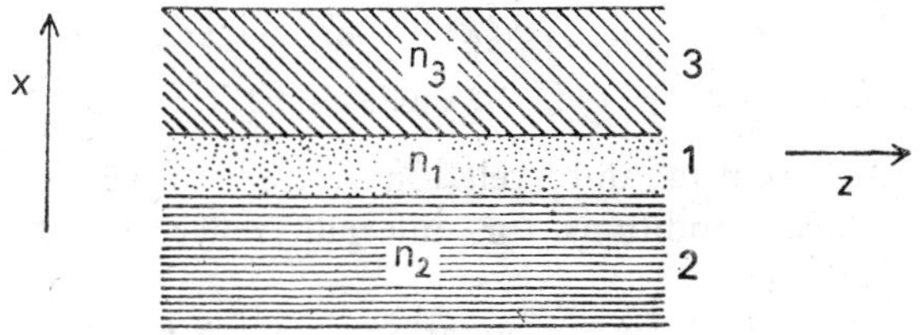

Fig. 8.5

Let us now discuss how a plane slab of a dielectric can be used as a wave guide.

Consider a plane dielectric sheet (1) sandwitched between two layers (2) and (3) as shown in Fig 8.5. The outer layers forming the cladding are referred to as substrate (2) and superstrate (3). Between these two there is the infinite sheet—a thin film of thickness of the order of the optical wave-length '*d*'. These are usually made of sputtered zinc oxide, gallium arsenide etc. We assume that the layers do not absorb the electromagnetic radiation incident on them, or absorb it very weakly, and their refractive indices satisfy the condition

$$n_1 > n_2 \geqslant n_3 \tag{8.51}$$

which meets the condition for internal reflection.

Let us consider a TE mode propagating in the z-direction. The field **E** will be transverse to this direction. Let its components be $\{0, E_y e^{i(\beta z-\omega t)}, 0\}$ where β is the propagation constant in the z-direction given by

$$\beta = k' \cos\theta = n_1 k \cos\theta \tag{8.52}$$

$$\left(\because \text{ If } \lambda_d \text{ is the wavelength in the dielectric, } k' = \frac{2\pi}{\lambda_d} = \frac{2\pi}{\lambda} n_1\right).$$

The wave equation

$$\left(\nabla^2 - \frac{1}{c^2}\frac{\partial^2}{\partial t^2}\right)\mathbf{E} = 0$$

reduces to

$$\frac{\partial^2 E_y}{\partial x^2} + (n^2k^2 - \beta^2)\, E_y = 0 \tag{8.53}$$

The fields must vanish at $x = \pm\infty$. The solution of this equation is

$$\begin{aligned} E_y &= Ae^{-\delta x} e^{i(\beta z-\omega t)} && \text{for } x \geqslant 0 \\ &= (B\cos Kx + C\sin Kx)e^{i(\beta z-\omega t)} && \text{for } 0 \geqslant x \geqslant -d \\ &= De^{\gamma x}\,.\,e^{i(\beta z-\omega t)} && \text{for } x \leqslant -d \end{aligned} \tag{8.54}$$

where $\delta^2 = \beta^2 - n_3{}^2k^2;\ K^2 = n_1{}^2k^2 - \beta^2;\ \gamma^2 = \beta^2 - n_2{}^2k^2,$

each one of these being a positive quantity. From the first of these relations, we have

$$\beta^2 > n_3^2k^2 \quad \text{i.e.} \quad n_1^2k^2\cos^2\theta > n_3^2k^2$$

or

$$n_1^2\cos^2\theta > n_3^2$$

and from the third relation $n_1^2\cos^2\theta > n_2^2$.

Thus, conditions for internal reflection at the two boundaries are satisfied.

Using Maxwell's equations, we can find the tangential components of the magnetic field **H**

$$\begin{aligned} H_z &= -\frac{i\delta}{\omega\mu_0}\left[Ae^{-\delta x}e^{i(\beta z-\omega t)}\right] && \text{for } x \geqslant 0 \\ &= \frac{iK}{\omega\mu_0}\left[-A\sin Kx + C\cos Kx\right]e^{i(\beta z-\omega t)} && \text{for } 0 \geqslant x \geqslant -d \\ &= \frac{i\gamma}{\omega\mu_0}\left[A\cos Kd - C\sin Kd\right]e^{\gamma(x+d)}e^{i(\beta z-\omega t)} && \text{for } x \leqslant -d. \end{aligned}$$

Since the tangential component of **H** has to satisfy boundary conditions, i.e. H_t must be continuous across the surface at $x = 0$ and $x = -d$, we have

$$-\delta A = KC \tag{8.55}$$

$$K[A \sin Kd + C \cos Kd] = \gamma [A \cos Kd - C \sin Kd] \tag{8.56}$$

i.e.
$$\delta A + KC = 0$$

and
$$(KA + \gamma C) \sin Kd + (KC - \gamma A) \cos Kd = 0.$$

For these to have a non-trivial solution for A and C, the determinant of the system must vanish, giving

$$\tan Kd = \frac{K(\gamma + \delta)}{K^2 - \gamma\delta}. \tag{8.57}$$

This equation, when solved graphically or numerically, will give the value of β^2 which, in turn, will give the value of the angle θ at which the energy must be coupled into the guide.

If $\gamma = 0$, there is a loss of total internal reflection at the lower boundary and the mode will no longer be guided.

PROBLEMS

8.1 ATM wave travels along the interface of two dielectrics of permittivities ε_1 and $-\varepsilon_2$. Find the dispersion equation.

8.2 Find the relation between the tangential components of an electric field and a magnetic field near a conductor.

8.3 What should be the maximum and minimum widths of a wave-guide of square cross-section, if it is to transmit waves in the TE_{01} mode only?

8.4 A plane electromagnetic wave is incident normally on a slab of dielectric material, the back surface of which is in contact with a perfect conductor. Show that if there is no reflection at the front surface, then $\varepsilon = \mu$ for the dielectric.

8.5 Find the number of resonances that may be there in a rectangular cavity with dimensions $a = 2$ cm, $b = 3$ cm, $d = 4$ cm within the frequency range $\nu = 5 \times 10^9$ Hz to $\nu = 10^{10}$ Hz.

8.6 Determine the cut-off thickness for the fundamental TE mode in the dielectric slab wave-guide in which the refractive index of the film is $n_1 = 1.59$, that of the substrate $n_2 = 1.53$, that of the superstrate $n_3 = 1$ and $k = 9.92$ rads/μm.

Chapter 9

Electromagnetic Radiation

In the last chapter we considered the electromagnetic waves and energy transported through wave-guides. In this chapter we shall be concerned with "radiation", i.e. the unguided transport of waves and energy through empty space. We shall show that any distribution of changing charge and current acts as a source of electromagnetic radiation. How is visible light produced? It results from the rapid adjustment of the charge and current distribution within the electron clouds of atoms. It is, indeed, true that the emission of radiation by atoms can only be described by quantum mechanics. It was actually in the theory of radiation that the breakdown of classical concepts first became evident. However, the classical theory presented in this chapter will prove to be of immense help in understanding the quantum theory of radiation.

9.1 Retarded Potentials

The relation of radiation fields to their sources can be easily found if the fields are expressed in terms of electromagnetic potentials A and Φ

$$\mathbf{E} = -\nabla \Phi - \frac{\partial \mathbf{A}}{\partial t}, \quad \mathbf{B} = \nabla \times \mathbf{A} \tag{9.1}$$

We have seen in Chapter 5, how the introduction of Lorentz condition in the Maxwell's equations yields the following two inhomogeneous equations for A and Φ

$$\nabla^2 \Phi - \frac{1}{c^2}\frac{\partial^2 \Phi}{\partial t^2} = -\frac{\rho}{\epsilon_0} \tag{9.2}$$

$$\nabla^2 \mathbf{A} - \frac{1}{c^2}\frac{\partial^2 \mathbf{A}}{\partial t^2} = -\mu_0 \mathbf{j} \tag{9.3}$$

The operator on the left-hand side is identical with that of the homogeneous wave equation, but the source terms j and ρ now appear on the right-hand side. It remains to solve these equations to obtain the expressions for A and Φ in terms of charge and current distribution.

We shall first obtain the solutions by analogy with the solutions obtained for steady state poblems in Chapter 1 and Chapter 5 in electrostatics and magnetostatics respectively. The solutions of the equations

$$\nabla^2\Phi = -\frac{\rho}{\epsilon_0} \quad \text{and} \quad \nabla^2 \mathbf{A} = -\mu_0 \mathbf{j}$$

that we found are

$$\Phi(\mathbf{r}) = \frac{1}{4\pi\epsilon_0}\int_V \frac{\rho(\mathbf{r})}{|\mathbf{r}-\mathbf{r}'|}\,d\tau$$

and

$$\mathbf{A}(\mathbf{r}) = \frac{\mu_0}{4\pi}\int_V \frac{\mathbf{j}(\mathbf{r})}{|\mathbf{r}-\mathbf{r}'|}\,d\tau. \tag{9.4}$$

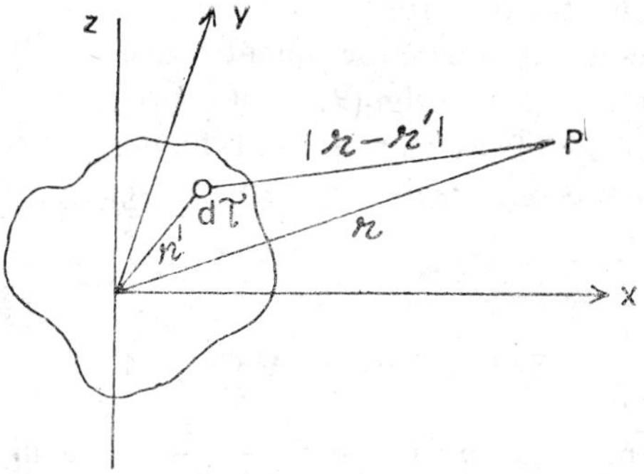

Fig. 9.1

In Fig. 9.1 $d\tau$ is a small volume element. The potentials are to be computed at a point P whose position vector is $\mathbf{r}$. This is done by integrating ρ and $\mathbf{j}$ throughout V by considering them as functions of $\mathbf{r}'$, the position vector of the volume element $d\tau$. The expressions (9.4) are valid if the charges are at rest and the currents are steady. Consider now a situation in which ρ changes with time. What is the potential at the point P at time t? If the charge distribution within the volume element $d\tau$ changes with time, the field observed at P at time t must have been "launched" by the element $d\tau$ at a time t' earlier than t. This is because the electric fields associated with these charges propagate with the finite velocity c and, hence, the time taken to travel the distance from $d\tau$ to P is $\frac{|\mathbf{r}-\mathbf{r}'|}{c}$. Therefore, the contribution to the potential at the point P at a time t, due to the charge in $d\tau$, does not depend on what the charge is at time t but on what it was at the time $t - \frac{|\mathbf{r}-\mathbf{r}'|}{c}$. This is the time at which the electric fields must have been propagated from the charges in $d\tau$ at $\mathbf{r}'$ in order to arrive at P at the time t.
Hence,

$$\Phi(\mathbf{r}, t) = \frac{1}{4\pi\epsilon_0}\int_V \frac{\rho\left(\mathbf{r}', t - \frac{|\mathbf{r}-\mathbf{r}'|}{c}\right)}{|\mathbf{r}-\mathbf{r}'|}\,d\tau \tag{9.5}$$

$$\mathbf{A}(\mathbf{r}, t) = \frac{\mu_0}{4\pi} \int_V \frac{\mathbf{j}\left(\mathbf{r}', t - \frac{|\mathbf{r} - \mathbf{r}'|}{c}\right)}{|\mathbf{r} - \mathbf{r}'|} d\tau. \tag{9.6}$$

These expressions for Φ and $\mathbf{A}$ are found on physical grounds. But are these the solutions of the inhomogeneous equations (9.2) and (9.3)?

To verify this, let us substitute (9.5) into (9.2). We write (9.5) as

$$\Phi(\mathbf{r}, t) = \frac{1}{4\pi\epsilon_0} \int_V \frac{\rho(\mathbf{r}', t - R'/c)}{R'} d\tau \tag{9.7}$$

where we have substituted R' for $|\mathbf{r} - \mathbf{r}'|$.

Since it is difficult to evaluate the potential at $R' = 0$, we divide the volume of integration in equation (9.7) into two regions. The first of these is a small sphere of radius r_0 around the point at which Φ is to be determined and the second region is rest of space. The potential is given by

$$\Phi = \Phi_1 + \Phi_2 \tag{9.8}$$

Hence $$\nabla^2\Phi = \nabla^2\Phi_1 + \nabla^2\Phi_2$$

In the region of the small sphere $\dfrac{R'}{c} = \dfrac{|\mathbf{r} - \mathbf{r}'|}{c}$ is negligible. Hence,

$$\nabla^2\Phi_1 = \frac{1}{4\pi\epsilon_0} \nabla^2 \int_V \frac{\rho(\mathbf{r}', t)}{R'} d\tau$$

$$= \frac{\rho}{4\pi\epsilon_0} \int_V \nabla^2 \left(\frac{1}{R'}\right) d\tau.$$

We have shown in Sec. 3.9 that

$$\nabla^2\Phi_1 = \frac{\rho}{4\pi\epsilon_0} \int_V \nabla^2 \left(\frac{1}{R'}\right) d\tau = -\rho/\epsilon_0 \tag{9.9}$$

This is independent of r_0 and, therefore, the sphere can be made as small as possible shrinking into a point.

Let us now find $\nabla^2\Phi_2$. We note that ∇^2 operates only on R'. In spherical polar coordinates,

$$\nabla^2\Phi_2 = \frac{1}{R'} \frac{\partial^2}{\partial R'^2} (R' \Phi_2) \tag{9.10}$$

$$= \frac{1}{4\pi\epsilon_0} \int_V \frac{1}{R'} \frac{\partial^2}{\partial R'^2} \rho \left(\mathbf{r}', t - \frac{R'}{c}\right) d\tau. \tag{9.11}$$

Now for any function $f(t - R'/c)$

$$\frac{\partial^2 f}{\partial R'^2} = \frac{1}{c^2} \frac{\partial^2 f}{\partial t^2} \tag{9.12}$$

$$\therefore \quad \nabla^2\Phi_2 = \frac{1}{4\pi\epsilon_0} \int_V \frac{1}{R' c^2} \frac{\partial^2}{\partial t^2} \rho \left(\mathbf{r}', t - \frac{R'}{c}\right) d\tau.$$

$$= \frac{1}{c^2}\frac{\partial^2}{\partial t^2}\left[\frac{1}{4\pi\epsilon_0}\int_V \frac{\rho\left(\mathbf{r}', t - \frac{R'}{c}\right)}{R'} d\tau'\right]$$

$$= \frac{1}{c^2}\frac{\partial^2 \Phi}{\partial t^2} \qquad (9.13)$$

$$\therefore \quad \nabla^2\Phi = \nabla^2\Phi_1 + \nabla^2\Phi_2 = -\frac{\rho}{\epsilon_0} + \frac{1}{c^2}\frac{\partial^2\Phi}{\partial t^2}$$

or
$$\nabla^2\Phi - \frac{1}{c^2}\frac{\partial^2\Phi}{\partial t^2} = -\frac{\rho}{\epsilon_0}.$$

We have thus proved that (9.5) is a solution of (9.2).

These potentials (9.5) and (9.6) are called **retarded potentials.** Note that the integrands $\rho(r')$ and $\mathbf{j}(\mathbf{r}')$ are to be evaluated at the retarded time $t - \frac{|\mathbf{r} - \mathbf{r}'|}{c}$, where $\mathbf{r}'$ is the radius vector at the retarded time. This concept is familiar in astronomy. The light of a star observed on the earth at a certain time gives information about the star at the much earlier time when the light was radiated from that star. The potentials are retarded by the time the light takes to travel the distance from the star to the earth and are to be calculated using the earlier values of $\mathbf{j}$ and ρ.

It is quite simple to find fields **E** and **B** once the potentials are found. But the computation of **A** and Φ is not as simple as it appears to be. Complications arise because retarded times are different for different parts of the sources. Only in certain simple situations, the fields can be determined by introducing some sensible approximations.

9.2 Radiation from an Oscillating Dipole

Let us calculate the radiation field produced by an electric oscillating dipole. This field has many important applications. Many practical radiating systems may be considered to be made up by putting together a large number of such dipoles. Besides, it contains many features useful in quantum theory of emission of radiation by atoms, molecules and nuclei. Consider two small spheres at the two ends of a short wire. Suppose the charge is transferred periodically from one sphere to the other, the time variation being harmonic, i.e.

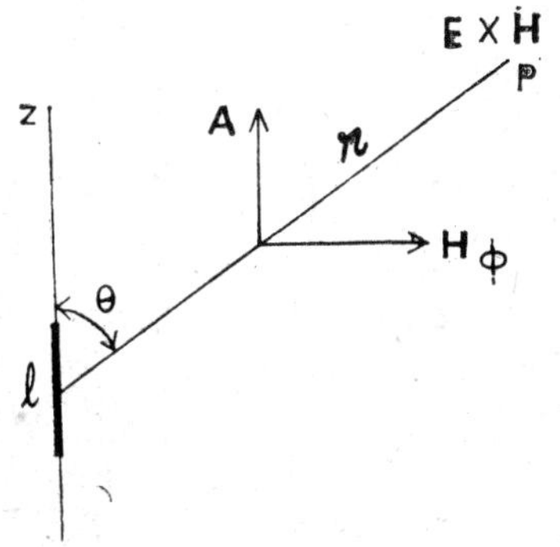

Fig. 9.2

$$q = q_0 \exp(i\omega t) \qquad (9.14)$$

where the amplitude of the oscillating charge and ω the angular frequency of oscillation. Let us assume that the wave-length of the radiation produced is large compared with the length of the wire 'l', i.e.

$$\lambda = \frac{2\pi c}{\omega} \gg l \quad \text{or} \quad \frac{2\pi}{\omega} = T \gg \frac{l}{c}.$$

This means the time l/c taken for a signal to propagate along the wire, from one end to the other, is very much less than that over which the source current changes appreciably. We may, therefore, take the current I to be the same at all points along its length. Now

$$I = \frac{dq}{dt} = i\omega q_0 \exp(i\omega t). \tag{9.15}$$

The charge oscillating between two spheres is equivalent to an oscillating dipole moment $\mathbf{p}$

$$|\mathbf{p}| = ql = q_0 l \exp(i\omega t) = p_0 \exp(i\omega t) = \frac{Il}{i\omega} \tag{9.16}$$

where $p_0 = q_0 l$.

Let us now determine: (i) how the radiation field of this dipole is distributed throughout space and (ii) the total power radiated.

The dipole is shown in Fig. 9.2. The wire lies along the z-axis. The origin of the coordinate system coincides with the centre of the wire. We are interested in the field values at the point P specified by the position vector $\mathbf{r}$. In order to determine the field we have first to find the retarded potentials. The vector potential $\mathbf{A}$ is given by

$$\mathbf{A}(\mathbf{r}, t') = \frac{\mu_0}{4\pi}\int_V \frac{\mathbf{j}\left(\mathbf{r}', t - \frac{|\mathbf{r} - \mathbf{r}'|}{c}\right)}{|\mathbf{r} - \mathbf{r}'|} d\tau.$$

The integration is over the volume occupied by the current. Now

$$\mathbf{j}\left(\mathbf{r}', t - \frac{|\mathbf{r} - \mathbf{r}'|}{c}\right) d\tau = I\left(\mathbf{r}', t - \frac{|\mathbf{r} - \mathbf{r}'|}{c}\right) \hat{\mathbf{e}}\, d.$$

Further, since the current is always along the z-axis, r' may be replaced by z. Hence,

$$\mathbf{A}(\mathbf{r}, t) = \frac{\mu_0}{4\pi}\int_{-l/2}^{l/2} \frac{I\left(z, t - \frac{|\mathbf{r} - \hat{\mathbf{e}}_z z|}{c}\right)}{|\mathbf{r} - \hat{\mathbf{e}}_z z|} \hat{\mathbf{e}}_z dz. \tag{9.17}$$

Since $r \gg l$, we may neglect $\hat{\mathbf{e}}_z z$ with respect to r and put the denominator in (9.17) as r and replace the time $t - \frac{|\mathbf{r} - \hat{\mathbf{e}}_z z|}{c}$ at which the current density in the wire is to be measured by the time $t - r/c$. Thus,

$$A(\mathbf{r}, t) = \frac{\mu_0}{4\pi}\int_{-l/2}^{l/2} \frac{I(z, t - r/c)}{r}\,\hat{\mathbf{e}}_z\, dz$$

$$= \frac{\mu_0}{4\pi}\,\hat{\mathbf{e}}_z l\,\frac{I(t - r/c)}{r} \qquad (9.18)$$

because, the current is the same at all points along the wire. This equation shows that the vector potential $\mathbf{A}$ is everywhere parallel to l as shown in Fig. 9.2. Thus the components of the vector potential are known.

$$A_x(\mathbf{r}, t) = 0; \qquad A_y(\mathbf{r}, t) = 0$$

and

$$A_z(\mathbf{r}, t) = \frac{\mu_0 l}{4\pi r}\, i\omega q_0 \exp\{i\omega(t - r/c)\}$$

$$= \frac{\mu_0}{4\pi r}\, i\omega p_0 \exp(i\omega t)\exp(-i\omega r/c)$$

$$= \frac{\mu_0}{4\pi r}\, i\omega\, p(t) e^{-ikr}$$

$$= \frac{\mu_0}{4\pi}\,\dot{p}(t)\,\frac{e^{-ikr}}{r}. \qquad (9.19)$$

The scalar potential Φ can easily be obtained from the Lorentz condition

$$\nabla \cdot \mathbf{A} = -\frac{1}{c^2}\frac{\partial \Phi}{\partial t}. \qquad (9.20)$$

Since only the z-component of $\mathbf{A}$ is non-zero

$$\frac{\partial A_z}{\partial z} = -\frac{1}{c^2}\frac{\partial \Phi}{\partial t}$$

$$\therefore \quad -\frac{1}{c^2}\frac{\partial \Phi}{\partial t} = \frac{\mu_0}{4\pi}\dot{p}(t)\frac{\partial}{\partial r}\left(\frac{e^{-ikr}}{r}\right)\cos\theta \quad \left(\because \frac{\partial r}{\partial z} = \cos\theta\right)$$

i.e. $$\frac{\partial \Phi}{\partial t} = -\frac{1}{4\pi\epsilon_0}\dot{p}(t)\cos\theta\,\frac{\partial}{\partial r}\left(\frac{e^{-ikr}}{r}\right)$$

Hence, $$\Phi = -\frac{1}{4\pi\epsilon_0}\, p(t)\cos\theta\,\frac{\partial}{\partial r}\left(\frac{e^{-ikr}}{r}\right). \qquad (9.21)$$

This can also be expressed as

$$\Phi = \frac{1}{4\pi\epsilon_0}\left\{\frac{z}{r^3}\, q\left(t - \frac{r}{c}\right) + \frac{z}{cr^2}\, I\left(t - \frac{r}{c}\right)\right\} \quad \text{(Problem 9.1)} \qquad (9.22)$$

We have determined $\mathbf{A}$ and Φ. We can now determine $\mathbf{E}$ and $\mathbf{B}$ using (9.1).

In spherical polar coordinates (which are sometimes convenient considering the symmetry of the problem) the components are:

$$A_r = \frac{\mu_0}{4\pi r} i\omega\, p_0 \cos\theta \exp\{i\omega(t - r/c)\} = \frac{\mu_0}{4\pi r} \dot{p} \cos\theta\, e^{-ikr} \quad (9.23)$$

$$A_\theta = \frac{\mu_0}{4\pi r} i\omega\, p_0 \sin\theta \exp\{i\omega(t - r/c)\} = \frac{\mu_0}{4\pi r} \dot{p} \sin\theta\, e^{-ikr}$$

$$A_\phi = 0.$$

Hence, the components of **E** and **H** are

$$E_r = -\frac{\partial\Phi}{\partial r} - \frac{\partial A_r}{\partial t} = \frac{ipk}{2\pi\epsilon_0 r} \cos\theta \left[1 - \frac{i}{kr}\right] \frac{e^{-ikr}}{r}$$

$$E_\theta = -\frac{1}{r}\frac{\partial\Phi}{\partial\theta} - \frac{\partial A_\theta}{\partial t} = -\frac{pk^2}{4\pi\epsilon_0} \sin\theta \left[1 - \frac{i}{kr}\left(1 - \frac{i}{kr}\right)\right] \frac{e^{-ikr}}{r}$$

$$E_\phi = 0 \quad (9.24)$$

$$H_r = 0, H_\theta = 0, H_\phi = -\frac{cpk^2}{4\pi} \sin\theta \left[1 - \frac{i}{kr}\right] \frac{e^{-ikr}}{r}. \quad (9.25)$$

The relations (9.24) and (9.25) are the **Hertz's relations** for the oscillating dipole.

We may imagine the space divided into two regions:

(i) The region in which $|r| \ll \lambda$ is known as the **near zone.** We shall see that the field in this case is exactly the same as that for an electrostatic dipole.

(ii) The region in which $|r| \gg \lambda$, is called the **radiation zone.** The fields in this region vary as $1/r$. We will not consider the region where $kr \simeq 1$ which is quite complicated.

Let us now examine the relations (9.24) and (9.25) in the near zone and the radiation zone.

(i) Near Zone: $|r| \ll \lambda$, i.e. $kr \ll 1$

$$E_r \simeq \frac{p\cos\theta}{2\pi\epsilon_0 r^3}, E_\theta \simeq \frac{p\sin\theta}{4\pi\epsilon_0 r^3}, E_\phi = 0 \quad (9.26)$$

$$H_r = 0, H_\theta = 0, H_\phi \simeq \frac{i\omega\, p\sin\theta}{4\pi r^2}.$$

These expressions are equivalent to the field of an electrostatic dipole. The ratio of magnetic to electric field in this zone is

$$\frac{\omega\mu_0 |\mathbf{H}|}{k|\mathbf{E}|} \simeq \frac{\omega\mu_0}{k} \epsilon_0 \omega r = kr \ll 1$$

This shows that the electric field dominates in this zone.

(ii) Radiation zone: $kr \gg 1$

$$E_r = 0, E_\theta = -\frac{pk^2}{4\pi\epsilon_0} \sin\theta \frac{e^{-ikr}}{r}, E_\phi = 0$$

$$H_r = 0, H_\theta = 0, H_\phi = -\frac{cpk^2}{4\pi} \sin\theta \frac{e^{-ikr}}{r}. \quad (9.27)$$

These results are arrived at neglecting the terms involving $1/r^2$ and its higher orders. We find from (9.27) that

$$E_\theta = \sqrt{\frac{\mu_0}{\epsilon_0}}\, H_\phi. \tag{9.28}$$

The directions of the fields are shown in Fig. 9.3.

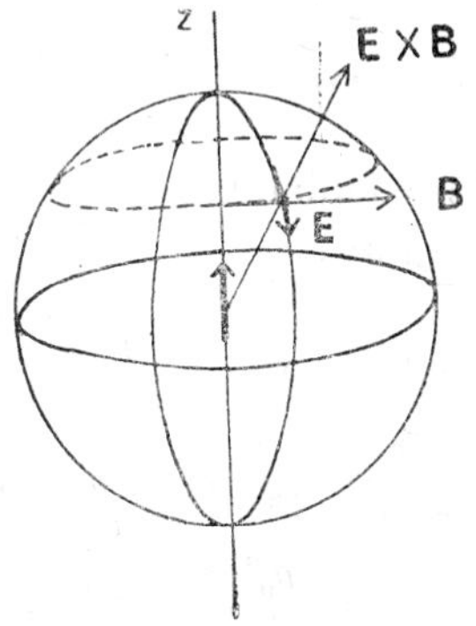

Fig. 9.3

The fields are mutually perpendicular. For the radiation zone $E_r = 0$, hence, the field distribution is spherical with the electric field as circles of longitude and the magnetic field as the circles of latitude. The field is maximum at the equator and zero at the poles. Since the magnetic field is transverse everywhere, the radiation from an oscillating electric dipole is generally TM. At large distances $E_r \to 0$ and the electric field also is transverse. Hence the waves are TEM.

The spatial distribution of the fields can be shown in a polar diagram (Fig. 9.4).

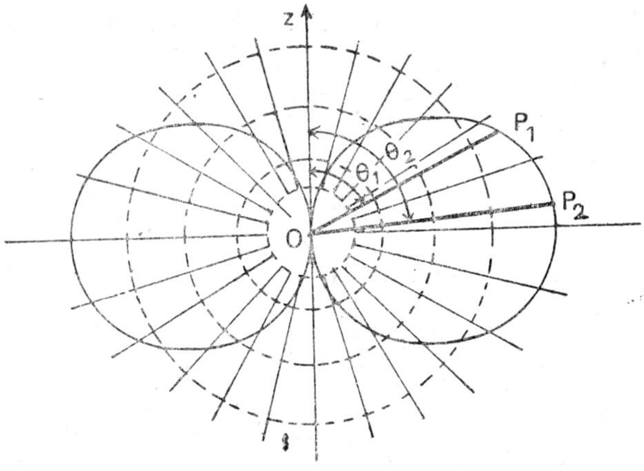

Fig. 9.4

The solid curved lines give the relative strength of the radiation field at different points on a sphere centred on the dipole. Along the axis the field is zero ($\theta = 0$). It is maximum in the equatorial plane. The field is independent of the azimuthal angle ϕ. The ratio of the electric fields at P_1 and P_2 with polar angles θ_1 and θ_2 is given by the ratio OP_1/OP_2.

Let us calculate the instantaneous power W crossing the surface of a sphere of radius, say, r, in the radiation zone.

The time-average Poynting vector $\langle \mathbf{N} \rangle$ is

$$\langle \mathbf{N} \rangle = \frac{1}{2} R_e(\mathbf{E} \times \mathbf{H}^*)$$

Using the values obtained for radiation zone in (9.27)

$$\langle \mathbf{N} \rangle = \frac{cp_0^2 k^4}{32\pi^2 \epsilon_0} \frac{\sin^2\theta}{r^2} \hat{\mathbf{e}}_n. \tag{9.29}$$

The total radiated power

$$W = \int_0^{2\pi} \int_0^{\pi} \frac{cp_0^2 k^4}{32\pi^2 \epsilon_0} \frac{\sin^2\theta}{r^2} r^2 \sin\theta \, d\theta \, d\phi$$

$$= \frac{cp_0^2 k^4}{32\pi^2 \epsilon_0} 2\pi \frac{4}{3} = \frac{cp_0^2 k^4}{12\pi\epsilon_0}$$

$$= \frac{cp_0^2}{12\pi\epsilon_0} \left(\frac{2\pi}{\lambda}\right)^4. \tag{9.30}$$

Thus, the power radiated varies directly as the square of the amplitude of the electric dipole and inversely as the fourth power of the wave length.

We now introduce the concept of **radiation resistance.** We know that when a current $I = I_0 e^{i\omega t}$ passes through a circuit containing a resistance R, the average rate of dissipation of energy is given by

$$\text{Rate of energy dissipated} = \frac{1}{2} R I_0^2. \tag{9.31}$$

The expression (9.30) for the power dissipated in the case of dipole can be written as

$$\frac{cp_0^2}{12\pi\epsilon_0} \left(\frac{2\pi}{\lambda}\right)^4 = \frac{p_0^2 \omega^4}{12\pi\epsilon_0 c^3}$$

$$= \frac{1}{12\pi\epsilon_0} \frac{I_0^2 l^2 \omega^2}{c^3} \quad \text{(see 9.16)}$$

$$= \frac{1}{12\pi\epsilon_0} \frac{l^2 4\pi^2}{\lambda^2} \frac{1}{c} I_0^2$$

$$= \frac{1}{2} \frac{2\pi}{3} \sqrt{\frac{\mu_0}{\epsilon_0}} \left(\frac{l}{\lambda}\right)^2 I_0^2. \tag{9.32}$$

Comparing this with (9.31) we find for the radiation resistance

$$R_r = \frac{2\pi}{3} \sqrt{\frac{\mu_0}{\epsilon_0}} \left(\frac{l}{\lambda}\right)^2$$

$$= 789 \left(\frac{l}{\lambda}\right)^2 \text{ ohms} \tag{9.33}$$

Note that the expression is valid provided $l \ll \lambda$.

9.3 Linear Antenna

The equations derived in the preceding sections apply only to systems with linear dimensions much smaller than the wave-length. The antennae used in radio or television transmission, generally, are not short compared to the wave-length of the radiation they transmit. Hence, the current in the antenna is not constant and the variation of its amplitude must be taken into account.

A simple antenna is a centre driven linear antenna shown in Fig. 9.5. The current is fed into the antenna (usually a very thin wire) via a coaxial cable transmission line. The antenna is assumed to be oriented along

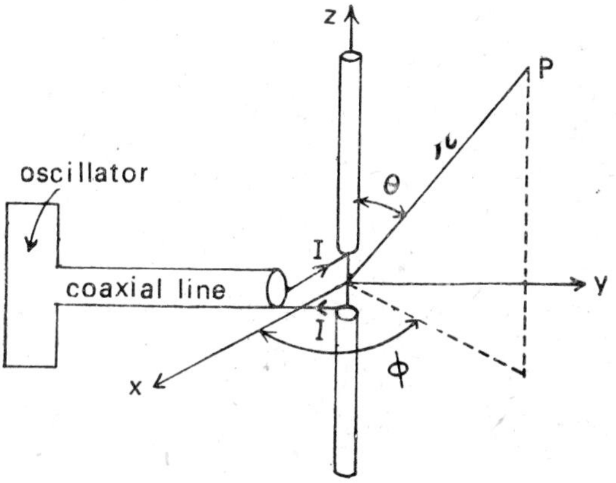

Fig. 9.5

the z-axis and has a length 'l'. The antenna is excited across a small gap at the mid-point which coincides with the origin. The current density is assumed to vary harmonically in time and space along the antenna. Standing waves are set up with the conditions that the current at each end is zero. To a good approximation the current density can be written as

$$\mathbf{j}\,(\mathbf{r}, t) = \hat{\mathbf{e}}_z\, I_0 \exp\,(i\omega t) \sin\left(\frac{kl}{2} - k\,|\,z\,|\right) \delta x \delta y. \qquad (9.34)$$

The delta functions assure that the current flows in z-direction only. One can easily see that this expression satisfies the required conditions. The current at the gap i.e. the input signal is

$$I(t) = I_0 \exp\,(i\omega t) \sin \frac{kl}{2}. \qquad (9.35)$$

The vector potential at a point P specified by the position vector $\mathbf{r}$ is

$$\mathbf{A}\,(\mathbf{r}, t) = \frac{\mu_0}{4\pi} \int \frac{\mathbf{j}\,(\mathbf{r}', t')}{|\,\mathbf{r} - \mathbf{r}'\,|}\, d\mathbf{r}', \text{ where } t' = t - \frac{|\,\mathbf{r} - \mathbf{r}'}{c}$$

$$= \frac{\mu_0}{4\pi} \hat{\mathbf{e}}_z I_0 \int_{-}^{l/2} \frac{e^{i\omega\left(t - \frac{|\mathbf{r}-\mathbf{r}'|}{c}\right)} \sin\left(\frac{kl}{2} - k|z|\right)}{|\mathbf{r}-\mathbf{r}'|} dr'$$

$$= \frac{\mu_0}{4\pi} \hat{\mathbf{e}}_z I_0 \int_{-l/2}^{l/2} \frac{e^{i\omega\left(t - \frac{|\mathbf{r}-\mathbf{z}|}{c}\right)} \sin\left(\frac{kl}{2} - k|z|\right)}{|\mathbf{r}-\mathbf{z}|} dz. \tag{9.36}$$

If the point P is far away, we can replace the denominator by r and in the numerator we may substitute

$$|\mathbf{r} - \mathbf{z}| = r - z\cos\theta$$

where θ is the angle the vector $\mathbf{r}$ make with the z-axis.

$$\therefore \quad \mathbf{A}(\mathbf{r}, t) = \frac{\mu_0}{4\pi} \hat{\mathbf{e}}_z \frac{I_0}{r} \exp(i\omega t) e^{-ikr} \int_{-l/2}^{l/2} \sin\left(\frac{kl}{2} - k|z|\right) e^{ikz\cos\theta}\, dz.$$

Integrating we get

$$\mathbf{A}(\mathbf{r}, t) = \frac{\mu_0}{2\pi} \hat{\mathbf{e}}_z I_0 \exp(i\omega t) \frac{e^{-ikr}}{kr} \left[\frac{\cos\left(\frac{kl}{2}\cos\theta\right) - \cos\left(\frac{kl}{2}\right)}{\sin^2\theta}\right]. \tag{9.37}$$

One can easily calculate from this the field $\mathbf{E}$ and $\mathbf{H}$. The average rate of power flow is given by

$$\langle \mathbf{N} \rangle = \frac{1}{2} R_e (\mathbf{E} \times \mathbf{H}^*) = \frac{\hat{\mathbf{e}}_n}{2c\mu_0} |E_0|^2$$

$$= \frac{1}{2} \sqrt{\frac{\mu_0}{\epsilon_0}}\, \hat{\mathbf{e}}_n \frac{I_0^2}{4\pi^2} \frac{1}{r^2} \left[\frac{\cos\left(\frac{kl}{2}\cos\theta\right) - \cos\left(\frac{kl}{2}\right)}{\sin\theta}\right]^2. \tag{9.38}$$

The average power radiated into unit solid angle

$$\left\langle \frac{dW}{d\Omega} \right\rangle = \frac{4\pi r^2 \langle \mathbf{N} \rangle}{4\pi} \cdot \hat{\mathbf{e}}_n = r^2 \langle \mathbf{N} \rangle \cdot \hat{\mathbf{e}}_n$$

$$= \sqrt{\frac{\mu_0}{\epsilon_0}} \frac{I_0^2}{8\pi^2} \left[\frac{\cos\left(\frac{kl}{2}\cos\theta\right) - \cos\left(\frac{kl}{2}\right)}{\sin\theta}\right]^2. \tag{9.39}$$

The angular distribution of the radiated power, therefore depends upon the value of $\frac{kl}{2}$. If we consider half-wave antenna, ($l = \lambda/2$), i.e. $kl = \pi$, we have

$$\left\langle \frac{dW}{d\Omega} \right\rangle = \sqrt{\frac{\mu_0}{\epsilon_0}} \frac{I_0^2}{8\pi^2} \left[\frac{\cos\left(\frac{\pi}{2}\cos\theta\right)}{\sin\theta}\right]^2. \tag{9.40}$$

The average total power $\langle W \rangle$ radiated by a half-wave antenna is by (9.38)

$$\langle W\rangle = \frac{1}{2}\sqrt{\frac{\mu_0}{\epsilon_0}}\frac{I_0^2}{4\pi^2}\frac{1}{r^2}\int_0^{2\pi}\int_0^{\pi}\left[\frac{\cos\left(\frac{\pi}{2}\cos\theta\right)}{\sin\theta}\right]^2 r^2\sin\theta\, d\theta\, d\phi$$

$$= \sqrt{\frac{\mu_0}{\epsilon_0}}\frac{I_0^2}{4\pi}\int_0^{\pi}\frac{\cos^2\left(\frac{\pi}{2}\cos\theta\right)}{\sin\theta}\, d\theta.$$

This integral can be evaluated numerically. Its value is found to be

$$\langle W\rangle \simeq 73\frac{I_0^2}{2} \tag{9.41}$$

The radiation resistance, therefore, comes out to be

$$R_r \simeq 73\ \Omega.$$

Comparing this value with the corresponding value for an electric dipole, we find that the half-wave antenna is a much more efficient radiator.

If l is equal to the integral number of half-wave lengths of the drivin oscillations

$$\left\langle\frac{dW}{d\Omega}\right\rangle = \sqrt{\frac{\mu_0}{\epsilon_0}}\frac{I_0^2}{8\pi^2}\left[\frac{\cos^2\left(\frac{m\pi}{2}\cos\theta\right)}{\sin^2\theta}\right] \quad \text{for odd values of } m \tag{9.42}$$

and

$$= \sqrt{\frac{\mu_0}{\epsilon_0}}\frac{I_0^2}{8\pi^2}\left[\frac{\sin^2\left(\frac{m\pi}{2}\cos\theta\right)}{\sin^2\theta}\right] \quad \text{for even values of } m \tag{9.43}$$

9.4 Lienard-Wiechert Potentials

Let us consider the application of the retarded potentials (9.5) and (9.6) to compute the radiation from a single charged particle, say, an electron, in arbitrary motion. Since the calculation of the potentials depends upon the position and velocity of the charge at the retarded time $t - \frac{|\mathbf{r} - \mathbf{r}'|}{c}$, we must know the details of the motion of the charge. In Fig. 9.6 we have shown a trajectory of the electron described by

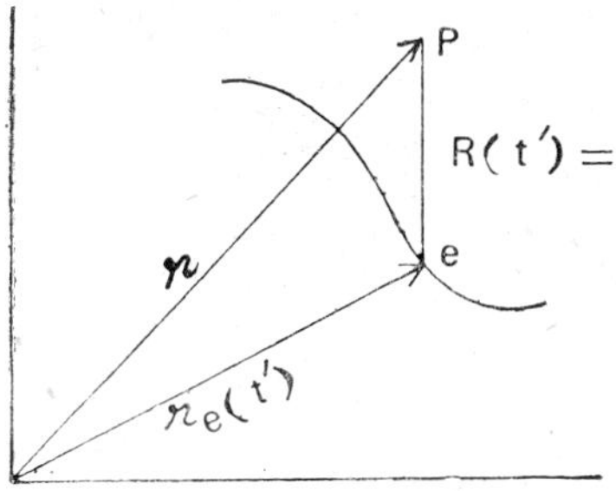

Fig. 9.6

by the radius vector $\mathbf{r}_e(t')$. The calculation of the potentials as given by the formulae (9.5) and (9.6) involves a retarded time integration over the entire volume containing the charge. Now we do not know how the charge is distributed geometrically within the electron. The only thing that we know is that it has a certain total charge. If we assume the electron to have zero physical extent we will land into difficulties. We, therefore, assume that the electron has a finite radius, but shall consider only those properties which are independent of the magnitude of the radius. For an electron we may express the retarded potential in terms of a delta-function. Thus

$$\Phi(r, t) = \frac{e}{4\pi\epsilon_0}\int_{-\infty}^{\infty}\frac{\delta\left\{t' - \left(t - \frac{|\mathbf{r} - \mathbf{r}_e t|}{c}\right)\right\}}{|\mathbf{r} - \mathbf{r}_e|}\, dt'. \tag{9.44}$$

If we put the integral in the form $\int_{-\infty}^{\infty} f(x)\,\delta(x - x')\,dx$, following a property of delta-function, its integral is readily found and is equal to $f(x')$.

We, therefore, introduce a now variable t'' such that

$$t'' = t' - t + \frac{|\mathbf{r} - \mathbf{r}_e(t')|}{c} \tag{9.45}$$

$$dt'' = dt' + \frac{1}{c}\frac{d}{dt'}|\mathbf{r} - \mathbf{r}_e(t')|\, dt' \tag{9.46}$$

Here we have taken $dt = 0$, because the observation is made at a fixed time t. Let the coordinates of the fixed point P be x_1, x_2, x_3 and those of the electron $x_{e_1}(t'), x_{e_2}(t'), x_{e_3}(t')$.

$$\text{Now } |\mathbf{r} - \mathbf{r}_e(t')| = \sqrt{\sum_i \left\{(x_i - x_{e_i}(t')\right\}^2}. \tag{9.47}$$

$$\therefore \quad \frac{1}{c}\frac{d}{dt'}\Big|\mathbf{r} - \mathbf{r}_e(t')\Big| = \frac{1}{c}\sum_i \frac{\partial}{\partial x_{e_i}}\Big|\mathbf{r} - \mathbf{r}_e(t')\Big|\frac{dx_{e_i}}{dt'} \tag{9.48}$$

Since $\frac{\partial}{\partial x_{e_i}}|\mathbf{r} - \mathbf{r}_e(t')/$are the components of the gradient of $|\mathbf{r} - \mathbf{r}_e(t')|$ and $\frac{dx_e}{dt'}$ are the components of $\frac{d\mathbf{r}_e}{dt}$ we can write

$$\frac{1}{c}\frac{d}{dt'}\Big|\mathbf{r} - \mathbf{r}_e(t')\Big| = \frac{1}{c}\,\mathrm{grad}_{re}\Big|\mathbf{r} - \mathbf{r}_e(t')\Big|\cdot\frac{d\mathbf{r}_e}{dt'}. \tag{9.49}$$

The gradient can readily be determined

$$\mathrm{grad}_{re}|\mathbf{r} - \mathbf{r}_e(t')| = -\frac{\mathbf{r} - \mathbf{r}_e(t')}{|\mathbf{r} - \mathbf{r}_e(t')|} = -\frac{\mathbf{R}}{|\mathbf{R}|} \tag{9.50}$$

where $\mathbf{r} - \mathbf{r}_e(t) = \mathbf{R}$.

We also know that

$$\frac{d\mathbf{r}_e}{dt} = \mathbf{u} \qquad \text{(the velocity of the electron)}$$

$$\therefore \quad \frac{1}{c}\frac{d}{dt'}\,|\,\mathbf{r} - \mathbf{r}_e(t')\,| = -\frac{1}{c}\frac{\mathbf{R}}{R}\cdot\mathbf{u} = -\frac{\boldsymbol{\beta}\cdot\mathbf{R}}{|\,\mathbf{R}\,|} \tag{9.51}$$

where $\boldsymbol{\beta} = \dfrac{\mathbf{u}}{c}$.

Therefore,

$$dt'' = dt'\left[1 - \frac{\boldsymbol{\beta}\cdot\mathbf{R}}{|\,\mathbf{R}\,|}\right]$$

i.e.

$$dt' = \frac{|\,\mathbf{R}\,|}{|\,\mathbf{R}\,| - \boldsymbol{\beta}\cdot\mathbf{R}}\,dt''. \tag{9.52}$$

Hence, the potential (9.44) can be expressed as

$$\Phi(\mathbf{r}, t) = \frac{e}{4\pi\epsilon_0}\int_{-\infty}^{\infty} \frac{\delta(t'')}{|\,\mathbf{R}(t')\,|}\,\frac{|\,\mathbf{R}(t')\,|}{|\,\mathbf{R}(t')\,| - \boldsymbol{\beta}(t)\cdot\mathbf{R}(t')}\,dt'' \tag{9.53}$$

$$= \frac{e}{4\pi\epsilon_0}\left[\frac{1}{|\,\mathbf{R}(t')\,| - \boldsymbol{\beta}(t')\cdot\mathbf{R}(t')}\right]_{t''=0}. \tag{9.54}$$

Since $t'' = 0$ implies $t' = t - R(t')/c$

$$\Phi(\mathbf{r}, t) = \frac{e}{4\pi\epsilon_0}\,\frac{1}{[R - \boldsymbol{\beta}\cdot\mathbf{R}]_{t'=t-\frac{R(t')}{c}}} \tag{9.55}$$

By similar arguments we find that the vector potential is given by

$$\mathbf{A}(\mathbf{r}, t) = \frac{\mu_0 e}{4\pi}\left[\frac{\mathbf{u}}{R - \boldsymbol{\beta}\cdot\mathbf{R}}\right]_{t'=t-\frac{R(t')}{c}} \tag{9.56}$$

The potentials Φ (9.55) and $\mathbf{A}$ (9.56) are called **Lienard-Wiechert potentials.** They are dependent on the velocity of the electron but independent of the extent of the charge, i.e. of any detailed electronic model.

9.5 Potentials for a Charge in Uniform Motion-Lorentz Formula

The fields of a moving charge can be found from the relations

$$\mathbf{E} = -\nabla\Phi - \frac{\partial\mathbf{A}}{\partial t}$$

$$\mathbf{B} = \nabla\times\mathbf{A}.$$

The evaluation of the fields will be in terms of the "retarded position" of the charge and its velocity, because the relation between the "present position" and the "retarded position" of the electron is not generally known. In the case of the uniform motion, however, it is possible to express the fields in terms of the present position of the charge at time t.

Let us use the Lienard-Wiechert potentials to find the fields of a charge moving with uniform velocity in a straight line.

If the observer is in the rest frame of the charge, i.e. if he is moving with the charge, the situation as observed by him is electrostatic and the scalar potential according to him is given by

$$\Phi(r) = \frac{e}{4\pi\epsilon_0 r}.$$

What will be the potential if the observer is at rest and the charge is moving? The expression for the potential can be obtained, as will be shown in the next chapter, using relativistic transformations. But relativity has its foundation in electrodynamics. In this section we shall show now Maxwell's equations lead to what is known as **Lorentz transformation.** We have already shown in the preceding section how potentials for a moving charge are obtained from Maxwell's equations. What remains now is to show how these potentials lead to Lorentz transformation. We shall follow essentially the arguments used by Lorentz.

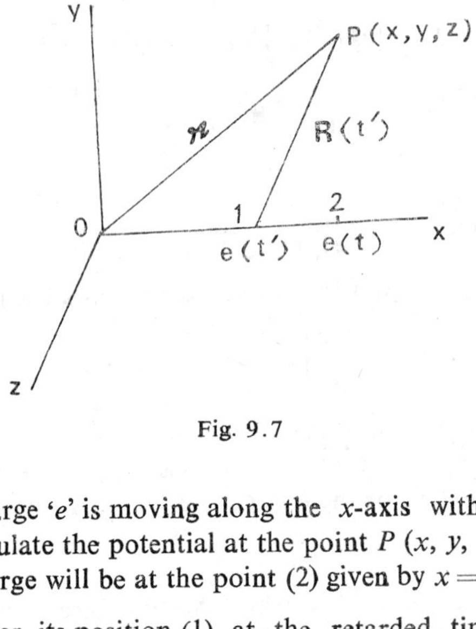

Fig. 9.7

Suppose a charge 'e' is moving along the x-axis with a velocity $u\hat{e}_x$. We want to calculate the potential at the point $P(x, y, z)$ at time t. At this time the charge will be at the point (2) given by $x = ut$, $y = 0$, $z = 0$. We need, however, its position (1) at the retarded time $t' = t - \dfrac{R(t')}{c}$ where $R(t')$ is the distance of the point P from the position of the charge at time t'. We find from the figure that

$$R(t') = |\mathbf{r} - \hat{\mathbf{e}}_x ut'|$$

where $\mathbf{r}$ is the position vector of the point P.

$$\therefore\ R(t') = \{(x - ut')^2 + y^2 + z^2\}^{1/2}$$

Since $\quad t' = t - R(t')/c$

$$R(t')^2 = c^2(t - t')^2 = (x - ut')^2 + y^2 + z^2$$

i.e. $\quad (c^2 - u^2)\,t'^2 - 2(c^2t - xu)\,t' - r^2 + c^2t^2 = 0$

Hence, $t' = \dfrac{(c^2t - xu) - \{(c^2t - xu)^2 + (c^2 - u^2)(r^2 - c^2t^2)\}^{1/2}}{(c^2 - u^2)}$

i.e. $\left(1 - \dfrac{u^2}{c^2}\right)t' = \left(t - \dfrac{xu}{c^2}\right) - \dfrac{1}{c}\left\{(x - ut)^2 + \left(1 - \dfrac{u^2}{c^2}\right)(y^2 + z^2)\right\}^{1/2}$

$$\therefore \quad c(t - t') - \frac{u}{c}(x - ut')$$

$$= \left\{(x - ut)^2 + \left(1 - \frac{u^2}{c^2}\right)(y^2 + z^2)\right\}^{1/2} \tag{9.57}$$

Now $\boldsymbol{\beta}\cdot\mathbf{R} = \dfrac{u}{c} R\cos\theta = \dfrac{u}{c}(x - ut')$.

Hence, $R - \boldsymbol{\beta}\cdot\mathbf{R} = c(t - t') - \dfrac{u}{c}(x - ut')$

$$= \left\{(x - ut)^2 + \left(1 - \frac{u^2}{c^2}\right)(y^2 + z^2)\right\}^{1/2}. \tag{9.58}$$

Therefore, Lienard-Wiechert potential is

$$\Phi(\mathbf{r}, t) = \frac{e}{4\pi\epsilon_0}\frac{1}{\left\{(x - ut)^2 + \left(1 - \dfrac{u^2}{c^2}\right)(y^2 + z^2)\right\}^{1/2}}$$

$$= \frac{e}{4\pi\epsilon_0}\frac{1}{\left(1 - \dfrac{u^2}{c^2}\right)^{1/2}}\frac{1}{\left\{\dfrac{(x - ut)^2}{1 - \dfrac{u^2}{c^2}} + y^2 + z^2\right\}^{1/2}}. \tag{9.59}$$

The vector potential **A** is

$$A(\mathbf{r}, t) = \frac{\mu}{4\pi}\frac{u}{\left(1 - \dfrac{u^2}{c^2}\right)^{1/2}\left\{\dfrac{(x - ut)^2}{1 - \dfrac{u^2}{c^2}} + y^2 + z^2\right\}^{\frac{1}{2}}} \tag{9.60}$$

If $u = 0$, we get our electrostatic formula for potential. This suggests, that in a moving coordinate system, the coordinates should transform as

$$x \to \frac{x - ut}{\sqrt{1 - u^2/c^2}}$$

$$y \to y \tag{9.61}$$

$$z \to z$$

This is **Lorentz transformation.**

9.6 Fields of an Accelerated Charge

Let us go back to the general case of a charge in arbitrary motion and the expression for the fields produced by it.

Let $x_\alpha = x_1, x_2, x_3$ be the coordinates of the point of observation P (Fig. 9.8) and $x_\alpha'(t') = x_1'(t'), x_2'(t'), x_3'(t')$ be the coordinates of the charge at time t' at which a signal propagated with velocity c is emitted at x'_α so as to arrive at x_α at time t.

$$\therefore \qquad R^2 = \Sigma\,(x_\alpha - x'_\alpha)^2.$$

Suppose, further that $x'_\alpha(t')$ is given.

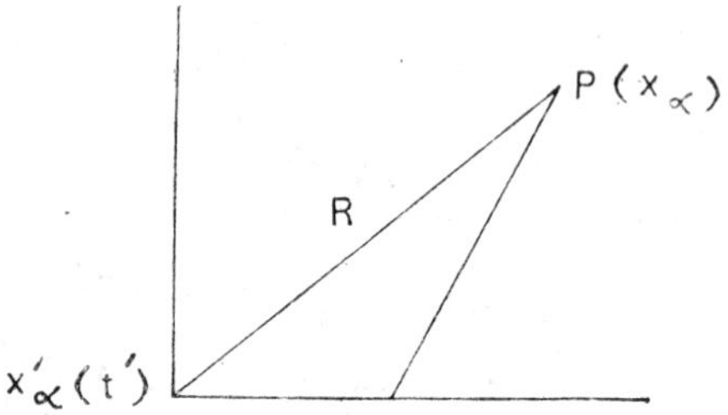

Fig. 9.8

The fields can be found from the relations

$$\mathbf{E} = -\nabla\Phi - \frac{\partial \mathbf{A}}{\partial t}$$

and

$$\mathbf{B} = \nabla \times \mathbf{A}.$$

The potentials are given by

$$\Phi = \frac{e}{4\pi\epsilon_0}\,\frac{1}{[R - \boldsymbol{\beta}\cdot\mathbf{R}]} = \frac{e}{4\pi\epsilon_0}\,\frac{1}{S} \tag{9.62}$$

$$\mathbf{A} = \frac{\mu_0}{4\pi}\,e\left[\frac{\mathbf{u}}{R - \boldsymbol{\beta}\cdot\mathbf{R}}\right] = \frac{e}{4\pi\epsilon_0}\,\frac{\mathbf{u}}{c^2 S}$$

where $S = (R - \boldsymbol{\beta}\cdot\mathbf{R})$.

The components of ∇ are partial derivatives at constant time t and not at constant time t'. Since the time variation with respect to t' is given, in order to compute the fields, we have to transform $\left.\frac{\partial}{\partial t}\right|_{x_\alpha}$ and $\nabla\,|_{x_\alpha}$ to expressions in terms of $\left.\frac{\partial}{\partial t'}\right|_{x_\alpha}$. This is necessary because in the case of an accelerated charge it is not possible, in general, to express the potentials in terms of the "present position" alone. Let us see how this can be done. From Fig. (9.8) we have

$$R[x_\alpha, x'_\alpha(t')] = [\Sigma(x_\alpha - x'_\alpha)^2]^{1/2}$$
$$= c\,(t - t'). \tag{9.63}$$

Since x'_α is given as a function of t', R is a function of x_α ant t'

$$R\,[x_\alpha, x'_\alpha(t')] = f\,(x_\alpha, t') = c\,(t - t'). \tag{9.64}$$

Now
$$\frac{\partial R}{\partial t} = c\left(1 - \frac{\partial t'}{\partial t}\right). \tag{9.65}$$

Also
$$\frac{\partial R}{\partial t} = \frac{\partial R}{\partial t'}\frac{\partial t'}{\partial t}. \tag{9.66}$$

But
$$\frac{\partial R}{\partial t'} = -\frac{\mathbf{R}\cdot\mathbf{u}}{R} \quad \text{(Can you prove this ?)}$$

Hence,
$$c\left(1 - \frac{\partial t'}{\partial t}\right) = -\frac{\mathbf{R}\cdot\mathbf{u}}{R}\frac{\partial t'}{\partial t}$$

or
$$\frac{\partial t'}{\partial t} = \frac{R}{R - \boldsymbol{\beta}\cdot\mathbf{R}} = \frac{R}{S}. \tag{9.67}$$

Therefore,
$$\frac{\partial}{\partial t} = \frac{R}{S}\frac{\partial}{\partial t'}. \tag{9.68}$$

Let us now transform the operator ∇. Because R is a function of x_α and t', we can write

$$\nabla R = \nabla_1 R + \frac{\partial R}{\partial t'}\nabla t' = \frac{\mathbf{R}}{R} - \frac{\mathbf{R}\cdot\mathbf{u}}{R}\nabla t' \tag{9.69}$$

where ∇_1 implies differentiation with respect to x_α at constant retarded time t'.

We have also from (9.64)

$$\nabla R = -c\nabla t'$$

$$\therefore \qquad -c\,\nabla t' = \frac{\mathbf{R}}{R} - \frac{\mathbf{R}\cdot\mathbf{u}}{R}\nabla t'$$

i.e.
$$\nabla t' = -\frac{\mathbf{R}}{Sc}. \tag{9.70}$$

Substituting this in (9.69) we see that we can write, in general, for ∇

$$\nabla = \nabla_1 - \frac{R}{Sc}\frac{\partial}{\partial t'}. \tag{9.71}$$

We have thus found the required transformation of the operator $\frac{\partial}{\partial t}$ (9.68) and ∇ (9.71).

We now compute $\mathbf{E}$ and $\mathbf{B}$

$$\mathbf{E} = -\frac{e}{4\pi\epsilon_0}\nabla\left(\frac{1}{S}\right) - \frac{e}{4\pi\epsilon_0}\frac{\partial}{\partial t}\left(\frac{\mathbf{u}}{Sc^2}\right)$$

$$= \frac{e}{4\pi\epsilon_0}\left[\frac{1}{S^2}\nabla_1 S - \frac{\mathbf{R}}{S^3c}\frac{\partial S}{\partial t'} - \frac{R}{S^2c^2}\dot{\mathbf{u}} + \frac{R\mathbf{u}}{c^2S^3}\frac{\partial S}{\partial t'}\right] \tag{9.72}$$

$$= \frac{e}{4\pi\epsilon_0}\left[\frac{\mathbf{R}}{s^2R} - \frac{\mathbf{u}}{cS^2} + \frac{\mathbf{R}}{S^3c}\left(\frac{\mathbf{R}\cdot\mathbf{u}}{R}\right) - \frac{R}{S^3c}\frac{u^2}{c} + \frac{R}{S^3c}\left(\frac{\mathbf{R}\cdot\mathbf{u}}{c}\right)\right]$$

$$-\frac{R\dot{u}}{S^2c^2}-\frac{R}{S^3c^2}\mathbf{u}\left(\frac{\mathbf{R}\cdot\mathbf{u}}{R}\right)+\frac{R}{S^3c^3}\mathbf{u}u^2-\frac{R}{c^2S^3}\mathbf{u}\left(\frac{\mathbf{R}\cdot\mathbf{u}}{c}\right) \tag{9.73}$$

$$\left(\because\quad \nabla_1 S=\frac{\mathbf{R}}{R}-\frac{u}{c}\right).$$

Rearranging and combining the terms, we have

$$\mathbf{E}=\frac{e}{4\pi\epsilon_0}\left[\frac{1}{S^3}\left(\mathbf{R}-\frac{R\mathbf{u}}{c}\right)\left(1-\frac{u^2}{c^2}\right)+\frac{1}{c^2S^3}\left\{\mathbf{R}\times\left(\left(\mathbf{R}-\frac{R\mathbf{u}}{c}\right)\times\dot{\mathbf{u}}\right)\right\}\right]. \tag{9.74}$$

Similarly

$$\mathbf{B}=\frac{e}{4\pi\epsilon_0c^2}\left[\frac{\mathbf{u}\times\mathbf{R}}{S^3}\left(1-\frac{u^2}{c^2}\right)+\frac{1}{cS^3}\frac{\mathbf{R}}{R}\times\left\{\mathbf{R}\times\left(\mathbf{R}-\frac{R\mathbf{u}}{c}\right)\times\dot{\mathbf{u}}\right\}\right]. \tag{9.75}$$

These last two relations lead to the result

$$\mathbf{B}=\frac{\mathbf{R}\times\mathbf{E}}{Rc}. \tag{9.76}$$

Thus the magnetic field **B** is always perpendicular to **R** and **E**.

If you examine the relation (9.74) carefully you will find that the field **E** is composed of two components. The first component given by the first term is a function of velocity **u**, while the second is a function of acceleration. We can, therefore, write

$$\mathbf{E}=\mathbf{E}_v+\mathbf{E}_a \tag{9.77}$$

where $\mathbf{E}_v$ is the velocity field and $\mathbf{E}_a$ the acceleration field. We further see that $\mathbf{E}_v \propto (1/R^2)$ while $\mathbf{E}_a \propto (1/\mathrm{R})$. If we compute the Poynting vector for the fields, we find that the contribution to this vector due to the two components is

$$\mathbf{N}_v \propto \frac{1}{R^4}$$

$$\mathbf{N}_a \propto \frac{1}{R^2}. \tag{9.78}$$

To find the energy radiated by the particle, we have to integrate the normal component of **N** over the surface of a sphere of radius R. Because the element of surface area involves R^2, the integral containing $\mathbf{N}_v$ varies as $1/R^2$ while that involving $\mathbf{N}_a$ remains finite. Therefore, for large R, the contribution due to $\mathbf{N}_v$ tends to zero while that due to $\mathbf{N}_a$ is finite. We conclude, therefore, that a particle moving with a uniform velocity cannot radiate energy. Energy can be radiated only by accelerated charges. This fact has important consequences and applications.

One may ask: "Since the accelerated charges radiate and acceleration may be positive or negative, should not deceleration also produce radiation?" Indeed, it does. For example, if a beam of electrons is projected into a block of material, the electrons are stopped and radiation is produced. This radiation is called **Bremsstrahlüng** (i.e. "braking radiation").

9.7 Radiation from an Accelerated Charged Particle at Low Velocity

If the velocity of the particle is so small that u/c can be neglected, then $S \simeq R$ and fields as obtained from (9.74) to (9.75) are:

$$\mathbf{E}_a = \frac{e}{4\pi\,\epsilon_0\,c^2R^3}\{\mathbf{R}\times(\mathbf{R}\times\dot{\mathbf{u}})\} \tag{9.79}$$

$$\mathbf{B}_a = \frac{e}{4\pi\,\epsilon_0\,c^3R^2}(\dot{\mathbf{u}}\times\mathbf{R}). \tag{9.80}$$

The Poynting vector that contributes to the radiation is

$$\mathbf{N}_a = \mathbf{E}_a\times\mathbf{H}_a = \mathbf{E}_a\times\frac{\mathbf{B}_a}{\mu_0} = \mathbf{E}_a\times\frac{1}{\mu_0 c}\left(\frac{\mathbf{R}\times\mathbf{E}_a}{R}\right).$$

Since $\mathbf{E}_a$ is perpendicular to $\mathbf{R}$

$$\mathbf{N}_a = \frac{1}{\mu_0 c}E_a^2\,\hat{\mathbf{e}}_n = \sqrt{\frac{\epsilon_0}{\mu_0}}\,E_a^2\,\hat{\mathbf{e}}_n \tag{9.81}$$

Now

$$\mathbf{E}_a = \frac{e}{4\pi\,\epsilon_0\,c^2R^3}\{\mathbf{R}\times(\mathbf{R}\times\dot{\mathbf{u}})\}$$

$$= \frac{e}{4\pi\,\epsilon_0\,c^2R^3}\{(\mathbf{R}\cdot\dot{\mathbf{u}})\,\mathbf{R}-(\mathbf{R}\cdot\mathbf{R})\,\dot{\mathbf{u}}\}$$

$$= \frac{e}{4\pi\,\epsilon_0\,c^2\,R^3}\{R\dot{u}\cos\theta\,\mathbf{R}-R^2\dot{\mathbf{u}}\}$$

where θ is the angle between $\mathbf{R}$ and $\dot{\mathbf{u}}$.

Therefore,

$$\mathbf{N}_a = \frac{1}{\mu_0 c}\,\frac{e^2}{16\pi^2\,\epsilon_0^2\,c^4\,R^6}\{R\dot{u}\cos\theta\,\mathbf{R}-R^2\dot{\mathbf{u}}\}^2\,\hat{\mathbf{e}}_n$$

$$= \frac{e^2}{16\pi^2\,\epsilon_0\,c^3\,R^6}\{R^4(\dot{u})^2\cos^2\theta+R^4(\dot{u})^2-2\,R^4(\dot{u})^2\cos^2\theta\}\,\hat{\mathbf{e}}_n$$

$$= \frac{e^2(\dot{u})^2}{16\pi^2\,\epsilon_0\,c^3\,R^2}(1-\cos^2\theta)\,\hat{\mathbf{e}}_n = \frac{e^2(\dot{u})^2}{16\pi^2\,\epsilon_0\,c^3\,R^2}\sin^2\theta\,\hat{\mathbf{e}}_n. \tag{9.82}$$

The Poynting vector gives us the energy flow per unit area per unit time. The power radiated per unit solid angle can be found by multiplying by R^2 which is the area per unit solid angle.

$$\therefore \quad \frac{dW}{d\Omega} = \frac{e^2(\dot{u})^2}{16\pi^2\,\epsilon_0\,c^3}\sin^2\theta \tag{9.83}$$

The angular distribution of energy, therefore, is just the $\sin^2\theta$ distribution (Fig. 9.9).

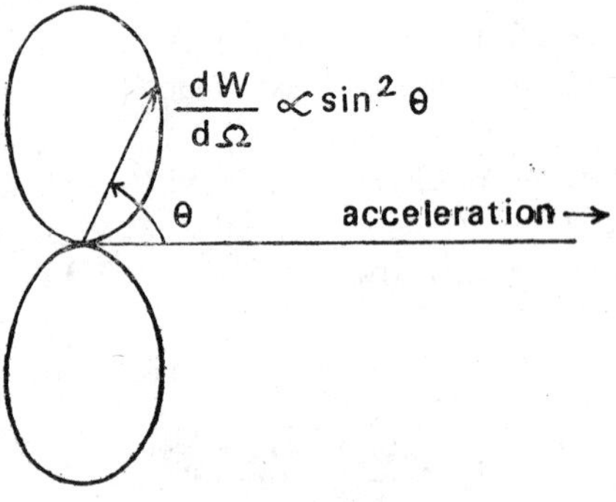

Fig. 9.9

To obtain the total radiated power, we have to integrate over the whole sphere

$$W = \frac{e^2(\dot{u})^2}{16\pi^2\,\epsilon_0\,c^3\,R^2}\int_0^{2\pi}\int_0^{\pi}(1-\cos^2\theta)\,R^2\sin\theta\,d\theta\,d\phi$$

$$= \frac{e^2(\dot{u})^2}{16\pi^2\,\epsilon_0\,c^3}\,2\pi\,\frac{4}{3} = \frac{e^2(\dot{u})^2}{6\pi\,\epsilon_0\,c^3}. \qquad (9.84)$$

This is known as **Larmor formula.**

9.8 Radiation when the Velocity and Acceleration of the Particles are Colinear

In this case, the radiation fields are:

$$\mathbf{E}_a = \frac{e}{4\pi\,\epsilon_0\,c^2\,S^3}\{\mathbf{R}\times(\mathbf{R}\times\dot{\mathbf{u}})\} \qquad (9.85)$$

$$\mathbf{B}_a = \frac{eR}{4\pi\,\epsilon_0\,c^3\,S^3}(\dot{\mathbf{u}}\times\mathbf{R}) \qquad (9.86)$$

$$\therefore \quad E_a^2 = \frac{e^2\,R^4\,(\dot{u})^2}{16\pi^2\,\epsilon_0^2\,c^4\,S^6}\sin^2\theta \qquad (9.87)$$

and

$$\mathbf{N}_a = \frac{1}{\mu_0 c}E_a^2\,\hat{\mathbf{e}}_n = \frac{e^2\,R^4\,(\dot{u})^2}{16\pi^2\,\epsilon_0\,c^3\,S^6}\sin^2\theta\,\hat{\mathbf{e}}_n. \qquad (9.88)$$

The amount of energy radiated into a unit solid angle at θ and measured during the interval dt is

$$-\,dW(\theta) = \frac{e^2\,R^6\,(\dot{u})^2}{16\pi^2\,\epsilon_0\,c^3\,S^6}\sin^2\theta\,dt. \qquad (9.89)$$

The left-hand side is preceded by negative sign because that is the energy lost by the electron in a time interval dt' during the emission of the signal.

The energy lost per unit time per unit solid angle is

$$\frac{dW}{d\Omega} = -\frac{dW(\theta)}{dt'} = \frac{e^2 R^6 (\dot{u})^2}{16\pi^2 \epsilon_0 c^3 S^6} \sin^2\theta \frac{dt}{dt'} \tag{9.90}$$

$$= \frac{e^2 R^6 (\dot{u})^2}{16\pi^2 \epsilon_0 c^3 S^6} \sin^2\theta \frac{S}{R}$$

where we have used (9.67)

$$\therefore \quad \frac{dW}{d\Omega} = \frac{e^2 R^5 (\dot{u})^2}{16\pi^2 \epsilon_0 c^3 S^6} \sin^2\theta = \frac{e^2 R^5 (\dot{u})^2}{16\pi^2 \epsilon_0 c^3 R^5 \left(1 - \frac{\boldsymbol{\beta}\cdot\mathbf{R}}{R}\right)^5} \sin^2\theta$$

$$= \frac{e^2 (\dot{u})^2 \sin^2\theta}{16\pi^2 \epsilon_0 c^3 (1 - \beta\cos\theta)^5}. \tag{9.91}$$

This gives the angular distribution of radiated energy.

If $\beta \ll 1$ i.e. $u \ll c$, we retrieve our formula (9.83) for low velocities. However, if $u \to c$, i.e. $\beta \to 1$, the radiated energy increases in the forward direction (Fig. 9.10), but not exactly in the forward direction (i.e. for $\theta = 0$).

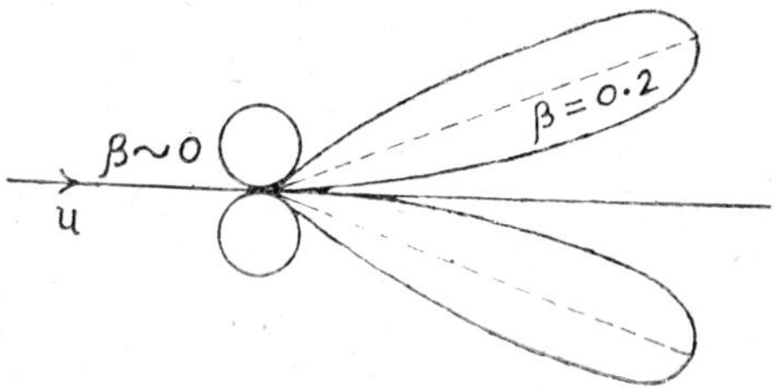

Fig. 9.10

9.9 Radiation from a Charged Particle Moving in a Circular Orbit

Let us now discuss an important case of a charge moving in a circle of radius ρ with an angular frequency ω

$$\therefore \quad u = \rho\omega, \dot{u} = \rho\omega^2$$

Fig. (9.11) shows the orbit of the particle in the yz-plane. The acceleration $\dot{\mathbf{u}}$ is directed towards the centre and, hence, is perpendicular to the velocity $\mathbf{u}$.

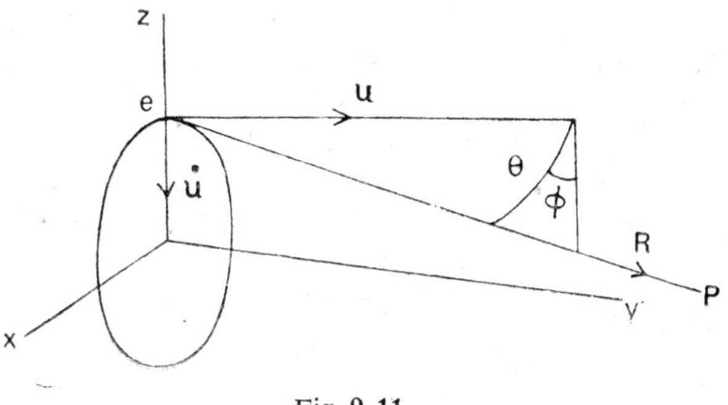

Fig. 9.11

Let θ be the angle between **u** and **R** and let ϕ be the azimuthal angle. Hence,

$$\mathbf{u}\cdot\mathbf{R} = uR\cos\theta \tag{9.92}$$

$$\dot{\mathbf{u}}\cdot\mathbf{R} = \dot{u}R\sin\theta\cos\phi. \tag{9.93}$$

The radiation field is given by

$$\mathbf{E}_a = \frac{e}{4\pi\,\epsilon_0\,c^2\,S^3}\left[\mathbf{R}\times\left\{\left(\mathbf{R}-R\frac{\mathbf{u}}{c}\right)\times\dot{\mathbf{u}}\right\}\right]$$

and the Poynting vector

$$\mathbf{N}_a = \frac{1}{\mu_0 c}E_a{}^2\,\hat{\mathbf{e}}_n$$

Now $\left[\mathbf{R}\times\left\{\left(\mathbf{R}-\frac{R\mathbf{u}}{c}\right)\times\dot{\mathbf{u}}\right]^2 = \left[\left(\mathbf{R}\cdot\dot{\mathbf{u}}\right)\left(\mathbf{R}-\frac{R\dot{\mathbf{u}}}{c}\right)-\left\{\mathbf{R}\cdot\left(\mathbf{R}-\frac{R\mathbf{u}}{c}\right)\right\}\dot{\mathbf{u}}\right]^2$

$$= \left(\mathbf{R}\cdot\dot{\mathbf{u}}\right)^2\left(\mathbf{R}-\frac{R\mathbf{u}}{c}\right)^2+\left\{R^2\dot{\mathbf{u}}-\frac{R}{c}\left(\mathbf{R}\cdot\mathbf{u}\right)\dot{\mathbf{u}}\right\}^2$$

$$-2\left(\mathbf{R}\cdot\dot{\mathbf{u}}\right)\left(\mathbf{R}-\frac{R\mathbf{u}}{c}\right)\left\{R^2\dot{\mathbf{u}}-\frac{R}{c}\left(\mathbf{R}\cdot\mathbf{u}\right)\dot{\mathbf{u}}\right\}$$

$$= -\left(\dot{\mathbf{u}}\cdot\mathbf{R}\right)^2R^2\left(1-\frac{u^2}{c^2}\right)+(\dot{u})^2R^4\left(1-\frac{\mathbf{R}\cdot\mathbf{u}}{Rc}\right)^2$$

$$= -(\dot{u})^2R^4\sin^2\theta\cos^2\phi+(\dot{u})^2R^4\frac{u^2}{c^2}\sin^2\theta\cos^2\phi$$

$$+(\dot{u})^2R^4-\frac{2\dot{u}^2R^4u\cos\theta}{c}+\frac{(\dot{u})^2R^4u^2\cos^2\theta}{c^2}$$

$$= (\dot{u})^2R^4\left[\left(1-\frac{u}{c}\cos\theta\right)^2-\left(1-\frac{u^2}{c^2}\right)\sin^2\theta\cos^2\phi\right] \tag{9.94}$$

where we have used the relations (9.92) and (9.93).

$$\therefore\quad \mathbf{N}_a = \frac{1}{\mu_0 c}\frac{e^2}{16\pi^2\,\epsilon_0{}^2\,c^4\,S^6}(\dot{u})^2R^4\left[\left(1-\frac{u}{c}\cos\theta\right)^2\right.$$

$$\left.-\left(1-\frac{u^2}{c^2}\right)\sin^2\theta\cos^2\phi\right]\hat{\mathbf{e}}_n. \tag{9.95}$$

The angular distribution of the radiated energy is given by

$$-\frac{dW(\theta)}{dt'}\,d\Omega = \left(\mathbf{N}_a\cdot\hat{\mathbf{e}}_n\right)R^2\frac{dt}{dt'}\,d\Omega$$

$$= \frac{e^2(\dot{u})^2}{16\pi^2\,\epsilon_0 c^3}\frac{\left[\left(1-\frac{u}{c}\cos\theta\right)^2-\left(1-\frac{u^2}{c^2}\right)\sin^2\theta\cos^2\phi\right]}{\left(1-\frac{u}{c}\cos\theta\right)^5}\,d\Omega \tag{9.96}$$

where we have used (9.67) and substituted for S as in (9.91).

Figure (9.12) shows the radiation pattern in the plane $\phi = 0$ for a given value of β. For large values of u the radiation is intense in the forward direction and becomes a sharp ray as u approaches c.

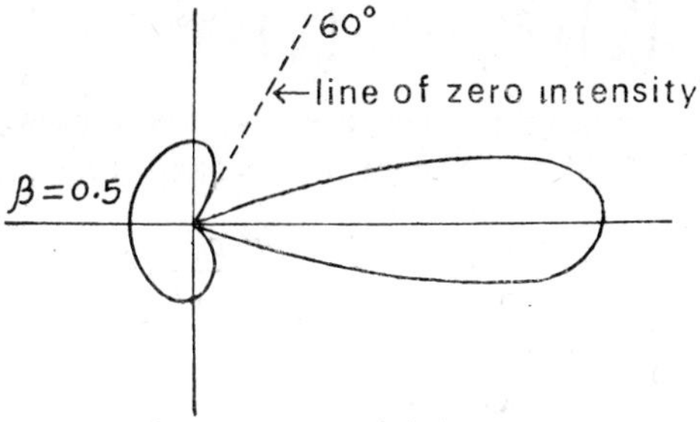

Fig. 9.12

Integration of (9.96) gives the total rate of radiation

$$-\frac{dW(\theta)}{dt'} = \frac{e^2(\dot{u})^2}{6\pi\epsilon_0 c^3} \frac{1}{\left[1 - \frac{u^2}{c^2}\right]^2} \tag{9.97}$$

9.10 Electric Quadrupole Radiation

Imagine two electric dipoles, each of amplitude p, aligned along a given straight line and oscillating exactly out of phase so that there is zero net dipole moment. The system, however, possesses quadrupole moment which varies with time and, hence, quadrupole radiation will be emitted.

The quadrupole field may be readily derived by considering it to be the superposition of two dipole fields. Fig. 9.13 shows the quadrupole.

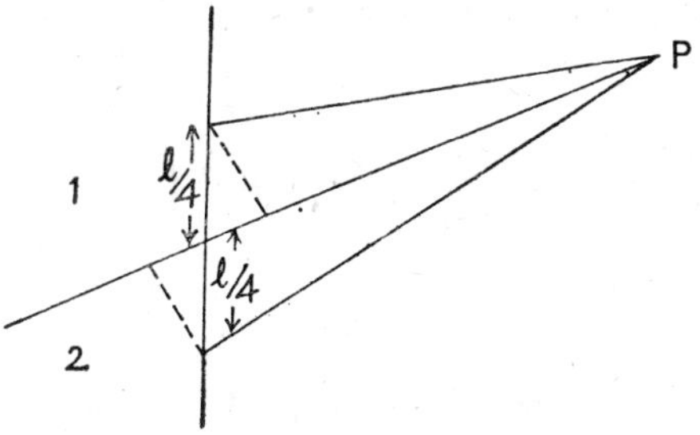

Fig. 9.13

The field due to a single dipole located at the origin, at a point P in the radiation zone is

$$E_\theta = -\frac{pk^2}{4\pi\epsilon_0} \sin\theta \frac{e^{-ikr}}{r} = -\frac{k^2 p_0}{4\pi\epsilon_0 r} e^{i\omega t'} \sin\theta \tag{9.98}$$

The dipoles are at distances $l/4$ and $-\ l/4$ from the origin. The dipole 1 is close to the point P than is the origin by approximately a distance $(l/4)\cos\theta$; and the dipole 2 is farther by the same amount. The phases of the two dipoles relative to the origin, therefore, are

$$\left(\frac{kl}{4}\right)\cos\theta \text{ and } -\left(\frac{kl}{4}\right)\cos\theta - \pi.$$

The phase factor $-\pi$ appears because the two dipoles are oscillating out of phase. The field component of the quadrupole, therefore, is

$$E_\theta = -\frac{k^2 p_0}{4\pi\epsilon_0 r}\sin\theta\, e^{i\left(\omega t' - \frac{kl}{4}\cos\theta\right)} - \frac{k^2 p_0}{4\pi\epsilon_0 r}\sin\theta\, e^{i\left\{\omega t' - \left(-\frac{kl}{4}\cos\theta - \pi\right)\right\}}$$

$$= -\frac{k^2 p_0 \sin\theta}{4\pi\epsilon_0 r}\left\{e^{-i\frac{kl}{4}\cos\theta} - e^{i\frac{kl}{4}\cos\theta}\right\} e^{i\omega t'}$$

$$= \frac{k^2 p_0 \sin\theta}{4\pi\epsilon_0 r}\, 2\, i \sin\left(\frac{kl}{4}\cos\theta\right) e^{i\omega t'}. \qquad (9.99)$$

Using $\qquad H_\phi = \sqrt{\frac{\epsilon_0}{\mu_0}}\, E_\theta$ we have

$$\langle \mathbf{N} \rangle = \frac{1}{2} R_e\,(\mathbf{E}\times\mathbf{H}^*) = \frac{1}{2}\sqrt{\frac{\epsilon_0}{\mu_0}}\,|\,E_0\,|^2\,\hat{\mathbf{e}}_n$$

$$= \frac{1}{2}\sqrt{\frac{\epsilon_0}{\mu_0}}\,\frac{k^4 p_0^2 \sin^2\theta}{16\pi^2\epsilon_0^2 r^2}\sin^2\left(\frac{kl}{4}\cos\theta\right)\cdot 4$$

and $\qquad \left\langle \frac{dW}{d\Omega} \right\rangle = \langle \mathbf{N} \rangle \cdot \hat{\mathbf{e}}_n r^2$

$$= \frac{\sqrt{\epsilon_0}\,k^4 p_0^2 \sin^2\theta}{8\sqrt{\mu_0}\,\pi^2\epsilon_0^2}\sin^2\left(\frac{kl}{4}\cos\theta\right)$$

$$= \frac{\omega^4 p_0^2}{8\pi^2\epsilon_0 c^3}\sin^2\theta\,\sin^2\left(\frac{kl}{4}\cos\theta\right). \qquad (9.100)$$

If $l = \lambda$, i.e. $kl = 2\pi$

$$\left\langle \frac{dW}{d\Omega} \right\rangle = \frac{\omega^4 p_0^2}{8\pi^2\epsilon_0 c^3}\sin^2\theta\,\sin^2\left(\frac{\pi}{2}\cos\theta\right). \qquad (9.101)$$

The quadrupole moment $Q = 2p_0 l$

$$\therefore \left\langle \frac{dW}{d\Omega} \right\rangle = \frac{\omega^4 Q^2}{8\pi^2\epsilon_0 c^3\, 4\, l^2}\sin^2\theta\,\sin^2\left(\frac{\pi}{2}\cos\theta\right)$$

$$= \frac{\omega^6 Q^2}{128\,\pi^4\epsilon_0 c^5}\sin^2\theta\,\sin^2\left(\frac{\pi}{2}\cos\theta\right). \qquad (9.102)$$

In the treatment above we have assumed $l = \lambda$. Let us now suppose l to be very small and as $l \to 0$, p_0 increases so that Q remains constant.

The dipole pair now reduces to a **point quadrupole** located at the origin. Since kl is very small we may write (9.99) as

$$E_\theta = 2i \frac{k^2 p_0}{4\pi \epsilon_0 r} \sin\theta \frac{kl}{4} \cos\theta \, e^{i\omega t'} \tag{9.103}$$

Hence,

$$\left\langle \frac{dW}{d\Omega} \right\rangle = \frac{k^6 p_0^{\,2} l^2 c}{128\pi^2 \epsilon_0} \sin^2\theta \cos^2\theta = \frac{\omega^6 p_0^{\,2} l^2}{128\pi^2 \epsilon_0 c^5} \sin^2\theta \cos^2\theta. \tag{9.104}$$

Evidently the maximum power is radiated at $\theta = 45°$

PROBLEMS

9.1 Show how (9.22) is obtained from (9.21).

9.2 Find the expression for the field produced by a charge moving with uniform velocity.

9.3 Obtain expressions for electric and magnetic fields due to a small current loop at points far away from the loop and show that Ampère's equivalence between magnetic dipole and current loops can be used in the calculation of radiation fields.

9.4 Show that a spherical shell uniformly charged and undergoing purely radial oscillations will not radiate.

9.5 A particle of mass m is moving in a circular orbit of radius 'a' under coulomb forces. Compute the intensity of radiation of the particle and express it in terms of the energy of the particle.

9.6 The charges $-e$, $2e$, $-e$ form a linear quadrupole oscillator. The positive charge $2e$ is stationary at the origin; one of the negative charges is at $z_1 = a \cos \frac{1}{2} \omega t$ and the other at $z_2 = -a \cos \omega t$. Find the fields at large distances and also the average rate at which energy is radiated.

Chapter 10

Relativistic Electrodynamics

The world around us has objective reality. It does not change when we alter our view point or our mode of description. One of the aims of physics, therefore, is to formulate laws that are entirely independent of observers or reference systems. The laws must be true for all observers.

Let us examine Newton's first law from this point of view. The law states:

"A body at rest remains at rest, and a body in motion will continue in motion with uniform velocity in a straight line as long as no outside force acts on the body.".

The law presupposes the existence of an observer with some means of measuring position and time. An observer A in a certain frame of reference determines that the motion of a body, not acted upon by any force, is uniform. Another observer B in a different frame, accelerated relative to the observer A, will find that the motion of the body is not uniform and, hence, one may conclude that Newton's first law is not true. Though Newton's law is not true for all reference systems, it is true for some '**privileged**' reference systems. These are known as "**inertial frames**" of reference. Inertial frames are those in which the law of inertia—Newton's first law—holds; or, we may say that a reference system in which a moving body not acted by any external force proceeds with constant velocity is an inertial system. Clearly, a system which is moving uniformly relative to an inertial system is also inertial. All frames moving with constant velocity relative to one another are inertial frames.

10.1 Galilean Transformation

We have in physics what is called the **principle of relativity** according to which all laws of nature are identical in all inertial systems of references. This means, the equations expressing the laws of nature retain their form in different inertial systems; that is, they are invariant with respect to transformation of coordinates and time from one system to another. Experiment has shown that the principle is valid.

Consider two inertial frames S and S' with parallel axes (Fig. 10.1). Suppose S' is moving along x-direction with a uniform velocity v relative to S. Any event occurring at a point P will be specified by an observer in S by the space coordinates (x, y, z) and time t. The same event will be

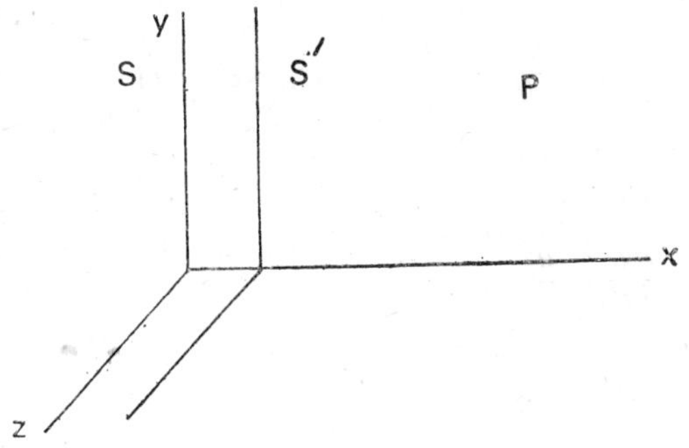

Fig. 10.1

described by an observer in frame S' by (x', y', z', t'). The transformation of coordinate of a point from one system to another is clearly

$$x' = x - vt,\ y' = y,\ z' = z,\ t' = t. \tag{10.1}$$

The last relation is a consequence of the implicit assumption of classical physics that time is absolute, that is independent of a frame of reference. The transformation (10.1) is known as **Galilean transformation.**

Our "common sense" leads us to the following assumptions: The length intervals and time intervals are the same for all inertial observers of the same events. That is, if the two rods have the same length when compared at rest, there is no reason why their length should change when they are in relative motion. Similarly, if two clocks are synchronized when at rest, their readings will agree even if they are put in relative motion. Thus let x_B and x_A be the end points of a rod as measured by an observer in the frame S. Note that the measurements of the two end points are made at the same time. The length of the rod, therefore, is

$$dl = x_B - x_A.$$

The corresponding measurements made by an observer in S' for whom the rod is moving with a velocity v, are x'_B and x'_A and its length is

$$dl' = x'_B - x'_A.$$

By our assumption we expect that $dl' = dl$. Let us see whether this expectation comes out to be true if Galilean transformation is used to transform the coordinates:

$$\begin{aligned} dl' &= x'_B - x'_A \\ &= (x_B - vt) - (x_A - vt) \\ &= (x_B - x_A) = dl. \end{aligned} \tag{10.2}$$

Thus the space-interval measurements are absolute, i.e. they are the same or all inertial observers according to the Galilean transformation.

From (10.1)

$$\frac{dx'}{dt'} = \frac{dx'}{dt} = \frac{dx}{dt} - v \; (\because \; t' = t)$$

$$\frac{dy'}{dt'} = \frac{dy}{dt}$$

$$\frac{dz'}{dt'} = \frac{dz}{dt}$$

and, hence,

$$\frac{d^2x'}{dt'^2} = \frac{d^2x}{dt^2}$$

$$\frac{d^2y'}{dt'^2} = \frac{d^2y}{dt^2}$$

$$\frac{d^2z'}{dt} = \frac{d^2z}{dt^2}.$$

We see that though their velocities vary, their accelerations are identical.

The Newton's equation of motion in the two systems is

$$\mathbf{F}_i = m\,\ddot{\mathbf{x}}_i = m\,\ddot{\mathbf{x}}_i' = \mathbf{F}_i' \text{ (where } i = 1, 2, 3). \tag{10.3}$$

The laws of classical mechanics, thus, are invariant under the Galilean transformation. Let us now examine the behaviour of laws of electromagnetism under Galilean transformation. We know that the following scalar wave-equation results from Maxwell's equation

$$\nabla^2\psi - \frac{1}{c^2}\frac{\partial^2\psi}{\partial t^2} = 0 \tag{10.4}$$

From Galilean transformation, we have

$$\frac{\partial}{\partial x} = \frac{\partial}{\partial x'} + \frac{1}{v}\frac{\partial}{\partial t'}$$

$$\therefore \quad \frac{\partial^2}{\partial x^2} = \frac{\partial^2}{\partial x'^2} + \frac{1}{v^2}\frac{\partial^2}{\partial t'^2} + \frac{2}{v}\frac{\partial^2}{\partial t'\partial x'}$$

and

$$\frac{\partial^2}{\partial y^2} = \frac{\partial^2}{\partial y'^2}, \; \frac{\partial^2}{\partial z^2} = \frac{\partial^2}{\partial z'^2}, \; \frac{\partial^2}{\partial t^2} = \frac{\partial^2}{\partial t'^2}.$$

Substituting in (10.4) we get

$$\nabla'^2\psi - \frac{1}{c^2}\frac{\partial^2\psi}{\partial t'^2} + \frac{1}{v^2}\frac{\partial^2\psi}{\partial t'^2} + \frac{2}{v}\frac{\partial^2\psi}{\partial t'\partial x'} = 0 \tag{10.5}$$

We see that the form of the wave equation is not preserved by the substitution (10.1). The electromagnetic situation, therefore, is different from the mechanical one and electrodynamics is not consistent with the assumption of Galilean invariance. This also can be seen if we remember that the electromagnetic waves, as found from the Maxwell's equations, propagate with a constant velocity c, while the velocity measured by the observer in the moving frame S' will be $c+v$ or $c-v$ depending on the direction of relative motion. The velocity of electromagnetic waves, therefore, is not invariant under the Galilean transformation and, hence, Maxwell's equations will change their form on transformation from one system to another. This situation places before us three possibilities and we have to choose one:

(1) Like Newtonian mechanics, the correct electromagnetic theory must be invariant under Galilean transformation. Since it is not, Maxwell's equations which form the basis of electrodynamic theory are not correct and must be modified.

(2) The Galilean transformation is applicable to mechanics, but there is a preferred unique inertial frame in which Maxwell's equations are valid. In other reference frames they must be suitably modified, and

(3) Maxwell's equations are valid and there is a transformation under which the laws of electrodynamics as well as those of mechanics are invariant, but it is not Galilean transformation. This implies that Newton's laws are not correct.

If the first possibility were correct, we ought to be able to perform experiments to show deviation from Maxwell's electrodynamics. Experiments have shown, on the contrary, that Maxwell's theory is amazingly successful. If the second alternative is accepted, we should be able to locate the "privileged" frame experimentally. The "privileged" frame in question is the classical "ether frame". The space is believed to be filled with a medium called "ether" through which the waves are supposed to propagate with a constant velocity 'c' and in which Maxwell's equations are valid. The privileged observer is the one at rest relative to ether.

The most famous experiment to locate the absolute frame—the ether frame—was the Michelson-Morley experiment. The findings of the experiment however, seem to rule out an ether frame. They could not find any evidence of a unique inertial frame.

Despite negative results of the Michelson-Morley experiments, the scientists were reluctant to give up the concept of the "ether" frame. To square the ether-theory with the results of Michelson-Morley experiment, they assumed that the ether is "dragged" along by moving bodies. This assumption would automatically give a null result for our interferometric experiment. However, Fizeau's experiments (1853) and more recent ones based on Mössbauer effect performed by Champeney, Isaak and Khan (1963) and similar experiments carried out by Jaseja, Jaran, Murray and Townes (1964) using ammonia maser gave null result, proving that any motion relative to some "absolute" reference frame can never be detected.

10.2 Postulates of Special Theory of Relativity

We have seen that the laws of electrodynamics are correct and do not need any modification. We have also seen that an ether hypothesis is untenable. Einstein, therefore, chose the third alternative. One of the important conclusions of Michelson-Morley experiment was that the speed of light is the same in all directions, i.e. it is isotropic and does not depend on the motion of the observer. Einstein assumed the speed of light to be the same in all inertial systems, independent of the relative motion of source and observer. Since Galilean principle requires the speed of light to depend on the relative motion of source and observer, Einstein concluded that Galilean transformation must be replaced and the laws of mechanics which are consistent with this transformation must be modified. Two ideas form the basis of Einstein's special theory of relativity, equivalence of inertial frames and the constancy of the speed of light. The two postulates are:

I. Physical laws are the same for all inertial systems, i.e. all inertial frames are equivalent. In other words, it is impossible by means of any physical measurements to pick up a "privileged" inertial frame. The state of "absolute" motion is impossible to find. We can only speak of relative motion.

II. The speed of light has the same value in all inertial systems.

It is necessary and worthwhile to comment on this second postulate. When a particle interacts with another particle, their interaction is described in mechanics by its potential energy. Interaction, therefore, depends only on the position of the particles. A change in the position of one particle influences the other particle **immediately**. This means the **signal of interaction** propagates instantaneously from one particle to another. In other words the velocity of propagation of the signal is assumed to be infinite. Galilean relativity is based on this assumption. Actually, a change occurring in a body begins to manifest itself in the other, only after the lapse of certain interval. The signal of interaction propagates with certain speed from one particle to another. This speed is finite and is the maximum speed of interaction. Clearly, there cannot be any motion with a speed greater than this, in nature. The principle of relativity demands that this speed—the speed of light—be the same in all inertial frames.

This second postulate was not tested when it was proposed, but has been decisively verified in recent years. For example, the speed of radiation emitted in the decay of π-meson as measured by Farley, Bailey and Picasso was $(2.9979 \pm 0.0003) \times 10^8$m/sec although the mesons were moving at a speed greater than $0.99975\,c$. The postulate, though simple, has forced the physicists to reconsider their ideas about space and time.

10.3 Lorentz Transformation

Our first concern now is to obtain the equation of transformation between two inertial systems which would replace Galilean transformation and which would keep the speed of light constant. We have then to check whether the laws of physics keep the same form under this new transformation. The laws that are not invariant under this transformation will have to be modified. The transformation we are going to obtain is known as **Lorentz transformation**. We have made a reference to this transformation in Sec. (9.5). We shall now show how the equations of transformation are derived directly from Einstein's postulates.

Consider two inertial systems Σ and Σ' (Fig. 10.2).

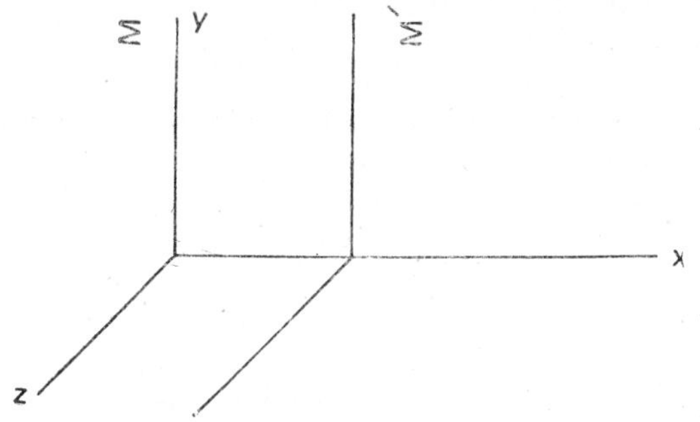

Fig. 10.2

The system Σ' moves relative to Σ at the constant speed v along the x-axis. According to Einstein we have to give up the concept of **absolute time** which is inherent in Galilean transformation. Time has only a relative meaning and must be treated on the same footing as space. At the Σ–time, $t = 0$ the origins of Σ and Σ' coincide. The x'-axis coincides with x–axis. Points which are at rest relative to Σ', will move with speed v relative to Σ.

The most general transformation between the two coordinate systems may be expressed as

$$x' = \alpha(x - vt), \; y' = Cy, \; z' = Dz,$$
$$t' = Ex + Ft \tag{10.6}$$

where the quantities α, C, D, E, and F are yet to be determined.

Let us assume that at time $t = 0$ an electromagnetic spherical wave leaves the point of origin of Σ. Since the speed of the propagation of the wave is the same in all directions and in each frame, the progress of the wave is described in the two frames by the equations

$$x^2 + y^2 + z^2 = c^2t^2 \tag{10.7}$$

and

$$x'^2 + y'^2 + z'^2 = c^2t'^2. \tag{10.8}$$

Therefore

$$x^2 + y^2 + z^2 - c^2t^2 = x'^2 + y'^2 + z'^2 - c^2t'^2 \tag{10.9}$$

If we assume that at $t = 0, x = 0$ and $z = 0$, then from (10.6) and (10.9), we have

$$y^2 = y'^2 \quad \therefore \quad C = 1$$

In the same way we can show that $D = 1$.

Substituting in (10.8) from (10.6), we have

$$\alpha^2(x - vt)^2 + y^2 + z^2 = c^2(Ex + Ft)^2$$

i.e.

$$(\alpha^2 - c^2E^2)x^2 - 2(\alpha^2 v + c^2EF)xt + (\alpha^2 v^2 - c^2F^2)t^2 + y^2 + z^2 = 0.$$

This must be identical with (10.7).

$$\therefore \quad \alpha^2 - c^2E^2 = 1,\ \alpha^2 v^2 - c^2F^2 = -c^2,\ \alpha^2 v + c^2EF = 0. \qquad (10.10)$$

In order to solve these equations we first eliminate α^2 and obtain

$$F(F + vE) = 1$$

$$c^2E(F + vE) = -v$$

Eliminating E we get

$$F^2 = \frac{1}{1 - \dfrac{v^2}{c^2}} \quad \therefore \quad F = \frac{1}{\sqrt{1 - \dfrac{v^2}{c^2}}} = \frac{1}{\sqrt{1 - \beta^2}}$$

where $\beta = \dfrac{v}{c}$.

Hence,
$$\alpha = \frac{1}{\sqrt{1 - \beta^2}},\ E = -\frac{v}{c^2\sqrt{1 - \beta^2}}.$$

Therefore, the equations of transformation (10.6) are

$$x' = \frac{x - vt}{\sqrt{1 - \beta^2}} = \gamma(t - vt)$$

$$y' = y \qquad (10.11)$$

$$z' = z$$

and
$$t' = \frac{1}{\sqrt{1 - \beta^2}}\left(t - \frac{vx}{c^2}\right) = \gamma\left(t - \frac{vx}{c^2}\right)$$

where
$$\gamma = \frac{1}{\sqrt{1 - \beta^2}}.$$

These are the **Lorentz transformation** equations. Note that for small value of v/c, they are approximated by the Galilean transformation equations.

The corresponding inverse transformation equations are

$$x = \frac{x' + vt'}{\sqrt{1 - \beta^2}}$$

$$y = y' \qquad (10.12)$$

$$z = z'$$

$$t = \frac{t' + \dfrac{vx^1}{c^2}}{\sqrt{1-\beta^2}}$$

The transformation equations for the velocity of a moving body can be found by taking the derivatives of (10.11) with respect to t and t'.

10.4 Some Consequences of Lorentz Transformation

We shall now discuss three important consequences of Lorentz transformation. These are particularly important when we consider electric and magnetic fields.

(i) The concept of simultaneity: Two events which are simultaneous to one observer in a reference system may not be simultaneous to an observer in the other system. Consider two systems Σ and Σ' of which the latter is moving with a velocity v relative to the former as shown in Fig. 10.3.

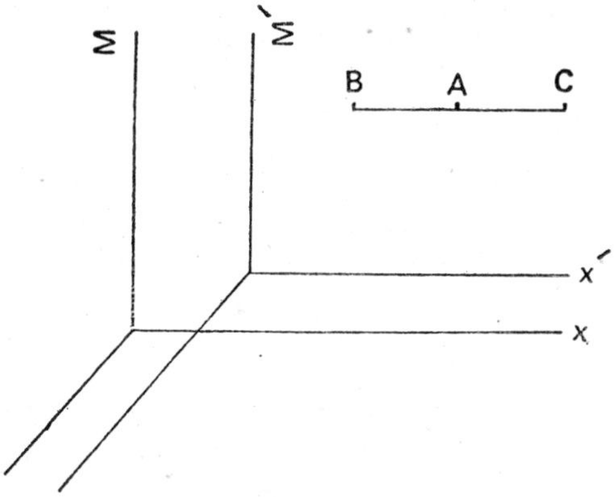

Fig. 10.3

Signals starting from a point A in Σ' in opposite directions parallel to x'-axis will reach points B and C equidistant from A at the same time, that is simultaneously. However, to an observer in Σ, since the point B approaches him while the points C recedes, the two signals will reach him at different times and the events will no longer be simultaneous.

(ii) FitzGerald contraction: Imagine a rigid rod with ends which are fixed in the frame Σ' at the points $(x_1', 0, 0)$ and $(x_2', 0, 0)$. At a particular time t in the frame Σ, the ends are seen to be at x_1 and x_2. From (10.11)

$$x_1' = \frac{x_1 - vt}{\sqrt{1-\beta^2}}$$

$$x_2' = \frac{x_2 - vt}{\sqrt{1-\beta^2}}$$

The length of the moving rod as measured on Σ is the distance between its ends at the same time t, i.e.

$$x_2 - x_1 = (x_2' - x_1')\sqrt{1-\beta^2}$$

or

$$l = l'\sqrt{1-\beta^2}. \tag{10.13}$$

The rod appears contracted by a factor $\sqrt{1-\beta^2}$. This is called **Fitz-Gerald contraction.** Thus, every body appears to be longest when at rest relatively to the observer. When it is moving, it appears contracted in the direction of motion by a factor $\sqrt{1-\beta^2}$. Since $y' = y$, $z' = z$ the apparent dimensions of the body are unaltered in the direction perpendicular to its motion.

(iii) Time dilatation: Suppose that a clock at a fixed point x' in the frame Σ' measures a time interval $(t_2'-t_1')$ between two events that occurred at x' at time t_1' and t_2'. To an observer in the frame Σ, the events appear to occur at time t_1 and t_2. By (10.12)

$$t_1 = \frac{t_1' + \dfrac{vx'}{c^2}}{\sqrt{1-\beta^2}}$$

$$t_2 = \frac{t_2' + \dfrac{vx'}{c^2}}{\sqrt{1-\beta^2}}$$

$$t_2 - t_1 = \frac{t_2' - t_1'}{\sqrt{1-\beta^2}} \tag{10.14}$$

The observer in Σ considers the time interval to be $t_2 - t_1$, which is larger than that measured by the clock in Σ'. This shows that the **moving clocks appear to run slow.**

10.5 Charges and Fields as Observed in Different Frames

(i) Charge: We know that electric charge is absolutely conserved. The protons and electrons in an atom are moving about, yet the atom preserves its electrical neutrality as the positive and negative charges exactly cancel. This means that the charge on each elementary particle is $\pm e$ whether or not the particle is moving relative to the observer. The charge on an elementary particle and, hence, the net charge carried by any macroscopic body is Lorentz invariant.

(ii) Charge density: The charge is Lorentz invariant but is charge density Lorentz invariant? To answer this question consider a series of positive charges equally spaced along a line, each with charge $+e$. Let the charges span a distance of unit length in the frame Σ in which the charges are at rest. The linear charge density as seen by an observer in Σ is Ne coulomb/m, if N is the number of charges in unit length. An observer in Σ' moving with a speed v parallel to the charges does not agree that the charges cover a unit length. Because of the FitzGerald contraction the charges are squashed up into a distance $\sqrt{1-\beta^2}$ and, hence, according to him the linear charge density is $\frac{N_e}{\sqrt{1-\beta^2}}$ coulomb/m.

(iii) Electric and magnetic forces: Let us now investigate the forces exerted by electric and magnetic fields, as viewed by observers in Σ and Σ'. Consider a long straight wire in the frame Σ in which free electrons move with a drift velocity v in the x direction Let n be the number of electrons per unit length. This is also the number of positive ions per unit length. Suppose a charge q is moving parallel to the wire at a speed v. This charge will experience a magnetic force $\mathbf{F} = q\,(\mathbf{v} \times \mathbf{B})$ in the Σ frame where $\mathbf{B}$ is the magnetic field produced by the current in the wire.

Now consider an observer in the frame Σ', moving with a velocity v in the x-direction. For this observer the charge q is at rest and, hence, there cannot be any magnetic force i.e. $\mathbf{F} = 0$. But all inertial frames are equivalent. The observer in Σ', therefore, must agree with the observer in Σ that there is a force on q though it is not a magnetic force. How does he interpret this force? For observer in Σ, the negative linear charge density in the wire, due to electrons is $\lambda^- = -ne$ and the positive charge density due to positive ions is $\lambda^+ = ne$. The net density

$$\lambda = \lambda^- + \lambda^+ = 0. \tag{10.15}$$

The current density in the wire is given by

$$\mathbf{J}_x^- = -ne\mathbf{v} = \lambda^-\mathbf{v} \text{ and } \mathbf{j}_x^+ = 0$$

$$\therefore \qquad \mathbf{j}_x = \mathbf{j}_x^+ + \mathbf{j}_x^- = \lambda^-\mathbf{v}. \tag{10.16}$$

The current produces a magnetic field $\mathbf{B}$ given by

$$\mathbf{B} = -\frac{\mu_0 ne\mathbf{v}}{2\pi r}$$

and this causes a force $\mathbf{F}$ to act on the charge

$$|\,\mathbf{F}\,| = |\,q(\mathbf{v} \times \mathbf{B})\,| = -\frac{\mu_0 nev^2 q}{2\pi r}. \tag{10.17}$$

Let us see how the force is evaluated by an observer in the frame Σ'. The positive charges which were at rest to the observer in Σ, are moving at a speed v in the opposite direction to the observer in Σ'. Taking the FitzGerald contraction into consideration, the unit length in Σ is equivalent to the length $\sqrt{1-\beta^2}$ in Σ' and the linear charge density due to the positive ions $\lambda^+ = \frac{ne}{\sqrt{1-\beta^2}}$. On the other hand electrons are at rest in this frame and they span a distance $\frac{1}{\sqrt{1-\beta^2}}$ in Σ'. (Note that the distance is of unit length in Σ.) Hence, the linear charge density is

$\lambda'^- = -ne\sqrt{1-\beta^2}$. The net charge density in Σ' therefore, is

$$\lambda' = \lambda'^- + \lambda'^+ = -ne\sqrt{1-\beta^2} + \frac{ne}{\sqrt{1-\beta^2}} = \frac{ne}{\sqrt{1-\beta^2}}\beta^2 \quad (10.18)$$

which is positive and not zero as found by the observer in Σ. The observer in Σ', therefore, does not think that the wire is electrically neutral. According to him there is an electrostatic field around the wire given by

$$E' = \frac{\lambda'}{2\pi\epsilon_0 r} = \frac{ne\beta^2}{2\pi\epsilon_0 r\sqrt{1-\beta^2}}.$$

Hence, the charge 'q' experiences a force given by

$$|\mathbf{F}'| = |q\mathbf{E}'| = \frac{qne\beta^2}{2\pi\epsilon_0 r\sqrt{1-\beta^2}}$$

$$= \frac{\mu_0 nev^2 q}{2\pi r\sqrt{1-\beta^2}} = \frac{|\mathbf{F}|}{\sqrt{1-\beta^2}} \quad \left(\because \; c^2 = \frac{1}{\epsilon_0\mu_0}\right). \quad (10.19)$$

Note that the force that was purely magnetic in Σ appears to be purely electrostatic in Σ'. There is, thus, disagreement between the observers regarding the origin of the forces. They also disagree regarding the magnitude of the force.

Although we have assumed that the set of charges constituting the current and the test charge q are moving with the same speed, the arguments are general and applicable to all currents.

One may ask: The arguments above show that a net positive charge density results in Σ' when the net charge density in Σ is zero. Then what about charge invariance? Does it mean that the total charge in Σ' is different than that in Σ? It is not. A complete circuit is electrically neutral. One segment of the wire may be positively charged and the other negatively charged because the direction of motion of electrons is different in different parts of the current.

10.6 The Lorentz Transformation as an Orthogonal Transformation

Suppose a frame of reference is rotated through a certain angle θ about z-axis. The coordinates of a point P in the rotated frame will not be the same as those in the original frame. The coordinates in the rotated frame are given by (Fig. 10.4).

$$x' = x\cos\theta + y\sin\theta$$

$$y' = -x\sin\theta + y\cos\theta \quad (10.20)$$

$$z' = z.$$

These are the linear combinations of the original coordinates. If th

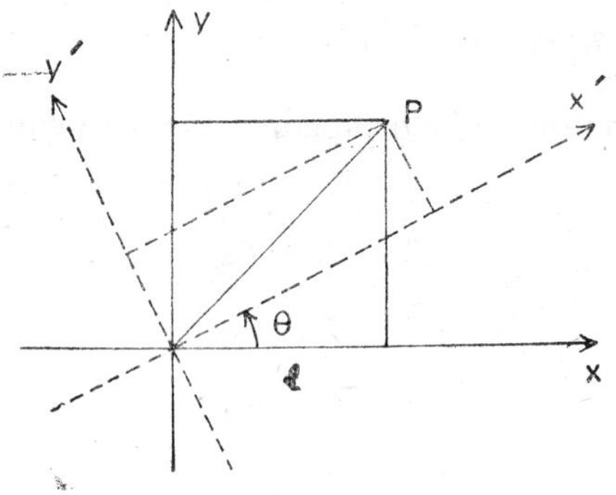

Fig. 10.4

point P is specified by a position vector r, its length in the original frame is given by

$$|\mathbf{r}|^2 = x^2 + y^2 + z^2.$$

It is easy to verify from (10.20) that

$$|\mathbf{r}|^2 = x^2 + y^2 + z^2 = x'^2 + y'^2 + z'^2. \tag{10.21}$$

The transformation (10.20) leaves the length of the vector unchanged. Such transformations are called **orthogonal transformations.**

A coordinate transformation is said to be *linear* if the new coordinates can be expressed as a linear combination of the old coordinates. Thus,

$$x_1' = a_{11}x_1 + a_{12}x_2 + a_{13}x_3$$
$$x_2' = a_{21}x_1 + a_{22}x_2 + a_{23}x_3$$
$$x_3' = a_{31}x_1 + a_{32}x_2 + a_{33}x_3$$

i.e.

$$x_i' = \sum_j a_{ij}\, x_j \tag{10.22}$$

is a linear transformation. This can be written in the matrix form as

$$\begin{bmatrix} x_1' \\ x_2' \\ x_3' \end{bmatrix} = \begin{bmatrix} a_{11} & a_{12} & a_{13} \\ a_{21} & a_{22} & a_{23} \\ a_{31} & a_{32} & a_{33} \end{bmatrix} \begin{bmatrix} x_1 \\ x_2 \\ x_3 \end{bmatrix} \tag{10.23}$$

If this transformation is orthogonal, then

$$\sum_i (x_i')^2 = \sum_k x_k^2$$

i.e.

$$\sum_j \sum_k \sum_i a_{ij}\, a_{ik}\, x_j\, x_k = \sum_k x_k^2. \tag{10.24}$$

This would be possible if

$$\sum_i a_{ij}\, a_{ik} = \begin{cases} 0 & \text{if } j \neq k \\ 1 & \text{if } j = k \end{cases}$$

i.e. if

$$\sum_i a_{ij}\, a_{ik} = \delta_{ik}. \tag{10.25}$$

For the transformation to be orthogonal the elements of the transformation matrix

$$\mathbf{A} = \begin{bmatrix} a_{11} & a_{12} & a_{13} \\ a_{21} & a_{22} & a_{23} \\ a_{31} & a_{32} & a_{33} \end{bmatrix}$$

have to satisfy this condition.

We have seen that the quadratic function $x^2 + y^2 + z^2 - c^2t^2$ has the same value in different inertial frames connected by the Lorentz transformation. The Lorentz transformation, like the rotational transformation, is a linear relation between different coordinate systems, but it mixes up time with the space coordinates. An event occurring at a particular time and place in a frame Σ is specified by four coordinates x, y, z and t.

By analogy with 3-dimensional vectors, we can interpret $s^2 = x^2 + y^2 + z^2 - c^2 t^2$ as the square of the "length" of the "four vector" s. The four "coordinates" are $x_\mu = x_1, x_2, x_3, x_4$ where

$$x_1 = x,\ x_2 = y,\ x_3 = z \text{ and } x_4 = ict. \tag{10.26}$$

This non-Euclidean four dimensional space was introduced by H. Minkowski and is also known as Minkowski space. The square of the length is given by

$$\begin{aligned} s^2 = x_\mu\, x_\mu &= x_1^2 + x_2^2 + x_3^2 + x_4^2 \\ &= x^2 + y^2 + z^2 - c^2t^2. \end{aligned} \tag{10.27}$$

In writing this equation, we have used the summation convention: Wherever a Greek suffix appears twice in a single term, the term is assumed to be summed from 1 to 4.

In the new notation, the Lorentz transformation equations are:

$$x_1' = \frac{x_1 + i\beta x_4}{\sqrt{1-\beta^2}};\ x_2' = x_2;\ x_3' = x_3; \tag{10.28}$$

$$x_4' = \frac{x_4 - i\beta x_1}{\sqrt{1-\beta^2}}.$$

The inverse transformation equations are

$$x_1 = \frac{x_1' - i\beta x_4'}{\sqrt{1-\beta^2}};\ x_2 = x_2';\ x_3 = x_3'; \tag{10.29}$$

$$x_4 = \frac{x_4' + i\beta x_1'}{\sqrt{1-\beta^2}}.$$

The matrix of Lorentz transformation is

$$
\mathbf{A} = \begin{bmatrix} \dfrac{1}{\sqrt{1-\beta^2}} & 0 & 0 & \dfrac{i\beta}{\sqrt{1-\beta^2}} \\ 0 & 1 & 0 & 0 \\ 0 & 0 & 1 & 0 \\ \dfrac{-i\beta}{\sqrt{1-\beta^2}} & 0 & 0 & \dfrac{1}{\sqrt{1-\beta}} \end{bmatrix}
$$

and we can easily verify that it satisfies the condition (10.25). Hence, Lorentz transformation is an orthogonal transformation.

10.7 Covariant Formulation of Electrodynamics

Covariance of equations is a mathematical property which corresponds to the existence of a relativity principle for physical laws expressed by these equations. Equations which do not change with the transformation are called **invariant**. The terms of such equations may not be invariant, but they transform according to identical transformation laws and are said to be **covariant**.

We have already seen in Sec. (9.5) that Maxwell's equations are covariant with respect to the Lorentz transformation. This signifies that the Maxwell's equations retain their form in any inertial frame of reference and, hence, the principle of relativity holds automatically. One may hasten to conclude, therefore, that the theory of relativity cannot introduce anything of importance in electrodynamics. This is far from being correct. A close relationship between certain physical quantities could be discovered only after the emergence of a relativistic approach to electromagnetic phenomena.

In order to investigate the relativistic behaviour of Maxwell's equations, it is necessary to write the equations in four dimensional form and for which we must know the transformation properties of differential operators. Applying the rules of partial differentiation, we get

$$
\frac{\partial}{\partial x_1'} = \frac{\partial x_1}{\partial x_1'}\frac{\partial}{\partial x_1} + \frac{\partial x_2}{\partial x_1'}\frac{\partial}{\partial x_2} + \frac{\partial x_3}{\partial x_1'}\frac{\partial}{\partial x_3} + \frac{\partial x_4}{\partial x_1'}\frac{\partial}{\partial x_4}
$$

$$
= \frac{1}{\sqrt{1-\beta^2}}\left(\frac{\partial}{\partial x_1} + i\beta\frac{\partial}{\partial x_4}\right) \tag{10.31}
$$

where we have used (10.29). Comparing this with (10.28), we see that the transformation law is the same as the one applicable to the coordinate x_1. The corresponding equations for other components $\dfrac{\partial}{\partial x_2}, \dfrac{\partial}{\partial x_3}, \dfrac{\partial}{\partial x_4}$ show that they all transform according to the Lorentz transformation.

We conclude that $\dfrac{\partial}{\partial x_\mu}$ is a four-vector. The scalar product of this vector is

$$\frac{\partial}{\partial x_\mu}\cdot\frac{\partial}{\partial x_\mu}=\frac{\partial^2}{\partial x_1^2}+\frac{\partial^2}{\partial x_2^2}+\frac{\partial^2}{\partial x_3^2}+\frac{\partial^2}{\partial x_4^2}$$

$$=\frac{\partial^2}{\partial x^2}+\frac{\partial^2}{\partial y^2}+\frac{\partial^2}{\partial z^2}-\frac{1}{c^2}\frac{\partial^2}{\partial t^2}=\Box^2. \tag{10.32}$$

We may remind you that this is the operator appearing in the wave-equation for waves moving at a speed c. It should also be clear that it is an invariant. This four-dimensional version of the operator is known as **d'Alembertian operator** and is denoted by the symbol $\Box^2$. The wave-equation can be expressed as

$$\nabla^2\psi-\frac{1}{c^2}\frac{\partial^2\psi}{\partial t^2}=\Box^2\psi=0. \tag{10.33}$$

The mathematical statement of charge conservation is contained in the equation of continuity

$$\operatorname{div}\mathbf{j}+\frac{\partial\rho}{\partial t}=0 \tag{10.34}$$

Let us see how this equation can be written in an invariant form so that it has the same form in all inertial frames.

In relativity theory, the current density and charge density are not distinct entities since the charge distribution that is static in one reference frame will appear to be a current distribution in a moving reference frame. We, therefore, introduce four-vector $\mathcal{J}_\mu=(\mathbf{j}, ic\rho)$ made up of the current density $\mathbf{j}$ and the charge density ρ, its components being $\mathcal{J}_1=j_1$, $\mathcal{J}_2=j_2$, $\mathcal{J}_3=j_3$, and $\mathcal{J}_4=ic\rho$. The first three of these are space like part of $\mathcal{J}_\mu$. We, thus, have

$$\frac{\partial\mathcal{J}_\mu}{\partial x_\mu}=\frac{\partial\mathcal{J}_1}{\partial x_1}+\frac{\partial\mathcal{J}_2}{\partial x_2}+\frac{\partial\mathcal{J}_3}{\partial x_3}+\frac{\partial(ic\rho)}{\partial(ict)}$$

$$=\operatorname{div}\mathbf{j}+\frac{\partial\rho}{\partial t}=0. \tag{10.35}$$

Representing the four-dimensional divergence operator by 'Div', we may express the equation of continuity in the four-dimensional form as

$$\operatorname{Div}\mathcal{J}=0. \tag{10.36}$$

We define the four vector potential $\mathcal{A}$ as

$$\mathcal{A}\equiv(\mathcal{A}_1,\mathcal{A}_2,\mathcal{A}_3,\mathcal{A}_4)\equiv\left(A_1,A_2,A_3,\frac{i\Phi}{c}\right)=\left(\mathbf{A},\frac{i\Phi}{c}\right). \tag{10.37}$$

The Lorentz condition is

$$\operatorname{div}\mathbf{A}+\frac{1}{c^2}\frac{\partial\Phi}{\partial t}=0$$

Now

$$\text{Div } \mathcal{A} = \frac{\partial \mathcal{A}_1}{\partial x_1} + \frac{\partial \mathcal{A}_2}{\partial x_2} + \frac{\partial \mathcal{A}_3}{\partial x_3} + \frac{\partial \mathcal{A}_4}{\partial x_4}$$

$$= \text{div } \mathbf{A} + \frac{\partial\left(\frac{i\Phi}{c}\right)}{\partial(ict)}$$

$$= \text{div } \mathbf{A} + \frac{1}{c^2}\frac{\partial \Phi}{\partial t} = 0. \tag{10.38}$$

Notice that the Lorentz condition is also invariant and can be written in four-dimensional form as

$$\text{Div } \mathcal{A} = 0, \text{ i.e. } \frac{\partial \mathcal{A}_\mu}{\partial x_\mu} = 0. \tag{10.39}$$

We know that in free space the potentials **A** and Φ satisfy the equations

$$\nabla^2 \mathbf{A} - \frac{1}{c^2}\frac{\partial^2 \mathbf{A}}{\partial t^2} = -\mu_0 \mathbf{j} \tag{10.40}$$

and

$$\nabla^2 \Phi - \frac{1}{c^2}\frac{\partial^2 \Phi}{\partial t^2} = -\frac{\rho}{\epsilon_0} \tag{10.41}$$

Using $\mathcal{J}$ and $\mathcal{A}$ as defined above, these two equations can be written together in the form of a single four-vector equation

$$\Box^2 \mathcal{A} = -\mu_0 \mathcal{J} \tag{10.42}$$

The space portion of this equation gives (10.40) and the fourth component gives (10.41).

The arguments above indicate that Maxwell's equations can be written in Lorentz invariant form without any modification and also that they satisfy the postulate about the consistency of the velocity of light.

10.8 The Electromagnetic Field Tensor

Let us now test whether the Lorentz force $\mathbf{F} = q(\mathbf{E} + \mathbf{v} \times \mathbf{B})$ is Lorentx invariant. For this, we have to investigate the transformation of the fields **E** and **B**. The vectors **E** and **B** are not four-vectors; but we can work out the transformation of the fields by expressing them in terms of potentials. We shall see how the six components $E_1, E_2, E_3, B_1, B_2, B_3$ can be used to define a tensor of rank two called the **electromagnetic field tensor.** In n-dimensional space, a tensor of the mth rank is a set of n^m quantities which transform as

$$T'_{abcd\cdots} = \sum_{i,j,k,l\ldots} \lambda_{ai}\lambda_{bj}\lambda_{ck}\lambda_{dl}\ldots T_{ijkl\cdots}$$

We are concerned with a four-dimensional space and need to consider only tensors of rank two which transform as

$$T'_{ij} = \sum_{k,l} \lambda_{ik} \lambda_{jl} T_{kl} \tag{10.43}$$

We know that

$$E_x = -\frac{\partial \Phi}{\partial x} - \frac{\partial A_x}{\partial t} \text{ etc.}$$

Since $\mathcal{A}_\mu = \left(A_1, A_2, A_3, A_4 = \frac{i\Phi}{c} \right)$ and $x_4 = ict$,

the above equation may be written as

$$E_1 = ci \left(\frac{\partial A_4}{\partial x_1} - \frac{\partial A_1}{\partial x_4} \right)$$

Similarly, $$E_2 = ci \left(\frac{\partial A_4}{\partial x_2} - \frac{\partial A_2}{\partial x_4} \right); E_3 = ci \left(\frac{\partial A_4}{\partial x_3} - \frac{\partial A_3}{\partial x_4} \right) \tag{10.44}$$

$$B_1 = \frac{\partial A_3}{\partial x_2} - \frac{\partial A_2}{\partial x_3}; B_2 = \frac{\partial A_1}{\partial x_3} - \frac{\partial A_3}{\partial x_1}; B_3 = \frac{\partial A_2}{\partial x_1} - \frac{\partial A_1}{\partial x_2}.$$

We define a set of quantities

$$F_{\mu\nu} = \frac{\partial A_\nu}{\partial x_\mu} - \frac{\partial A_\mu}{\partial x_\nu} \tag{10.45}$$

so that

$$\begin{aligned} E_1 &= ci\, F_{14};\ E_2 = ci\, F_{24};\ E_3 = ci\, F_{34} \\ B_1 &= F_{23};\ B_2 = F_{31};\ B_3 = F_{12}. \end{aligned} \tag{10.46}$$

We can now write the tensor $\{F\}$ in terms of its elements $F_{\mu\nu}$ as

$$\{F\} = \begin{Bmatrix} F_{11} & F_{12} & F_{13} & F_{14} \\ F_{21} & F_{22} & F_{23} & F_{24} \\ F_{31} & F_{32} & F_{33} & F_{34} \\ F_{41} & F_{42} & F_{43} & F_{44} \end{Bmatrix} = \begin{Bmatrix} 0 & B_3 & -B_2 & -\frac{i}{c}E_1 \\ -B_3 & 0 & B_1 & -\frac{i}{c}E_2 \\ B_2 & -B_1 & 0 & -\frac{i}{c}E_3 \\ \frac{i}{c}E_1 & \frac{i}{c}E_2 & \frac{i}{c}E_3 & 0 \end{Bmatrix} \tag{10.47}$$

The tensor is clearly anti-symmetric, i.e.

$$F_{\mu\nu} = -F_{\nu\mu}$$

Let us see how this tensor can be used to express Maxwell's equations. Consider the equation

$$\frac{\partial F_{\lambda\mu}}{\partial x_\nu} + \frac{\partial F_{\mu\nu}}{\partial x_\lambda} + \frac{\partial F_{\nu\lambda}}{\partial x_\mu} = 0 \tag{10.48}$$

If we assign to λ, μ, ν the values 1, 2, 3 in this order or in any other order, the equation (10.48) reduces to

$$\frac{\partial F_{12}}{\partial x_3}+\frac{\partial F_{23}}{\partial x_1}+\frac{\partial F_{31}}{\partial x_2}=0$$

i.e.
$$\frac{\partial B_3}{\partial x_3}+\frac{\partial B_1}{\partial x_1}+\frac{\partial B_2}{\partial x_2}=0$$

or
$$\nabla\cdot\mathbf{B}=0$$

which is one of the Maxwell's equations. If we assign to one of the indices the value 4, say $\lambda = 2$, $\mu = 3$, $\nu = 4$, we have from equation (10.48)

$$\frac{\partial F_{23}}{\partial x_4}+\frac{\partial F_{34}}{\partial x_2}+\frac{\partial F_{42}}{\partial x_3}=0$$

i.e.
$$\frac{\partial B_1}{\partial(ict)}+\frac{1}{ic}\frac{\partial E}{\partial x_2}-\frac{1}{ic}\frac{\partial E_2}{\partial x_3}=0$$

or
$$\frac{\partial B_1}{\partial t}+\frac{\partial E_3}{\partial x_2}-\frac{\partial E_2}{\partial x_3}=0$$

or
$$(\text{curl }\mathbf{E})_1+\frac{\partial B_1}{\partial t}=0$$

which is the x-component of $\nabla\times\mathbf{E}+\dfrac{\partial\mathbf{B}}{\partial t}=0$.

Thus, we see that the equation (10.48) also represents the other homogeneous Maxwell's equation.

The other two inhomogeneous Maxwell's equations may readily be obtained from the equation

$$\frac{\partial F_{\mu\nu}}{\partial x_\nu}=\mu_0 j_\mu. \tag{10.49}$$

For example, if $\mu = 1$, equation (10.49) takes the form

$$\frac{\partial F_{11}}{\partial x_1}+\frac{\partial F_{12}}{\partial x_2}+\frac{\partial F_{13}}{\partial x_3}+\frac{\partial F_{14}}{\partial x_4}=\mu_0 j_1$$

i.e.
$$0+\frac{\partial B_3}{\partial x_2}-\frac{\partial B_2}{\partial x_3}+\frac{1}{ic}\frac{\partial E_1}{\partial(ict)}=\mu_0 j_1$$

or
$$(\text{curl }\mathbf{B})_1-\frac{1}{c^2}\frac{\partial E_1}{\partial t}=\mu_0 j_1,$$

and, in general,

$$\text{curl }\mathbf{B}-\frac{1}{c^2}\frac{\partial\mathbf{E}}{\partial t}=\mu_0\,\mathbf{j}.$$

For $\mu = 4$, we have from (10·49)

$$\frac{\partial F_{41}}{\partial x_1}+\frac{\partial F_{42}}{\partial x_2}+\frac{\partial F_{43}}{\partial x_3}+\frac{\partial F_{44}}{\partial x_4}=\mu_0 j_4$$

i.e. $$-\frac{1}{ic}\frac{\partial E_1}{\partial x_1}-\frac{1}{ic}\frac{\partial E_2}{\partial x_2}-\frac{1}{ic}\frac{\partial E_3}{\partial x_3}+0=\mu ic\rho$$

or $$\frac{\partial E_1}{\partial x_1}+\frac{\partial E_2}{\partial x_2}+\frac{\partial E_3}{\partial x_3}=\text{div }\mathbf{E}=\frac{\rho}{\epsilon_0}.$$

Thus the four Maxwell's equations are represented by only two equations involving operations on the components of the field tensor.

10.9 Transformation of the Fields

We shall now work out the transformation of the fields. Let us evaluate E_x' in the Σ' frame in terms of $\mathbf{E}$ and $\mathbf{B}$ in the Σ frame.

$$E_x'=-\frac{\partial \Phi'}{\partial x'}-\frac{\partial A_x'}{\partial t'}$$

Because $\mathcal{A}_\mu\equiv\left(A_1, A_2, A_3, \frac{i\Phi}{c}\right)$

and $\frac{\partial}{\partial x_\mu}\equiv\left(\frac{\partial}{\partial x_1}, \frac{\partial}{\partial x_2}, \frac{\partial}{\partial x_3}, \frac{\partial}{\partial (ict)}\right)$

$$E_x'=ci\,F_{14}'=ci\left\{\frac{\partial A_4'}{\partial x_1'}-\frac{\partial A_1'}{\partial x_4'}\right\}$$

Since A_μ, $\frac{\partial}{\partial x_\mu}$ are Lorentz covariant

$$A_4'=\frac{1}{\sqrt{1-\beta^2}}(A_4-i\beta A_1)=\gamma(A_4-i\beta A_1)$$

$$\frac{\partial}{\partial x_4'}=\gamma\left(\frac{\partial}{\partial x_4}-i\beta\frac{\partial}{\partial x_1}\right)$$

$$A_1'=\gamma(A_1+i\beta A_4);\quad \frac{\partial}{\partial x_1'}=\gamma\left(\frac{\partial}{\partial x_1}+i\beta\frac{\partial}{\partial x_4}\right).$$

$$\therefore\quad E_x'=ci\left[\gamma\left(\frac{\partial}{\partial x_1}+i\beta\frac{\partial}{\partial x_4}\right)\{\gamma(A_4-i\beta A_1)\}\right.$$
$$\left.-\gamma\left(\frac{\partial}{\partial x_4}-i\beta\frac{\partial}{\partial x_1}\right)\{\gamma(A_1+i\beta A_4)\}\right]$$
$$=ci\gamma^2\left[\frac{\partial A_4}{\partial x_1}+i\beta\frac{\partial A_4}{\partial x_4}-i\beta\frac{\partial A_1}{\partial x_1}+\beta^2\frac{\partial A_1}{\partial x_4}\right.$$
$$\left.-\frac{\partial A_1}{\partial x_4}+i\beta\frac{\partial A_1}{\partial x_1}-i\beta\frac{\partial A_4}{\partial x_4}-\beta^2\frac{\partial A_4}{\partial x_1}\right]$$
$$=ci\gamma^2(1-\beta^2)\left(\frac{\partial A_4}{\partial x_1}-\frac{\partial A_1}{\partial x_4}\right)=ci\left(\frac{\partial A_4}{\partial x_1}-\frac{\partial A_1}{\partial x_4}\right)=E_x.$$

The other components can be calculated in the same way. We, thus, have

$$E_x'=E_x;\qquad B_x'=B_x;$$

$$E'_y = \gamma (E_y - c\beta B_z); \quad B'_y = \gamma \left(B_y + \frac{\beta}{c} E_z\right) \tag{10.50}$$

$$E'_z = \gamma (E_z + c\beta B_y); \quad B'_z = \gamma \left(B_z - \frac{\beta}{c} E_y\right).$$

Thus, the transverse components of **E** and **B** are modified, but the components in the direction of motion are unaffected.

We see also that electric and magnetic fields are relative. The field **E** or **B** may be present in one frame and be zero in the other frame.

10.10 Field due to a Point Charge in Uniform Motion

We have already shown in Sec. 9.6, how the expression for the electric field of a moving charge can be obtained using Liènard-Wiechert potentials. Let us now see what kind of expression we arrive at from relativity theory.

Suppose the charge is stationary in the system Σ. In this system the magnetic field $\mathbf{B} = 0$ and the electric field is given by

$$\mathbf{E} = \frac{q}{4\pi\epsilon_0} \frac{\mathbf{r}}{|\mathbf{r}|^3}. \tag{10.51}$$

What will be the field as it appears to an observer in the frame Σ'? For convenience we assume the charge to be at the origin of Σ and the field is calculated at the instant $t = 0$, that is, when the two origins coincide. From (10.50)

$$E'_1 = E_1; \; E'_2 = \gamma E_2; \; E'_3 = \gamma E_3$$

$$(\because \quad B_2 = B_3 = 0 \text{ in the static frame } \Sigma) \tag{10.52}$$

The coordinates at $t = 0$, are related by the Lorentz transformation

$$x_1 = \gamma x'_1; \; x_2 = x'_2, \; x_3 = x'_3.$$

Hence, the distance r from the origin to the observation point P is

$$r = \sqrt{x_\mu x_\mu} = \sqrt{\gamma^2 x'^2_1 + x'^2_2 + x'^2_3} \tag{10.53}$$

The components of the electric field in Σ', therefore, are

$$\mathbf{E}'_1 = E_1 = \frac{q}{4\pi\epsilon_0} \frac{x_1}{r^3} = \frac{q}{4\pi\epsilon_0} \frac{\gamma x'_1}{(\gamma^2 x'^2_1 + x'^2_2 + x'^2_3)^{3/2}}$$

$$E'_2 = \gamma E_2 = \frac{q}{4\pi\epsilon_0} \frac{\gamma x_2}{r^3} = \frac{q}{4\pi\epsilon_0} \frac{\gamma x'_2}{(\gamma^2 x'^2_1 + x'^2_2 + x'^2_3)^{3/2}}$$

$$E'_3 = \gamma E_3 = \frac{q}{4\pi\epsilon_0} \frac{\gamma x_3}{r^3} = \frac{q}{4\pi\epsilon_0} \frac{\gamma x'_3}{(\gamma^2 x'^2_1 + x'^2_2 + x'^2_3)^{3/2}}$$

In general,

$$\mathbf{E}' = \frac{q\gamma}{4\pi\epsilon_0} \frac{\mathbf{r}'}{(\gamma^2 x'^2_1 + x'^2_2 + x'^2_3)^{3/2}}. \tag{10.54}$$

Now $\qquad x'_1 = r' \cos\theta$

where θ is the angle that r' makes with the x–axis

and $$x'^2_1 + x'^2_2 + x'^2_3 = r'^2$$

$\therefore$ $$x'^2_2 + x'^2_3 = r'^2 \sin^2\theta$$

Hence $$\gamma^2 x'^2_1 + x'^2_2 + x'^2_3 = \gamma^2 r'^2 \cos^2\theta + r'^2 \sin^2\theta$$

$$= r'^2\gamma^2\left(\cos^2\theta + \frac{\sin^2\theta}{\gamma^2}\right)$$

$$= r'^2\gamma^2 (1 - \beta^2 \sin^2\theta) \tag{10.55}$$

$$\mathbf{E}' = \frac{q}{4\pi\epsilon_0} \frac{\mathbf{r}'(1-\beta^2)}{r'^3 (1-\beta^2\sin^2\theta)^{3/2}}. \tag{10.56}$$

This is equivalent to the expression for **E** obtained in Sec. 9.6. This is yet another illustration of the fact that Maxwell's equations are relativistically correct.

We see from equation (10.56) that the electric field is no longer spherically symmetric. For $\theta = 0$, i.e. along the line of motion the field is

$$\mathbf{E} = \frac{q}{4\pi\epsilon_0}\frac{\mathbf{r}}{r^3}(1-\beta^2) \tag{10.57}$$

which means the Coulomb field is reduced by a factor $(1-\beta^2)$. For $\theta = \frac{\pi}{2}$, i.e. perpendicular to the line of motion

$$\mathbf{E} = \frac{q}{4\pi\epsilon_0}\frac{\mathbf{r}}{r^3}\frac{1}{\sqrt{1-\beta^2}} \tag{10.58}$$

that is, the field is enhanced by a factor $\dfrac{1}{\sqrt{1-\beta^2}}$

10.11 Lagrangian Formulation of the Motion of a Charged Particle in an Electromagnetic Field

The Lagrangian method of classical mechanics can be applied to the motion of a charged particle in an electromagnetic field if a suitable Lagrangian function is devised. We shall first illustrate the method for a non-relativistic case and then give a more general relativistic approach.

In a static electric field, Lagrangian can be expressed in a conventional manner as the difference between the kinetic and potential energies.

$$L = T - U = \frac{1}{2}mv^2 - q\Phi$$

If a magnetic field is present, the Lagrangian must be modified suitably. Since the magnetic field depends upon the velocity of the moving charges and because L is a scalar function, the term to be added for to modify L is a scalar product of **v** and the vector potential **A** which describes the field. We assume the Lagrangian to be

$$L = \frac{1}{2} mv^2 + q\mathbf{v}\cdot\mathbf{A} - q\Phi \tag{10.59}$$

Let us obtain the equation of motion. We have from (10.59)

$$\frac{\partial L}{\partial v_i} = mv_i + qA_i \; (i = 1, 2, 3,) \tag{10.60}$$

or vectorially

$$\sum_i \hat{\mathbf{e}}_i \frac{\partial L}{\partial v_i} = \mathbf{p} + q\mathbf{A} \tag{10.61}$$

where $\mathbf{p} = m\mathbf{v}$ is the linear momentum and $\mathbf{p} + q\mathbf{A}$ is called the *generalized momentum*. We have also

$$\frac{\partial L}{\partial x_i} = q \frac{\partial}{\partial x_i} (\mathbf{v}\cdot\mathbf{A}) - q \frac{\partial \Phi}{\partial x_i} \tag{10.62}$$

or vectorially

$$\sum_i \hat{\mathbf{e}}_i \frac{\partial L}{\partial x_i} = q \sum_i \hat{\mathbf{e}}_i \frac{\partial}{\partial x_i} (\mathbf{v}\cdot\mathbf{A}) - q \sum_i \hat{\mathbf{e}}_i \frac{\partial \Phi}{\partial x_i}$$

$$= q \operatorname{grad} (\mathbf{v}\cdot\mathbf{A}) - q \operatorname{grad} \Phi. \tag{10.63}$$

Using the identity

$$\operatorname{grad} (\mathbf{v}\cdot\mathbf{A}) = (\mathbf{v}\cdot\operatorname{grad}) \mathbf{A} + (\mathbf{A}\cdot\operatorname{grad}) \mathbf{v} + \mathbf{A} \times \operatorname{curl} \mathbf{v} + \mathbf{v} \times \operatorname{curl} \mathbf{A}$$

We have for constant $\mathbf{v}$

$$\sum_i \hat{\mathbf{e}}_i \frac{\partial L}{\partial x_i} = q (\mathbf{v}\cdot\operatorname{grad}) \mathbf{A} + q \mathbf{v} \times \operatorname{curl} \mathbf{A} - q \operatorname{grad} \Phi \tag{10.64}$$

The Lagrangian equation of motion is

$$\frac{d}{dt} \frac{\partial L}{\partial v_i} = \frac{\partial L}{\partial x_i}$$

Therefore, from (10.61) and (10.64) we have

$$\frac{d}{dt} (\mathbf{p} + q\mathbf{A}) = \frac{d\mathbf{p}}{dt} + q \frac{d\mathbf{A}}{dt}$$

$$= q (\mathbf{v}\cdot\operatorname{grad}) \mathbf{A} + q\mathbf{v} \times \operatorname{curl} \mathbf{A} - q \operatorname{grad} \Phi. \tag{10.65}$$

The time derivative of $\mathbf{A}$ is

$$\frac{d\mathbf{A}}{dt} = \frac{\partial \mathbf{A}}{\partial t} + \sum_i \frac{\partial \mathbf{A}}{\partial x_i} \frac{dx_i}{dt}$$

$$= \frac{\partial \mathbf{A}}{\partial t} + \left(\sum_i v_i \cdot \frac{\partial}{\partial x_i} \right) \mathbf{A}$$

$$= \frac{\partial \mathbf{A}}{\partial t} + (\mathbf{v}\cdot\operatorname{grad}) \mathbf{A}$$

∴ Equation (10.65) becomes

$$\frac{d\mathbf{p}}{dt} = -q\frac{\partial \mathbf{A}}{\partial t} + q\mathbf{V} \times \text{curl } \mathbf{A} - q \text{ grad } \Phi$$

i.e.
$$\mathbf{F} = q\left(-\text{ grad } \Phi - \frac{\partial \mathbf{A}}{\partial t}\right) + q\,\mathbf{v} \times \text{curl } \mathbf{A}$$

$$= q\,\mathbf{E} + q\,\mathbf{v} \times \text{curl } \mathbf{A}$$

$$= q\,(\mathbf{E} + \mathbf{v} \times \mathbf{B}). \tag{10.66}$$

This is the Lorentz force equation.

For a relativistic particle in an electromagnetic field we assume the Lagrangian to be

$$L = -m_0c^2\sqrt{1-\beta^2} + q\mathbf{v}\cdot\mathbf{A} - q\Phi \tag{10.67}$$

where m_0 is the mass of the particle as measured in the frame of reference in which the particle is at rest.

In Minkowski space, the position vector of a point is

$$\mathscr{X} = (x_1, x_2, x_3, ict) \tag{10.68}$$

The differential of $\mathscr{X}$ is also a four-vector

$$d\mathscr{X} = (dx_1, dx_2, dx_3, icdt) \tag{10.69}$$

We know that four-dimensional element of length is invariant under Lorentz transformation

$$ds = \sqrt{dx_\mu\, dx_\mu} = \sqrt{dx_1^2 + dx_2^2 + dx_3^2 - c^2\, dt^2}. \tag{10.70}$$

We now introduce a quantity $d\tau$ such that

$$d\tau = \sqrt{dt^2 - \frac{1}{c^2}(dx_1^2 + dx_2^2 + dx_3^2)}$$

$$= \sqrt{-\frac{1}{c^2}(dx_1^2 + dx_2^2 + dx_3^2 - c^2\, dt^2)}$$

$$= \frac{i}{c}\sqrt{dx_\mu\, dx_\mu}. \tag{10.71}$$

We can also write $d\tau$ as

$$d\tau = dt\sqrt{1 - \frac{1}{c^2}\left\{\left(\frac{dx_1}{dt}\right)^2 + \left(\frac{dx_2}{dt}\right)^2 + \left(\frac{dx_3}{dt}\right)^2\right\}}$$

$$= dt\sqrt{1 - \frac{v^2}{c^2}} = dt\sqrt{1-\beta^2} = \frac{dt}{\gamma} \tag{10.72}$$

This quantity, obviously is invariant and is called the element of **proper time** in Minkowski space. The vector

$$\mathscr{U} = \frac{d\mathscr{X}}{d\tau} = \left(\frac{dx_1}{d\tau}, \frac{dx_2}{d\tau}, \frac{dx_3}{d\tau}, ic\frac{dt}{d\tau}\right)$$

$$= \left(\gamma \frac{dx_1}{dt}, \gamma \frac{dx_2}{dt}, \gamma \frac{dx_3}{dt}, ic\gamma\right)$$

$$= (\gamma v_1, \gamma v_2, \gamma v_3, ic\gamma) \tag{10.73}$$

is a four-vector velocity.

Since $\mathcal{A} = \left(\mathbf{A}, \frac{i\Phi}{c}\right)$

The Lagrangian (10.67) may be expressed as

$$L = -\frac{m_0c^2}{\gamma} + \frac{q\mathcal{U}\cdot\mathcal{A}}{\gamma} = \frac{1}{\gamma}(-m_0c^2 + q\mathcal{U}_\mu \mathcal{A}_\mu) \tag{10.74}$$

The equations of motion are obtained by using this Lagrangian in the principle of least action.

$$\delta\int_{t_1}^{t_2} Ldt = 0$$

or

$$\delta\int_{\tau_1}^{\tau_2} \gamma L d\tau = 0 \tag{10.75}$$

i.e.

$$\delta\int_{\tau_1}^{\tau_2} [-m_0c^2 + q\mathcal{U}_\mu \mathcal{A}_\mu]\, d\tau$$

$$= \delta\int_{\tau_1}^{\tau_2} [-m_0c^2\, d\tau + q\mathcal{A}_\mu\, dx_\mu] = 0\left(\because\ \mathcal{U}_\mu = \frac{d\mathcal{X}_\mu}{d\tau}\right). \tag{10.76}$$

Performing the variation,

$$\int_{\tau_1}^{\tau_2} [-m_0c^2\delta\,(d\tau) + q\mathcal{A}_\mu\delta\,(dx_\mu) + q\delta\mathcal{A}_\nu\, dx_\nu] = 0 \tag{10.77}$$

Now $\delta\,(d\tau) = \frac{\partial\tau}{\partial x_\mu}\,\delta(dx_\mu) = \frac{\partial\tau}{\partial x_\mu}\, d\,(\delta x_\mu)$

Since $d\tau = \frac{i}{c}\sqrt{dx_\mu\, dx_\mu}$ (see 10.71)

$$\frac{\partial\tau}{\partial x_\mu} = \frac{i}{c}\frac{dx_\mu}{\sqrt{dx_\mu\, dx_\mu}} = -\frac{1}{c^2}\frac{dx_\mu}{d\tau}$$

$$\therefore \quad \delta\,(d\tau) = -\frac{1}{c^2}\frac{dx_\mu}{d\tau}\, d\,(\delta x_\mu) = -\frac{1}{c^2}\mathcal{U}_\mu\, d(\delta x_\mu). \tag{10.78}$$

Also $$\delta\mathcal{A}_\nu = \frac{\partial\mathcal{A}_\nu}{\partial x_\mu}\,\delta x_\mu \tag{10.79}$$

Hence, the equation (10.77) may be written as

$$\int_{\tau_1}^{\tau_2}\left[\{m_0\mathcal{U}_\mu + q\mathcal{A}_\mu\}\, d(\delta x_\mu) + q\frac{\partial\mathcal{A}_\nu}{\partial x_\mu}\,\delta x_\mu\, dx_\nu\right] = 0 \tag{10.80}$$

Integration of the first term gives

$$\int_{\tau_1}^{\tau_2} \{m_0 \mathcal{U}_\mu + q\mathcal{A}_\mu\}\, d(\delta x_\mu) = \Big[\{m_0 \mathcal{U}_\mu + q\mathcal{A}_\mu\}\,\delta x_\mu\Big]_{\tau_1}^{\tau_2}$$

$$- \int_{\tau_1}^{\tau_2} \frac{\partial}{\partial x_\nu} \{m_0 \mathcal{U}_\mu + q\mathcal{A}_\mu\}\,\delta x_\mu\, dx_\nu$$

$$= -\int_{\tau_1}^{\tau_2} \frac{\partial}{\partial x_\nu} \{m_0 \mathcal{U}_\mu + q\mathcal{A}_\mu\}\,\delta x_\mu\, dx_\nu.$$

The first term vanishes because the variation of the coordinates must vanish at the end points.

Therefore, equation (10.80) becomes

$$\int_{\tau_1}^{\tau_2} \left[-\frac{\partial}{\partial x_\nu} \{m_0 \mathcal{U}_\mu + q\mathcal{A}_\mu\} + q\frac{\partial \mathcal{A}_\nu}{\partial x_\mu} \right] \delta x_\mu\, dx_\nu = 0$$

i.e.

$$\int_{\tau_1}^{\tau_2} \left[m_0 \frac{\partial \mathcal{U}_\mu}{\partial x_\nu} + q\frac{\partial \mathcal{A}_\nu}{\partial x_\nu} - q\frac{\partial \mathcal{A}_\nu}{\partial x_\mu} \right] \delta x_\mu\, dx_\nu = 0$$

or

$$\int_{\tau_1}^{\tau_2} \left[m_0 \frac{\partial \mathcal{U}_\mu}{\partial x_\nu} - q\left(\frac{\partial \mathcal{A}_\nu}{\partial x_\mu} - \frac{\partial \mathcal{A}_\mu}{\partial x_\nu} \right) \right] \delta x_\mu\, dx_\nu = 0$$

and using (10.45)

$$\int_{\tau_1}^{\tau_2} \left[m_0 \frac{\partial \mathcal{U}_\mu}{\partial x_\nu} - qF_{\mu\nu} \right] \delta x_\mu\, dx_\nu = 0. \tag{10.81}$$

Now

$$\mathcal{U}_\nu = \frac{dx_\nu}{d\tau} \text{ and } \frac{\partial \mathcal{U}_\mu}{\partial x_\nu}\, dx_\nu = d\,\mathcal{U}_\mu = \frac{d\mathcal{U}_\mu}{d\tau}\, d\tau$$

Therefore, (10.81) changes to

$$\int_{\tau_1}^{\tau_2} \left[m_0 \frac{d\mathcal{U}_\mu}{d\tau} - qF_{\mu\nu}\mathcal{U}_\nu \right] \delta x_\mu\, d\tau = 0 \tag{10.82}$$

Since δx_μ is an arbitrary variation, it follows that

$$m_0 \frac{d\mathcal{U}_\mu}{d\tau} - qF_{\mu\nu}\,\mathcal{U}_\nu = 0 \tag{10.83}$$

which is the desired covariant equation of motion. The space portion of this equation is obtained by giving μ the values 1, 2, 3. Thus

$$m_0 \frac{d\mathcal{U}_1}{d\tau} = qF_{1\nu}\mathcal{U}_\nu = qF_{11}\mathcal{U}_1 + qF_{12}\mathcal{U}_{12} + qF_{13}\mathcal{U}_3 + qF_{14}\mathcal{U}_4$$

$$= \gamma q v_2 B_3 - \gamma q v_3 B_2 + q\gamma E_1 \qquad \text{[using (10.46) and (10.73)]}$$

i.e.

$$\frac{d}{dt}(m_0 \gamma \mathcal{U}_1) = \gamma q\,[E_1 + v_2 B_3 - v_3 B_2].$$

The general vectorial equation is

$$\frac{d\mathcal{P}}{dt} = q\,[\mathbf{E} + \mathbf{v} \times \mathbf{B}] \tag{10.84}$$

where we have used the symbol $\mathcal{P}$ for $m_0\gamma\mathbf{v}$ which is the momentum

$$\mathcal{P} = m_0\gamma\mathbf{v} = m\mathbf{v}. \tag{10.85}$$

Utilizing the first three components of (10.83), we have seen that the equation of motion retains its form as in (10.66). Let us now consider the fourth component.

$$m_0 \frac{d\mathcal{U}_4}{d\tau} = qF_{41}\mathcal{U}_1 + qF_{42}\mathcal{U}_2 + qF_{43}\mathcal{U}_3 + qF_{44}\mathcal{U}_4$$

$$= -\frac{\gamma q}{ic} E_1 v_1 - \frac{\gamma q}{ic} E_2 v_2 - \frac{\gamma q}{ic} E_3 v_3$$

$$\therefore \quad \gamma \frac{d}{dt}(m_0 i\gamma c) = -\frac{\gamma q}{ic} \mathbf{E}\cdot\mathbf{v}$$

i.e.

$$\frac{d}{dt}(m_0\gamma c^2) = q\,\mathbf{E}\cdot\mathbf{v}. \tag{10.86}$$

What is the physical interpretation of this equation? The right-hand side gives the rate at which the work is done on the particle by the electric field, which, in fact, is equal to the rate of change of the kinetic energy T with time. Hence,

$$q\mathbf{E}\cdot\mathbf{v} = \frac{dT}{dt} = \frac{d}{dt}(m_0\gamma c^2)$$

and

$$\int_{t_1}^{t_2} \frac{dT}{dt}\, dt = \Big[m_0\gamma c^2\Big]_{t_1}^{t_2}.$$

If the particle is at rest at $t = t_1$, the value of γ at this time is 1.

$$\therefore \qquad T = m_0\gamma c^2 - m_0 c^2 = mc^2 - m_0 c^2 \tag{10.87}$$

where $m = m_0\gamma$.

The quantity m_0c^2 is called the *rest energy*. The total energy W is given by

$$W = mc^2 = T + m_0c^2. \tag{10.88}$$

The total energy is sometimes expressed in terms of momentum $\mathcal{P}$

$$W = m_0\gamma c^2 \qquad \therefore \quad \frac{W^2}{c^2} = m_0^2\gamma^2c^2$$

Now $\quad \mathcal{P}^2 = m_0^2\gamma^2v^2 \qquad$ (see 10.85)

$$\therefore \quad \frac{W^2}{c^2} = \mathcal{P}^2 + m_0^2\gamma^2(c^2 - v^2)$$

or

$$W = [m_0^2c^4 + \mathcal{P}^2c^2]^{1/2}. \tag{10.89}$$

10.12 Radiation from Relativistic Particles

We have shown in the preceding chapter that only charges undergoing acceleration could produce radiation and the rate of radiation is given by Larmor formula

$$-W = \frac{e^2|\dot{\mathbf{v}}|^2}{6\pi\epsilon_0 c^3}. \tag{10.90}$$

If the velocity of the moving charges is much smaller than c the formula is exact in the frame at rest with respect to charges. We shall now investigate the formula for the particles moving with velocities comparable with the velocity of light.

We know that

$$\mathcal{U} = (\gamma\mathbf{v},\, i\gamma c).$$

From this, by differentiating with respect to τ, we obtain the four-vector acceleration

$$\frac{d\mathcal{U}}{d\tau} = \left\{\gamma^2\dot{\mathbf{v}} + \frac{\gamma^4\mathbf{v}\,(\mathbf{v}\cdot\dot{\mathbf{v}})}{c^2},\, \frac{\gamma^4 i\,(\mathbf{v}\cdot\dot{\mathbf{v}})}{c}\right\} \tag{10.91}$$

where $\dot{\mathbf{v}} = \frac{d\mathbf{v}}{dt}$, $\gamma d\tau = dt$.

We may generalize the result (10.90) to relativistic energies by replacing $\dot{\mathbf{v}}$ by the four-vector acceleration

Now $$\left(\frac{d\mathcal{U}}{d\tau}\right)^2 = \left[\gamma^4\,(\dot{\mathbf{v}}\cdot\dot{\mathbf{v}}) + \frac{2\gamma^6(\mathbf{v}\cdot\dot{\mathbf{v}})^2}{c^2} + \frac{\gamma^8\,(\mathbf{v}\cdot\mathbf{v})\,(\mathbf{v}\cdot\dot{\mathbf{v}})^2}{c^4} - \frac{\gamma^8(\mathbf{v}\cdot\dot{\mathbf{v}})^2}{c^2}\right]$$

$$= \left[\gamma^4\,(\dot{\mathbf{v}}\cdot\dot{\mathbf{v}}) + \frac{2\gamma^6(\mathbf{v}\cdot\dot{\mathbf{v}})^2}{c^2} - \frac{\gamma^6(\mathbf{v}\cdot\dot{\mathbf{v}})^2}{c^2}\right]$$

$$= \gamma^4\left[(\dot{\mathbf{v}}\cdot\dot{\mathbf{v}}) + \frac{\gamma^2(\mathbf{v}\cdot\dot{\mathbf{v}})^2}{c^2}\right] \tag{10.92}$$

$$-W = \frac{e^2}{6\pi\epsilon_0 c^3}\,\frac{1}{(1-\beta^2)^2}\left[\dot{\mathbf{v}}\cdot\dot{\mathbf{v}} + \frac{(\mathbf{v}\cdot\dot{\mathbf{v}})^2}{c^2(1-\beta^2)}\right] \tag{10.93}$$

The particular cases that we may consider here are:

1. When $\mathbf{v}$ and $\dot{\mathbf{v}}$ are colinear

$$-W = \frac{e^2}{6\pi\epsilon_0 c^3(1-\beta^2)^2}\Bigg]|\dot{\mathbf{v}}|^2 + \frac{v^2\,|\dot{\mathbf{v}}|^2}{c^2\,(1-\beta^2)}\Bigg]$$

$$= \frac{e^2}{6\pi\epsilon_0 c^3}\,\frac{|\dot{\mathbf{v}}|^2}{(1-\beta^2)^3} = \frac{e^2}{6\pi\epsilon_0 c}\,\frac{\dot{\beta}^2}{(1-\beta^2)^3}. \tag{10.94}$$

2. When $\mathbf{v} \perp \dot{\mathbf{v}}$

$$-W = \frac{e^2}{6\pi\epsilon_0 c^3}\,\frac{|\dot{\mathbf{v}}|^2}{(1-\beta^2)^2} = \frac{e^2}{6\pi\epsilon_0 c}\,\frac{\dot{\beta}^2}{(1-\beta^2)^2} \tag{10.95}$$

PROBLEMS

10.1 Show that two successive Lorentz transformations in the same direction are equivalent to a single Lorentz transformation with a velocity

$$v = \frac{v_1 + v_2}{1 + \dfrac{v_1 v_2}{c^2}}$$

10.2 A space craft travels directly away from the earth with a velocity $v = 0.8\,c$. A radio signal is transmitted from the craft back to the earth, after it reaches a distance 8 light-years from the earth, (i) How much time has elapsed on the space craft since departure when the signal is sent from it? (ii) From the point of view of an observer in the space craft how much time appears to have elapsed on the earth when the signal is sent?

10.3 Two oppositely directed space ships move with identical velocity of $0.7c$ as measured by an observer on the earth. What is the velocity of one space ship as observed from the other?

10.4 Show that the product $\mathbf{E}.\mathbf{B}$ is invariant to Lorentz transformation.

10.5 In a frame of reference Σ the electric and magnetic fields are perpendicular to one another. What is the velocity of the frame Σ' relative to Σ if the observer in it finds that there is (i) only an electric field? (ii) only a magnetic field?

10.6 A charge q is at rest in a frame Σ. Determine, applying Lorentz transformation, the potentials $\mathbf{A}$ and Φ in a frame Σ' moving with a velocity $\mathbf{v}$. Hence, find $\mathbf{E}$ and $\mathbf{B}$.

10.7 A cosmic ray proton moving perpendicularly to a magnetic field of magnitude 10^{-5}T, approaches the earth. If the energy of the proton is 10^{15} eV, what is the magnitude of the electric field in the proton's rest frame?

10.8 A charge q is uniformly distributed over the surface of a spherical shell of radius 'a'. If the sphere moves with a velocity v along the x-axis, calculate the total electric and total magnetic energy in the space exterior to the sphere. Show that when $v \ll c$ the result reduces to the static value $\dfrac{q^2}{8\pi\,\varepsilon_0\,a}$.

10.9 Find the path of a charged particle moving in a uniform field $\mathbf{E}$. Discuss the case of a slow particle.

10.10 The frequency of vibration of an oscillator is ω_0. What would be the frequency if the oscillator is placed in a magnetic field?

Chapter 11

Scattering and Dispersion

11.1 Scattering of Radiation

We have obtained in Chapter 9 expressions for the radiation fields in terms of the dipole moment $p(t)$. We did not specify, however, the cause of the time dependence of $p(t)$. Consider an electromagnetic wave incident on a system of charged particles. Due to the interaction of the wave with the charges, the latter are set into motion periodic in time. There will, thus, be accelerations and, hence, the system will radiate. We may consider the entire process as taking place in two steps: The energy is absorbed by the particles and is then re-emitted into space. We describe this process as a scattering of the original incident radiation by the charged particles.

Consider first a plane monochromatic linearly polarized wave incident on a particle carrying a charge 'q'. The wave is given by

$$\mathbf{E} = \mathbf{E}_0 e^{i(\mathbf{k}_0 \cdot \mathbf{r} - \omega_0 t)}. \tag{11.1}$$

The electric field will exert a force on q given by

$$\mathbf{F} = q\mathbf{E}. \tag{11.2}$$

We assume that the velocity acquired by the particle as a result of this interaction is much smaller than c ($v \ll c$). Our treatment, therefore, will be non-relativistic and we may ignore the second term in the expression for the Lorentz force

$$\mathbf{F} = q(\mathbf{E} + \mathbf{v} \times \mathbf{B}) \tag{11.3}$$

Thus, we have

$$\mathbf{F} = m\ddot{\mathbf{r}} = q\mathbf{E} = q\mathbf{E}_0 e^{i(\mathbf{k}_0 \cdot \mathbf{r} - \omega_0 t)}. \tag{11.4}$$

Now $\qquad \mathbf{p}(t) = q\mathbf{r}\,(t)$

$$\therefore \qquad \ddot{\mathbf{p}}(t) = q\,\ddot{\mathbf{r}}\,(t) = \frac{q^2}{m}\,\mathbf{E}(t). \tag{11.5}$$

The time average of the power radiated per unit solid angle is

$$\left\langle \frac{dW}{d\Omega} \right\rangle = \frac{q^2(\dot{u})^2}{16\,\pi^2 \epsilon_0 c^3} \sin^2\theta \qquad \text{(see 9.83)}$$

$$= \frac{\langle \ddot{p}^2 \rangle \sin^2\theta}{16\,\pi^2 \epsilon_0 c^3}$$

$$= \frac{q^4 E^2_0 \sin^2\theta}{16\,\pi^2 m^2 \epsilon_0 c^3}\,. \tag{11.6}$$

We define **differential scattering cross-section** as

$$\frac{d\sigma}{d\Omega} = \frac{\text{scattered energy per unit solid angle per unit time}}{\text{Incident energy per unit area per unit time}}$$

Hence,
$$\frac{d\sigma}{d\Omega} = \frac{\left\langle \dfrac{dW}{d\Omega} \right\rangle}{\langle \text{incident flux} \rangle}$$

$$= \frac{\left\langle \dfrac{dW}{d\Omega} \right\rangle}{\sqrt{\dfrac{\epsilon_0}{\mu_0}}\, E_0^{\,2}} \qquad \text{(see 9.81)}$$

$$= \left(\frac{q^2}{4\pi\epsilon_0 mc^2} \right)^2 \sin^2\theta \tag{11.7}$$

The angle θ is the angle between the induced dipole moment **p** and the direction of the outgoing radiation, i.e. it is the angle between the electric field **E** and the unit vector $\hat{\mathbf{e}}_r$. Fig. 11.1 shows this angle.

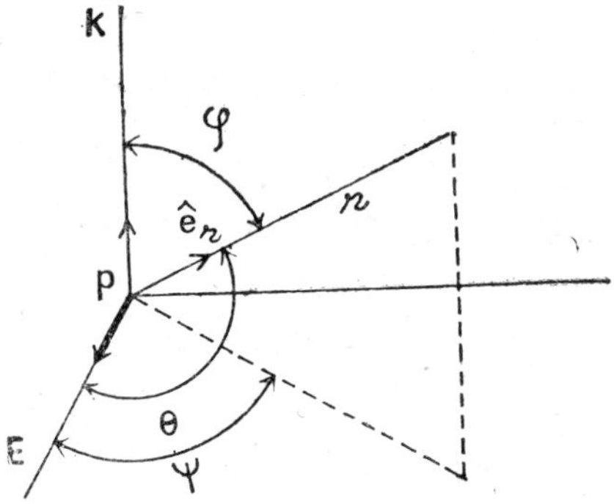

Fig. 11.1

Let us now consider a more general case. Suppose that the incident light is unpolarized. What would be the angular distribution of scattered light? To find this, we must average over all possible azimuthal orientations of the field **E**, i.e. we must average over the angle ψ in Fig. 11.1. We see from the figure that

$$\cos\theta = \cos\psi \sin\varphi$$

i.e.
$$\sin^2\theta = 1 - \cos^2\psi \sin^2\varphi$$

We have now to average over ψ. Let $\overline{\sin^2\theta}$ represent the average of $\sin^2\theta$ over all angles ψ

$$\therefore \qquad \overline{\sin^2\theta} = 1 - \overline{\cos^2\psi}\,\sin^2\varphi$$

$$= 1 - \frac{1}{2}\sin^2\varphi = \frac{1}{2}(1 + \cos^2\varphi).$$

Therefore, in the case of unpolarized radiation

$$\left\langle\frac{d\sigma}{d\Omega}\right\rangle_{\text{unpolarized}} = \left(\frac{q^2}{4\pi\epsilon_0 mc^2}\right)^2 \overline{\sin^2\theta} = \left(\frac{q^2}{4\pi\epsilon_0 mc^2}\right)^2 \frac{1+\cos^2\varphi}{2}. \tag{11.8}$$

We can see from the figure that φ is the angle between the directions of the incident radiation and the scattered radiation and, hence, is known as **the scattering angle.**

The total cross-section is obtained by integrating over the entire solid angle

$$\langle\sigma\rangle_{\text{unpolarized}} = \int \frac{d\sigma}{d\Omega}\,d\Omega = \left(\frac{q^2}{4\pi\epsilon_0 mc^2}\right)^2 \int \frac{1+\cos^2\varphi}{2}\sin\varphi\,d\varphi\,d\psi$$

$$= \frac{8\pi}{3}\left(\frac{q^2}{4\pi\epsilon_0 mc^2}\right)^2. \tag{11.9}$$

This scattering of an unpolarized plane electro-magnetic wave by a free charged particle is known as **Thomson scattering.**

If the charged particle is an electron $q = e$

and
$$\langle\sigma\rangle = \frac{8\pi}{3}\left(\frac{e^2}{4\pi\epsilon_0 mc^2}\right)^2 = \frac{8\pi}{3}r_0^2 \tag{11.10}$$

where $r_0 = \dfrac{e^2}{4\pi\epsilon_0 mc^2}$ is called the **classical electron radius.** The classical distribution of charge totalling 'e' must have a radius of this order if its self-energy is to equal its electron mass. For scattering by a free electron, Thomson scattering cross-section is

$$\sigma_T = 0.665 \times 10^{-28}\ \text{m}^2$$

11.2 Radiation Damping

We have so far discussed problems in electrodynamics in which we considered either the electromagnetic radiation produced by the accelerated charges or the influence of an external electromagnetic field on the motion of a particle. The two facts were considered as independent and unrelated. However, when the charges are accelerated by the electromagnetic field in which they are placed, electromagnetic radiation field is produced and this in turn is bound to influence the subsequent motion of the particle and, hence, for a correct treatment this influence must be taken into account. The reaction of the radiation on the motion of the particle is called **radiation reaction** or **radiation damping**.

Let us see how to include the reactive effects of radiation in the equation of motion of a charged particle. The Newtonian equation of motion for a particle of mass m and charge e acted on by a conservative external force $\mathbf{F}_e$ is

$$m\dot{\mathbf{v}} = \mathbf{F}_e. \tag{11.11}$$

In writing this equation we have neglected the emission of radiation. Since the particle is accelerated, it will emit radiation and to account for the reaction of this radiation field on the particle, we have to modify the equation (11.11) by adding a reaction force $\mathbf{F}_r$

i.e.
$$m\dot{\mathbf{v}} = \mathbf{F}_e + \mathbf{F}_r \tag{11.12}$$

But what is this force $\mathbf{F}_r$ equal to? The energy radiated per unit time by a charge 'e' moving with a speed v, which is much less than c and which undergoes acceleration $\mathbf{a} = \dot{\mathbf{v}}$ as given by (9.84) is

$$W = \frac{e^2(\dot{\mathbf{v}})^2}{6\pi\epsilon_0 c^3}. \tag{11.13}$$

For the conservation of energy it is necessary that the work done by the force $\mathbf{F}_r$ on the particle in a time interval, say, t_1 to t_2, be equal to the negative of the energy radiated in this interval, i.e.

$$\int_{t_1}^{t_2} \mathbf{F}_r \cdot \mathbf{v}\, dt = -\frac{e^2}{6\pi\epsilon_0 c^3}\int_{t_1}^{t_2} \dot{\mathbf{v}} \cdot \dot{\mathbf{v}}\, dt. \tag{11.14}$$

We can integrate the right-hand side by parts. We have,

$$\int_{t_1}^{t_2} \mathbf{F}_r \cdot \mathbf{v}\, dt = -\frac{e^2}{6\pi\epsilon_0 c^3}\Big[\mathbf{v} \cdot \dot{\mathbf{v}}\Big]_{t_1}^{t_2} + \frac{e^2}{6\pi\epsilon_0 c^3}\int^{t_2} \mathbf{v} \cdot \ddot{\mathbf{v}}\, dt.$$

If the motion is periodic or if the time interval is short, then the state of the system will be approximately the same at t_2 as at t_1, and we may neglect the integrated term. We, thus, have the approximate relation

$$\int_{t_1}^{t_2} \left(\mathbf{F}_r - \frac{e^2}{6\pi\epsilon_0 c^3}\ddot{\mathbf{v}}\right) \cdot \mathbf{v}\, dt \simeq 0 \tag{11.15}$$

which shows that

$$\mathbf{F}_r = \frac{e^2}{6\pi\epsilon_0 c^3}\ddot{\mathbf{v}} = m\tau\ddot{\mathbf{v}} \tag{11.16}$$

where, for convenience, we have put

$$\tau = \frac{e^2}{6\pi\epsilon_0 m c^3}. \tag{11.17}$$

The modified equation (11.12), therefore, can be written as

$$m\dot{\mathbf{v}} = \mathbf{F}_e + m\tau\ddot{\mathbf{v}}$$

or
$$m(\dot{\mathbf{v}} - \tau\ddot{\mathbf{v}}) = \mathbf{F}_e. \tag{11.18}$$

This equation is known as **Abraham-Lorentz equation of motion.**

If now the external force $\mathbf{F}_e$ is a linear restoring force of the type

$\mathbf{F}_e = -m\gamma r$*, we can write the equation (11.18) in the form

$$m\ddot{\mathbf{r}} - m\tau\dddot{\mathbf{r}} + m\gamma\mathbf{r} = 0. \tag{11.19}$$

If the radiation damping term is small, the equation reduces to

$$m\ddot{\mathbf{r}} + m\gamma\mathbf{r} = 0 \tag{11.20}$$

and its solution is

$$\mathbf{r} = \mathbf{r}_0 e^{-i\gamma^{1/2}t} = \mathbf{r}_0 e^{-i\omega_0 t} \tag{11.21}$$

where we have put $\omega_0^2 = \gamma$. We have, therefore,

$$\dot{\mathbf{r}}(t) = -i\omega_0\mathbf{r}(t)$$

and
$$\dddot{\mathbf{r}}(t) = i\omega_0^3\mathbf{r}(t) = -\omega_0^2\dot{\mathbf{r}}(t) \tag{11.22}$$

Using this approximate relation, we can write (11.19) as

$$m\ddot{\mathbf{r}} + m\tau\omega_0^2\dot{\mathbf{r}} + m\gamma\mathbf{r} = 0$$

or
$$m[\ddot{\mathbf{r}} + l\dot{\mathbf{r}} + \gamma\mathbf{r}] = 0 \tag{11.23}$$

where
$$l = \tau\omega_0^2 = \frac{e^2\gamma}{6\pi\epsilon_0 mc^3}. \tag{11.24}$$

11.3 Dispersion in Dilute Gases

We shall now discuss the propagation of electromagnetic waves in a dilute gas in which the mutual interactions between the constituent particles are neglected. As the wave passes through the gas, the electrons in the molecules will be displaced from their equilibrium position and the molecules will be polarized. We will neglect the difference between the applied electric field and the local field since the gas has low density. We assume that the electrons are bound by a linear restoring force and that the damping as shown in the preceding section, is proportional to the velocity.

The equation of motion for an electron (αth) is

$$m[\ddot{\mathbf{r}}_\alpha + l_\alpha\dot{\mathbf{r}}_\alpha + \gamma_\alpha\mathbf{r}_\alpha] = e\mathbf{E} \tag{11.25}$$

where l_α measures the damping force. This can be written as

$$\ddot{\mathbf{r}}_\alpha + l_\alpha\dot{\mathbf{r}}_\alpha + \gamma_\alpha\mathbf{r}_\alpha = \frac{e}{m}\mathbf{E}_0 \exp(-i\omega t) \tag{11.26}$$

The steady-state solution of this equation is

$$\mathbf{r}_\alpha(t) = \frac{\left(\frac{e}{m}\right)\mathbf{E}_0}{(\omega_\alpha^2 - \omega^2) - il_\alpha\omega} \exp(-i\omega t) \tag{11.27}$$

The dipole moment resulting from this displacement of the electron is

*Not to be confused with γ used in the preceding chapter.

$$\mathbf{p}_\alpha = e\mathbf{r}_\alpha(t) = \frac{\left(\frac{e^2}{m}\right)\mathbf{E}}{(\omega_\alpha^2 - \omega^2) - il_\alpha\omega}. \tag{11.28}$$

Suppose there are N electrons per unit volume in the gas and a fraction f_α of these have the characteristic resonance frequency ω_α. The total dipole moment per unit volume is then

$$\mathbf{P} = \sum_\alpha Nf_\alpha p_\alpha = \mathbf{E}\sum_\alpha \frac{\left(\frac{e^2}{m}\right)Nf_\alpha}{(\omega_\alpha^2 - \omega^2) - il_\alpha\omega} \tag{11.29}$$

where f_α known as oscillator strengths satisfy the sum rule $\sum_\alpha f_\alpha = 1$.

Since $\mathbf{P} = \epsilon_0 \chi \mathbf{E}$ (see 2.21), the electric susceptibility χ is given by

$$\chi = \sum_\alpha \frac{\left(\frac{e^2}{m\epsilon_0}\right)Nf_\alpha}{(\omega_\alpha^2 - \omega^2) - il_\alpha\omega} \tag{11.30}$$

and the dielectric constant ϵ

$$\epsilon = 1 + \chi = 1 + \sum_\alpha \frac{\left(\frac{e^2}{m\epsilon_0}\right)Nf_\alpha}{(\omega_\alpha^2 - \omega^2) - il_\alpha\omega} \tag{11.31}$$

We know that the refractive index n is equal to $\sqrt{\epsilon}$

$$\therefore \qquad n^2 = \epsilon = 1 + \sum_\alpha \frac{\left(\frac{e^2}{m\epsilon_0}\right)Nf_\alpha}{(\omega_\alpha^2 - \omega^2) - il_\alpha\omega}. \tag{11.32}$$

The equation (11.32) is the **dispersion relation** for a dilute gas. The damping constants l_α are generally small, and, therefore n^2 (or ϵ) is approximately real for most frequencies. For values of $\omega < \omega_\alpha$ all terms in the sum (11.32) are positive and ϵ is greater than one; on the other hand, for value of $\omega > \omega_\alpha$, more and more negative terms occur in the sum and finally ϵ is less than unity. In the neighbourhood of ω_α, the behaviour is peculiar. The real part vanishes and the term is imaginary and large. The variation of the real and imaginary parts of n^2 is shown in Fig. 11.2.

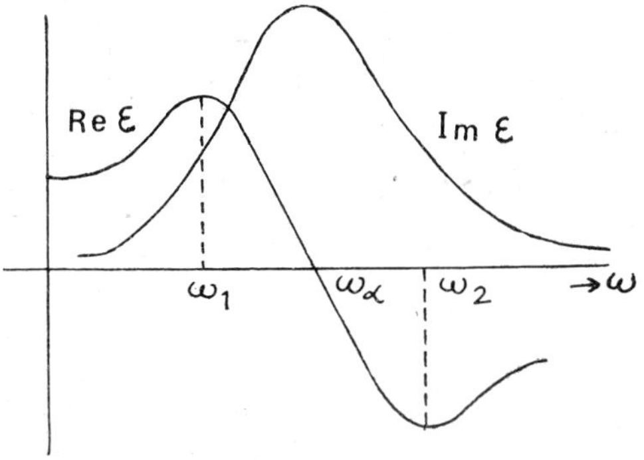

Fig. 11.2

When $\frac{dn}{d\omega} > 1$, the dispersion is said to be **normal**. When $\frac{dn}{d\omega} < 1$, the ordering of prismatic colours is reversed and the dispersion is said to be **anomalous**. We see from Fig. 11.2 that anomalous dispersion occurs between two frequencies ω_1 and ω_2 in the region in which $\frac{dn}{d\omega} < 1$. Everywhere else it is normal. If $I_m \epsilon > 0$, there is dissipation of energy from the electromagnetic wave into the medium. Hence, the regions where $I_m\epsilon$ is large, are called regions of *resonant absorption*. If $I_m\epsilon < 0$, the energy is given to the wave by the medium resulting into the amplification of waves as in the case of masers and lasers.

11.4 Dispersion in Liquids and Solids

When the medium is not dilute, the local field is not equal to the external field. We have to replace the electric field vector **E** in (11.29) by

$$\mathbf{E}_{eff} = \mathbf{E} + \frac{\mathbf{P}}{3\epsilon_0}.$$

We, thus, have

$$\mathbf{P} = \left(\mathbf{E} + \frac{\mathbf{P}}{3\epsilon_0}\right) \sum_{\alpha} \frac{N f_\alpha \left(\frac{e^2}{m}\right)}{(\omega_\alpha^2 - \omega^2) - i l_\alpha \omega} \tag{11.33}$$

Since

$$\mathbf{D} = \epsilon_0 \mathbf{E} + \mathbf{P} = \epsilon \mathbf{E},$$

$$\frac{\mathbf{P}}{\mathbf{E} + \frac{\mathbf{P}}{3\epsilon_0}} = \frac{\epsilon - \epsilon_0}{\epsilon + 2\epsilon_0} 3\epsilon_0$$

$\therefore$

$$\frac{\epsilon - \epsilon_0}{\epsilon + 2\epsilon_0} = \sum_{\alpha} \frac{N f_\alpha \left(\frac{e^2}{m}\right)}{3\epsilon_0\{(\omega_\alpha^2 - \omega^2) - i l_\alpha \omega\}} \tag{11.34}$$

or

$$\frac{\frac{\epsilon}{\epsilon_0} - 1}{\frac{\epsilon}{\epsilon_0} + 2} = \frac{n^2 - 1}{n^2 + 2} = \sum \frac{N f_\alpha \left(\frac{e^2}{m}\right)}{3\epsilon_0\{(\omega_\alpha^2 - \omega^2) - i l_\alpha \omega\}} \tag{11.35}$$

We see that on the right-hand side, except for N all other quantities are independent of density. The number of electrons per unit volume N is proportional to the number of atoms per unit volume, that is, proportional to the density. Hence, for a given frequency, we have

$$\frac{\frac{\epsilon}{\epsilon_0} - 1}{\frac{\epsilon}{\epsilon_0} + 2} \frac{1}{\rho} = A \text{ (const.)}. \tag{11.36}$$

This is analogous to the Clausius-Mossoti equation (2.72). When written in terms of n^2, i.e.

$$\frac{n^2-1}{n^2+2}\frac{1}{\rho}=A \tag{11.37}$$

it is known as **Lorentz-Lorenz** formula. The quantity A is known as the **atomic refractivity** of the medium. If the substance is molecular with molecular weight W, the formula can be written as

$$\frac{n^2-1}{n^2+2}\frac{W}{\rho}=A_m \; (\because \quad N=\frac{\rho N_a}{W} \text{ where } N_a \text{ is Avogadro's number})$$

The quantity A_m is known as **molecular refractivity**.

11.5 Media Containing Free Electrons

If the media contain free electrons, their motion is random and there is no current flow. When the field is applied, the electrons acquire an additional velocity component and a current results. The electrons collide with the atoms of the material and are scattered. A damping, therefore, results which clearly depends upon the velocity of the electrons. The equation of motion of an electron can be written in the form

$$m\dot{\mathbf{v}}_\alpha + ml_\alpha \mathbf{v}_\alpha = e\mathbf{E}_0 \exp(-i\omega t) \tag{11.38}$$

where we have assumed that the field varies harmonically with time. You can verify that the solution of this equation is

$$\mathbf{v}_\alpha = \frac{e\mathbf{E}_0 \exp(-i\omega t)}{m(l_\alpha - i\omega)}. \tag{11.39}$$

Therefore, the current density $\mathbf{j}$ is

$$\mathbf{j} = \sum_\alpha e\mathbf{v}_\alpha = Ne\mathbf{v} = \frac{Ne^2\mathbf{E}_0 \exp(-i\omega t)}{m(l-i\omega)} \tag{11.40}$$

Here we have assumed that there are N electrons per unit volume moving with the common velocity $\mathbf{v}$.

Since $\quad \mathbf{j} = \sigma\mathbf{E}$

$$\sigma = \frac{Ne^2}{m(l-i\omega)}. \tag{11.41}$$

Let us find the damping constant l for copper using this formula. For copper $N \simeq 8\times 10^{28}$ atom/m^3 and at normal temperature the low frequency conductivity is $\sigma \simeq 6\times 10^7$ mho m^{-1}. Therefore,

$$l = \frac{Ne^2}{m\sigma} = \frac{8\times 10^{28}\times 1.6\times 1\cdot 6\times 10^{-38}}{9\times 10^{-31}\times 6\times 10^7}$$

$$= 3\times 10^{13}\ \mathrm{s}^{-1}$$

Therefore, for the frequencies of the order of 10^{11} or less, the conductivities of the metals are essentially real and do not depend upon the frequency. For infrared frequency and beyond, the conductivity is complex and varies with frequency according to the formula (11.41).

PROBLEMS

11.1 Find the differential cross-section for the scattering of an elliptically polarized wave by a free charged particle.

11.2 Obtain the expression for the differential cross-section for the scattering of an elliptically polarized wave of frequency ω by an oscillator of mass m, charge e, and natural frequency ω_0.

11.3 When a linearly polarized plane electromagnetic wave is incident on an atom, an electron bound harmonically to the atom experiences a damping linearly proportional to its velocity. Calculate the differential and total cross-sections. How does the angular distribution of scattered radiation differ from that for Thomson scattering?

11.4 A dipole with moment $\mathbf{p}$ oscillating with a frequency ω is placed at the origin of a coordinate system. A particle with polarizability β is placed at a point specified by the radius vector $\mathbf{r}$ ($\mathbf{r} \perp \mathbf{p}$) where $|\mathbf{r}|$ is much smaller than the radiation wavelength λ. Find the radiation intensity of electromagnetic wave for the system.

Chapter 12

Plasma Physics

If we increase the temperature of a gas above a certain value, the kinetic energy of the atoms grows to such an extent that the collision of two atoms may split them up into electrons and positively charged ions. Generally, plasma can be assumed to be a mixture of three components: free electrons, positive ions and neutral atoms (or molecules). The particles interact between themselves via the simplest mechanism, namely, the Coulomb electrostatic forces. The "organization" of such a system, therefore, is entirely of a different nature from that existing in the solid or liquid states, which are organised by the action of short range inter-crystalline or cohesive forces.

Plasma which is known as the fourth state, is the most widespread state of matter in the universe. The ionosphere is a plasma envelope surrounding the earth's atmosphere. The so-called radiation belt outside the ionosphere contains plasma formations. The sun and the stars can be regarded as lumps of hot plasma. The studies in plasma phenomena, therefore, are of great interest and a physicist has to understand the mechanism of various processes occurring in a plasma.

12.1 Quasi-neutrality of a Plasma

One of the definitive properties of plasma is its quasineutrality (according to Schottky). The electric forces which bind the opposite forces in a plasma provide for its quasineutrality, that is, its tendency to balance positive and negative space charge in each macroscopic volume element. Any separation of charges when electrons shift with respect to ions, gives rise to electric fields which act in the direction of restoring neutrality, However, in a small volume of plasma, quasineutrality may be violated if the electric field generated by an excess of particles of one sign is too weak to have a significant effect in the motion of particles; but the space charge densities adjust themselves so that the major part of the plasma is shielded from the field. For a given temperature and concentration, the quasineutrality is described by a characteristic linear parameter δ. Within a volume with the linear size x, if $x \gg \delta$, then the concentrations of the opposite charges in this volume are equal and the neutrality is maintained; if $x \ll \delta$, the separation of charges has no significance on the motion of particles and the neutrality is isolated. The electric field is screened off at a distance δ.

The characteristic length δ can be estimated in the following way: Suppose the charges are completely separated in a volume element with the linear size δ. The potential energy of a charged particle in this volume is of the order of the energy of the thermal motion of particles, i.e. kT, where T is the plasma temperature. The electric field in a volume satisfies the Poisson equation

$$\text{div}\,\mathbf{E} = \frac{\rho}{\epsilon_0} \tag{12.1}$$

where ρ is the charge density. If the linear dimensions of the volume are of the order δ and the concentration of the charged particles is N_0, we have

$$\text{div}\,\mathbf{E} \simeq \frac{E}{\delta} \simeq \frac{N_0 e}{\epsilon_0}$$

$$\therefore \qquad \mathbf{E} \simeq \frac{N_0 e \delta}{\epsilon_0}. \tag{12.2}$$

Hence, the plasma potential in the volume of charge separation is

$$V = E\delta \simeq \frac{N_0 e \delta^2}{\epsilon_0} \tag{12.3}$$

and the potential energy of a particle is

$$U = eV \simeq \frac{N_0 e^2 \delta^2}{\epsilon_0} \simeq kT. \tag{12.4}$$

This gives

$$\delta \simeq \left(\frac{\epsilon_0 kT}{N_0 e^2}\right)^{1/2}. \tag{12.5}$$

The parameter δ can be derived also from the analysis of electric field screening in a plasma. Suppose that a test charge 'q' is introduced into a plasma. Electrons in the plasma will tend to move closer to the charge, whereas positive ions will tend to move away from it. In a state of statistical equilibrium, the spatial distribution of electrons and ions in the vicinity of the test charge is given by the Boltzmann formula $N \propto \exp\left(-\frac{U}{kt}\right)$ where U is the potential energy. Thus the electron density is

$$N_e = N_0 \exp\left(e\frac{U - U_0}{kT}\right) \tag{12.6}$$

and the positive ion density

$$N_i = N_0 \exp\left(-e\frac{U - U_0}{kT}\right) \tag{12.7}$$

where U is the local potential, U_0 the plasma potential and N_0 is the electronic density as well as the positive ion density when $U = U_0$.

The potential U must satisfy the Poisson equation

$$\frac{1}{r^2}\frac{d}{dr}\left(r^2\frac{dU}{dr}\right) = -\frac{N_i e - N_e e}{\epsilon_0}$$

$$= \frac{2N_0 e}{\epsilon_0}\sinh\left(e\frac{U-U_0}{kT}\right) \qquad (12.8)$$

This is a non-linear equation. However if $kT > eU$ we can find its approximate solution by putting

$$\sinh\left(\frac{eU}{kT}\right) \simeq \frac{eU}{kT}$$

The equation (12.8) reduces to

$$\frac{1}{r^2}\frac{d}{dr}\left(r^2\frac{dU}{dr}\right) = \frac{2N_0 e^2}{\epsilon_0 kT}(U - U_0) \qquad (12.9)$$

and its solution is

$$U = U_0 + \frac{q}{4\pi\epsilon_0 r}\exp\left(-\frac{r}{\delta}\right) \qquad (12.10)$$

where r is the distance from the charge q and

$$\delta = \left(\frac{\epsilon_0 kT}{2N_0 e^2}\right)^{1/2}. \qquad (12.11)$$

Debye first introduced the concept of characteristic length in his study of strong electrolytes. Later, it was introduced in plasma physics. The parameter δ is known as **Debye length** and indicates the shortest distance at which electrons moving at random in the plasma, screen the coulomb field of the probe particle.

Langmuir who was the pioneer of the plasma science defined plasma as an ionised gas in which the Debye length is small compared to the size of the volume occupied by the gas.

12.2 Plasma Behaviour in Magnetic Field

Since the charged particles are always present in the plasma and the plasma is very frequently exposed to the effects of electric and magnetic fields, it would be worthwhile first to examine the interaction of the particles and these fields. Application of a constant electric field is not of any particular interest because it leads to the formation of a thin sheath of space charge which, as stated above, shields the major part of the plasma. A constant magnetic field, however, gives rise to some interesting effects which we shall discuss in this section. Let us consider the following cases.

Case I: Charged particles in homogeneous magnetic fields ($\mathbf{E} = 0$)

Suppose the lines of force of the magnetic field point perpendicularly out of the plane of the paper. Let a particle with mass m and charge q be shot into the magnetic field with a velocity $\mathbf{v}$ in a direction perpendicular to the lines of force as shown in Fig. 12.1. The Lorentz force

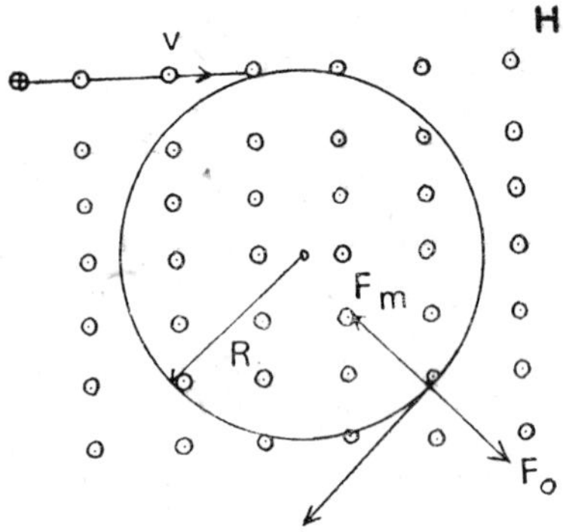

Fig. 12.1

$\mathbf{F} = q(\mathbf{v} \times \mathbf{B})$ acting on the particle is always at right angles to **v** as well as to **B** and, hence, causes the particle to trace a circular path in the plane of the paper. The force **F** must in every instant be in equilibrium with the centrifugal force. Therefore,

$$qvB = \frac{mv^2}{R} \tag{12.12}$$

where R is the radius of the orbit and is known as the **Larmor radius** of the orbit. From this, we find the angular frequency

$$\omega_c = \frac{v}{R} = \frac{qB}{m}. \tag{12.13}$$

This is known as the **cyclotron frequency**, its independence of v provides the physical mechanism for the design of low energy cyclotrons.

If we shoot the particle into the magnetic field under an angle other than a right angle with respect to its lines of force, we resolve **v** into two components: $v_{\parallel}$ parallel to the field and $v_{\perp}$ in the plane perpendicular to **B**. Since $v_{\parallel}$ is unaffected by the field, equation (12.12) may be written as

$$qv_{\perp}B = \frac{mv_{\perp}^2}{R}$$

i.e.

$$R = \frac{mv_{\perp}}{qB}. \tag{12.14}$$

The resulting curve will be a spiral as shown in Fig. 12.2, because the centre of the circular path with radius R will be moving on the line of force at uniform velocity $v_{\parallel}$. The particle, therefore, is gyrating about the line of force. In a sense the particle is "locked" to the field line.

Case II: Charged particles in crossed electric and magnetic fields

Suppose an additional homogeneous electric field **E** is superimposed

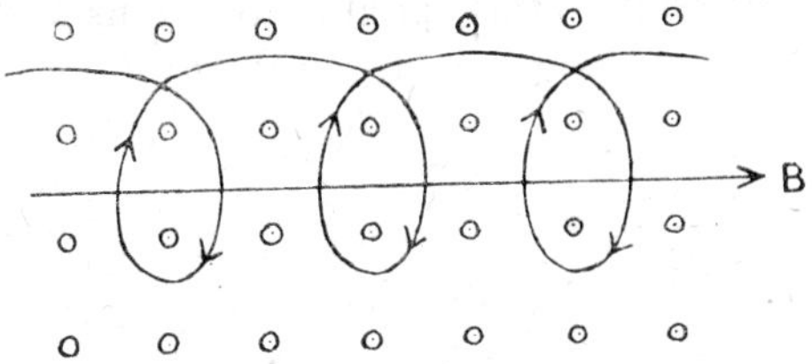

Fig. 12.2

on the homogeneous magnetie field **B** (**E** $\perp$ **B**). The trajectory is shown in Fig. 12.3. Three forces are now acting at every point along the path. The magnetic force $F_m = qvB$, the centrifugal force $F_c = \dfrac{mv^2}{R}$ and the electric force $F_e = qE$. At point A the magnetic and the electric forces act in the same sense and the trajectory is more strongly curved; at point B the forces partly balance each other and, therefore, the radius of curvature is larger. The resultant path is such as if the particle were uniformly rotating round a centre moving with a drift velocity $\mathbf{v}_D$. The expression for v_D can be found if we write

$$\mathbf{v} = \mathbf{v}_D + \mathbf{v}' \tag{12.15}$$

$$\therefore \quad \mathbf{F} = q(\mathbf{E} + \mathbf{v} \times \mathbf{B})$$

$$= q(\mathbf{E} + \mathbf{v}_D \times \mathbf{B} + \mathbf{v}' \times \mathbf{B}). \tag{12.16}$$

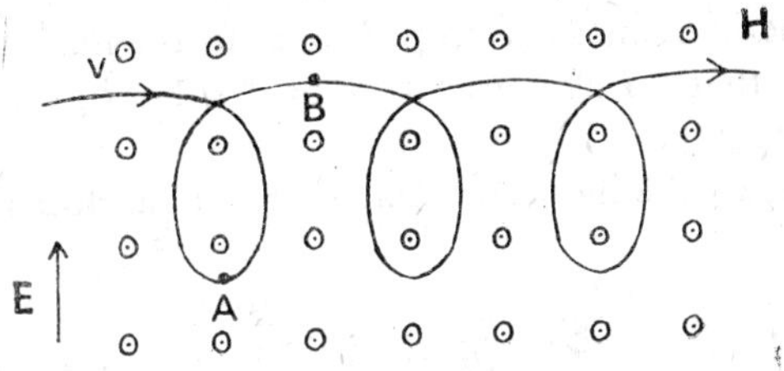

Fig. 12.3

If we choose
$$\mathbf{v}_D = \frac{\mathbf{E} \times \mathbf{B}}{B^2}, \tag{12.17}$$

the first two terms on the right hand side of (12.16) cancel each other and the force is given by

$$\mathbf{F} = q\mathbf{v}' \times \mathbf{B}$$

which is what we studied in Case I.

Case III: Charged particles in non-homogeneous magnetic field

The non-homogeneity of the magnetic field is expressed by the density of lines of force (Fig. 12.4). Where the field is large, the lines of forces are more densely spaced, the curvature of the trajectory is more marked and vice versa. The result of this variation of radius of curvature during particle motion is also to induce a particle drift.

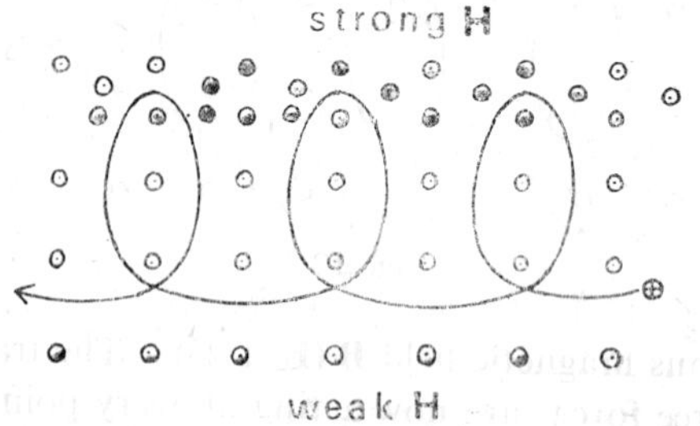

Fig. 12.4

A very important and useful property of gyrating charged particles is the magnetic moment μ which is defined as

μ = current due to the particle gyration $\times$ area enclosed by that current

$$= \frac{qv_\perp}{2\pi R}\,\pi R^2 = \frac{qv_\perp R}{2}$$

$$= \pi R^2 B\left(\frac{q^2}{2\pi m}\right) = \left(\frac{q^2}{2\pi m}\right)\Phi \qquad (12.18)$$

$$\left(\because \quad B = \frac{mv_\perp}{qR} \text{ and flux } \Phi = \pi R^2 B\right).$$

Thus, the magnetic moment is proportional to the flux Φ.

We can show that for slow variation of B either in space or time, μ is essentially constant.

By Faraday's law the induced e.m.f. $\mathcal{E}$ around a closed path is given by

$$\mathcal{E} = \frac{d\Phi}{dt} = \pi R^2 \frac{dB}{dt}. \qquad (12.19)$$

When the charge in the field is small, the work done upon a particle of charge q in one orbit is $q\mathcal{E}$ and this must be equal to $\frac{dW}{dt}$ multiplied by the time taken by the particle to go through one orbit, W being the work done. That is

$$q\mathcal{E} = \frac{2\pi R}{v}\frac{dW}{dt} = \frac{2\pi R}{v}\frac{dW}{dB}\frac{dB}{dt} = q\pi R^2 \frac{dB}{dt} \quad \text{(see 12.19)} \qquad (12.20)$$

$$\therefore \quad \frac{dW}{dB} = \frac{qRv}{2} = \mu \qquad (12.21)$$

Now $$W = \frac{1}{2}mv^2 = \mu B \quad \text{(from 12.18)} \qquad (12.22)$$

$$\therefore \quad \frac{dW}{dB} = B\frac{d\mu}{dB} + \mu \qquad (12.23)$$

Comparing this with (12.21) we find that $\frac{d\mu}{dB} = 0$, i.e. μ is independent of B.

The effects of such field variations, therefore, may be treated by noting the consequences of requiring the constancy of μ. For μ to be constant, the flux within particle orbit must remain constant. Thus, in a magnetic field that varies slowly with position, as the magnetic field lines converge, the particle orbit is compressed in such a way as to maintain Φ constant. The particle, therefore, is seen to move along the surface of a flux tube (Fig. 12.5).

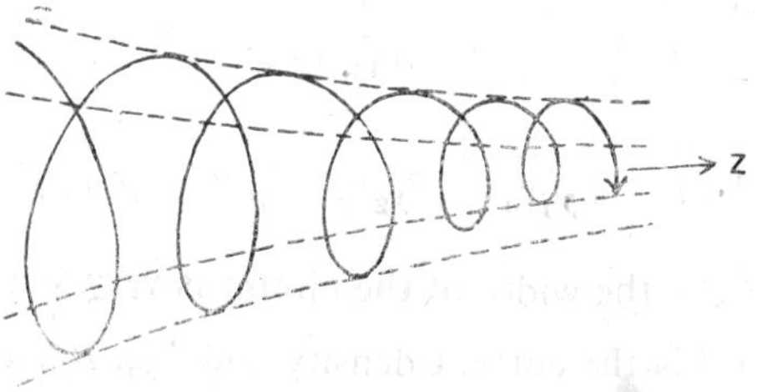

Fig. 12.5

12.3 Plasma as a Conducting Fluid—Magnetohydrodynamics

In principle, the properties of a plasma can be accounted for by considering the individual contributions made by all the particles and their interactions. In practice, however, this procedure is difficult and one has to take recourse to a macroscopic approach.

In the macroscopic treatment a plasma is considered to be divided into small volumes, each of which is large compared to the average spacings of the individual particles it contains, yet small compared to any distance over which the macroscopic properties vary appreciably. One associates with this volume the average values of velocity, magnetic field, density, temperature, pressure, conductivity etc. appropriate to that volume and the behaviour of the plasma is deduced in terms of the action of these volumes. This technique is generally used in the treatment of electromagnetic and hydrodynamic phenomena. In this and the following sections, we shall treat a conducting fluid—in particular plasma—subject simultaneously to the laws of electromagnetics and hydrodynamics. The treatment is known as **magnetohydrodynamics** or **hydromagnetics**.

With a view to obtain the equations of hydrodynamics that a plasma, regarded as a classical fluid, must obey, we shall first calculate the condition that a magnetic field **B** has to satisfy at a current sheet of infinite extent and a very small thickness 'd'. Figure 12.6 shows a sheet carrying a current I in the z-direction. It may easily be seen that the associated magnetic field **B** acts in the y-direction. Imagine a small rectangular circuit c in the sheet with one side of length y_0 in front of the sheet and the other behind it. By Ampère's law

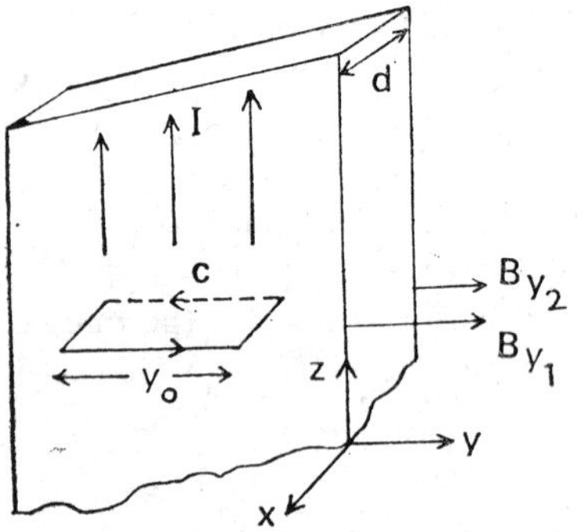

Fig. 12.6

$$\oint \mathbf{B}\cdot dl = By_1 y_0 - By_2 y_0 = \mu_0 I = \mu_0 j y_0 d \tag{12.24}$$

($\because$ the width of the circuit is very small)

where j amp m^{-2} is the current density and By_1, By_2 are the magnitudes of B in front of the sheet and behind it.

Hence, $$By_1 - By_2 = \mu_0 jd = \mu_0 j_s \tag{12.25}$$

where $j_s = jd$ is called the surface current density.

We may remind you that in Chapter 1, we showed that the electrostatic field **E** changes across a charge layer by an amount

$$E_1 - E_2 = \frac{\sigma}{\epsilon_0}. \tag{12.26}$$

The force per unit area on the current sheet can be determined by evaluating the integral

$$\mathbf{F} = \int_{-d/2}^{d/2} \mathbf{j} \times \mathbf{B} dx = -\hat{\mathbf{e}}_x \int_{-d/2}^{d/2} jB_y dx$$

($\because$ j is along z-direction)

$$= -\hat{\mathbf{e}}_x j \left[\int_{-d/2}^{0} By_2 dx + \int_{0}^{d/2} By_1 dx\right]$$

$$= -\hat{\mathbf{e}}_x j \frac{d}{2}(By_1 + By_2)$$

$$= -\hat{\mathbf{e}}_x (By_1 + By_2)\frac{(By_1 - By_2)}{2\mu_0} \quad \text{(From (12.25))}$$

$$= \frac{\hat{\mathbf{e}}_x}{2\mu_0}(B^2_{y_2} - B^2_{y_1}). \tag{12.27}$$

If the only current source is the sheet itself $By_1 = By_2$ giving $\mathbf{F} = 0$.

Consider now a situation in which other current sources are present and $By_1 > By_2$. The force now acts in the $-x$ direction. By analogy with hydrostatics we can consider the side where the magnetic field is higher as a region of higher pressure and write the expression for force as

$$\mathbf{F} = (\mathcal{P}_2{}^M - \mathcal{P}_1{}^M)\,\hat{\mathbf{e}}_x \tag{12.28}$$

where $\mathcal{P}^\mu$ is the magnetic pressure given by

$$\mathcal{P}^M = \frac{B^2}{2\mu_0} \tag{12.29}$$

This concept of magnetic pressure was introduced for the first time by Faraday. He thought of tubes of force as elastic filaments under tension in the direction of the field and compressed in the transverse direction. These ideas were later translated into mathematical language by Maxwell in terms of a stress tensor. Electromagnetic fields treated in this manner together with the hydrodynamics of a conducting fluid (liquid or gas) form the basis of magnetohydrodynamics.

Let us consider the behaviour of a conducting fluid in an electromagnetic field. If $\mathbf{j}$ is the current density within the fluid and $\mathcal{P}$ the fluid pressure, the force acting on unit volume of the fluid is

$$\mathbf{F} = \mathbf{j} \times \mathbf{B} - \nabla \mathcal{P} \tag{12.30}$$

This is generally known as the fundamental equation of magnetohydrodynamics.

Using the relation $\nabla \times \mathbf{B} = \mu_0 \mathbf{j}$, we can write (12.30) as

$$\begin{aligned}
\mathbf{F} &= -\frac{\mathbf{B}}{\mu_0} \times (\nabla \times \mathbf{B}) - \nabla \mathcal{P} \\
&= -\nabla\left(\frac{B^2}{2\mu_0}\right) + \frac{1}{\mu_0}\left(\mathbf{B}\cdot\nabla\right)\mathbf{B} - \nabla \mathcal{P} \\
&= \frac{1}{\mu_0}\left(\mathbf{B}\cdot\nabla\right)\mathbf{B} - \nabla\left(\mathcal{P} + \frac{B^2}{2\mu_0}\right).
\end{aligned} \tag{12.31}$$

This, in a way, justifies the suggestion made above (12.29), that $\frac{B^2}{2\mu_0}$ may be regarded as the magnetic pressure. Note that $\frac{B^2}{2\mu_0}$ accounts only for the part of the effect of the force $\mathbf{j} \times \mathbf{B}$. The remaining contribution comes from the term $\frac{1}{\mu_0}\left(\mathbf{B}\cdot\nabla\right)\mathbf{B}$. You, perhaps, wonder why this term did not appear in equation (12.28). This is because in the current sheet example, $\mathbf{B}$ being unidirectional, div $\mathbf{B} = 0$ guarantees that $\mathbf{B}$ does not change along the field direction and since the space variation occurs only at right angles to $\mathbf{B}$ it follows that $(\mathbf{B}\cdot\nabla)\,\mathbf{B} = 0$.

For equilibrium $\mathbf{F} = 0$, and, hence, if the first term vanishes as in the case of current sheet

$$\mathscr{P} + \frac{B^2}{2\mu_0} = \text{constant} \tag{12.32}$$

is the condition for static equilibrium.

12.4 Magnetic Confinement-Pinch Effect

We find from (12.30) that since the condition for equilibrium is $\mathbf{F} = 0$

$$\nabla \cdot \mathscr{P} = \mathbf{j} \times \mathbf{B}. \tag{12.33}$$

The scalar product with **B** and **J** gives

$$\mathbf{B} \cdot \nabla \mathscr{P} = \mathbf{B} \cdot (\mathbf{j} \times \mathbf{B}) = 0 \tag{12.34}$$

and

$$\mathbf{j} \cdot \nabla \mathscr{P} = \mathbf{j} \cdot (\mathbf{j} \times \mathbf{B}) = 0. \tag{12.35}$$

These equations indicate that both **B** and **J** lie on surfaces of constant pressure. If these surfaces happen to be **closed,** then since (according to equation (12.34) and (12.35)) no **B**-lines or **j**-lines can cross them, they may be viewed as made up from a winding of **B** lines and of **j**-lines as shown in Fig. 12.7. In these isobaric surfaces, the pressure increases from outside towards the axis, the force $\mathbf{j} \times \mathbf{B}$ also points towards the axis. This means the plasma is contained by the $\mathbf{j} \times \mathbf{B}$ force. This is known as **magnetic confinement**.

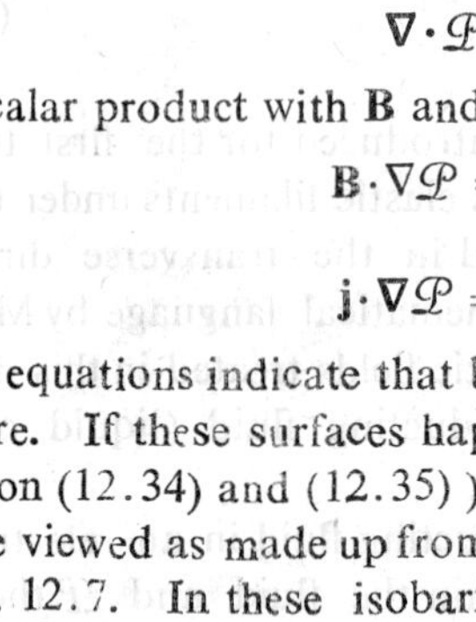

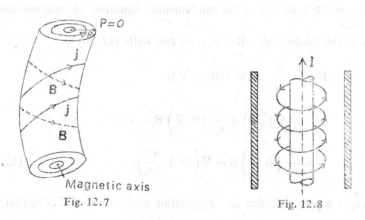

Fig. 12.7

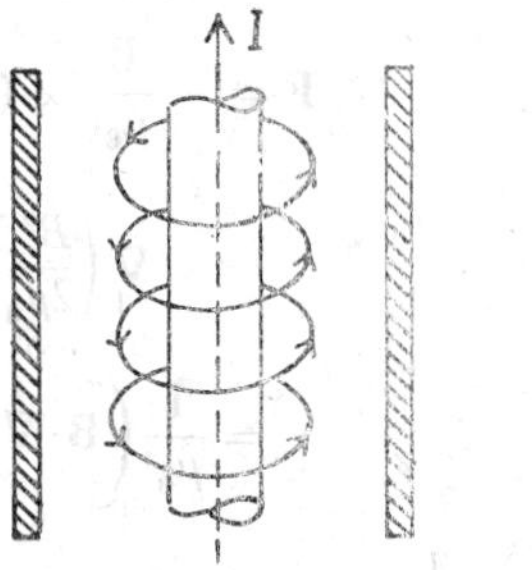

Fig. 12.8

A simple example may explain how the magnetic field can affect the dynamical state of a plasma. We consider a cylindrical plasma with an electric current flowing through it axially (Fig. 12.8). Associated with this column there will be azimuthal **B**-field lines and a force $\mathbf{j} \times \mathbf{B}$ directed towards the axis which tends to "pinch" or squeeze the plasma. Under the action of this force the plasma contracts until the compressing electrodynamic force is compensated by the increased kinetic pressure $p = NkT$. This phenomenon is called "**pinch effect**".

12.5 Instabilities

Experience shows that there is no stable equilibrium pinch. That is, a slight deviation from the equilibrium state tends to increase and results into the disintegration of the form of plasma. Plasma instabilities are important in a wide range of physical situations in which plasmas play a part.

Figure 12.9 shows what is called a *kink instability* which occurs when a linear pinch is bent due to some perturbation. The magnetic lines of force attenuate, i.e. the field is weakened on the outside of the curve, and they grow denser, i.e. the field is intensified, on its inside. The stronger field causes stronger magnetic pressure in the bend which is conducive to further bending resulting ultimately into the breaking up of the linear pinch.

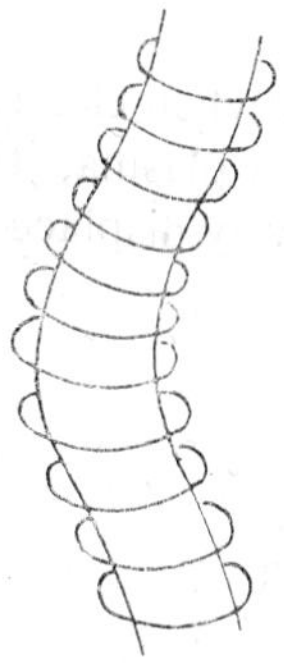

Fig. 12.9

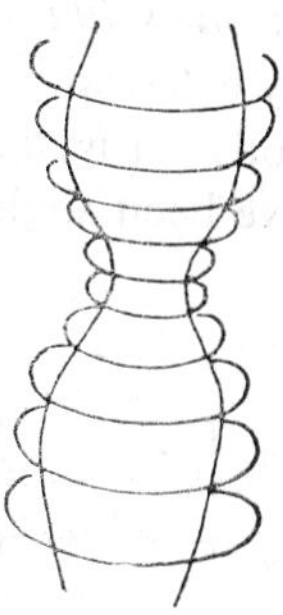

Fig. 12.10

The so-called *sausage instability* is shown in Fig. 12.10. This instability occurs if the plasma columns is pinched, by chance.

Since the field **B** is inversely proportional to r, the pressure $\frac{B^2}{2\mu_0}$ is greater in the squeezed region, i.e. at the neck which causes further pinching ultimately disrupting the column.

Apart from these instabilities, several other kinetic instabilities exist which are caused either by chance or by artificial deviation of the distribution of particles from Maxwellian distribution.

In order to get over the difficulty of instability various schemes for containing plasma are proposed. One such system is what is known as **tokamak**—an axially symmetric toroidal system (Fig. 12.11) in which the plasma is contained by the magnetic field of the current which flows along its axis. A very powerful magnetic field is applied parallel to the current which suppresses the instability.

12.6 Plasma Waves

In a plasma there can develop and propagate a variety of oscillations and waves in a very wide frequency range. We shall briefly consider two of these:

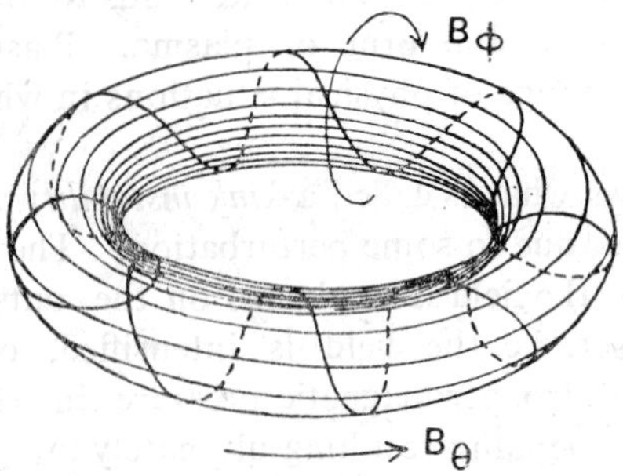

Fig. 12.11

1. *Plasma-electron oscillations*

High frequency electron oscillations result when electrons are displaced with respect to ions. Fig. 12.12 represents a plasma, the electrons and ions of which have been slightly but collectively displaced relative to each

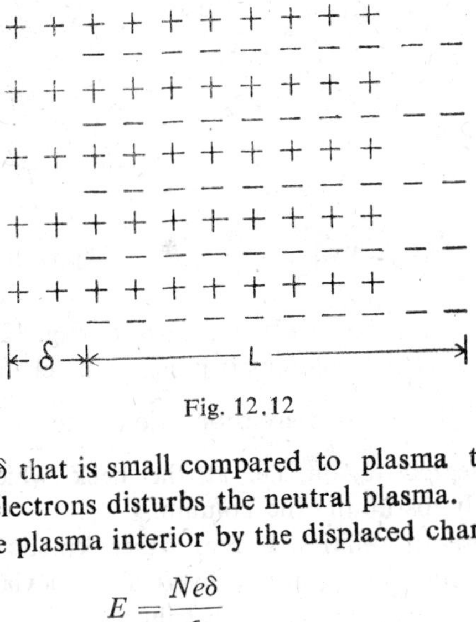

Fig. 12.12

other a distance δ that is small compared to plasma thickness L. The displacement of electrons disturbs the neutral plasma. The electric field **E** produced in the plasma interior by the displaced charges is given by

$$E = \frac{Ne\delta}{\epsilon_0} \tag{12.36}$$

where N is the charge density. The force per unit area acting on the electron in the plasma interior is $LNeE = \dfrac{N^2e^2\delta L}{\epsilon_0}$. The mass per unit area upon which this force acts is LmN, where m is the electron mass. The equation of motion for each electron, therefore is

$$LmN\frac{d^2\delta}{dt^2} + \frac{N^2e^2L\delta}{\epsilon_0} = 0 \tag{12.37}$$

This equation has the solution

$$\delta = A\exp\left\{i\left(\frac{Ne^2}{m\epsilon_0}\right)^{1/2}t\right\} = A\exp\,(i\omega_p t) \tag{12.38}$$

where

$$\omega_p = 2\pi\nu_p = \left(\frac{Ne^2}{m\epsilon_0}\right)^{1/2} \tag{12.39}$$

$$\therefore \qquad \nu_p = \frac{1}{2\pi}\left(\frac{Ne^2}{m\epsilon_0}\right)^{1/2} \tag{12.40}$$

is the oscillation frequencey commonly called the **plasma frequency**.

These oscillations can propagate in the plasma as longitudinal electrostatic waves. There can also be generated waves of a much lower frequency due to the longitudinal oscillations of ions—the so called **ion sound**; but we shall not discuss them here.

2. *Hydromagnetic or Alfvèn waves*

It can be shown that the stress due to a magnetic field is equivalent to an isotropic magnetic pressure $\frac{B^2}{2\mu_0}$ and a tension $\frac{B^2}{\mu_0}$ along the magnetic field lines as shown in Fig. 12.13. Stretching the tube increases the tension. The situation is analogous to that of a stretched elastic string. We know that by plucking the string transverse waves can be set up which propagate along the string. Should it not be possible, then, to create transverse waves propagating along the magnetic field lines by "plucking" a tube of force? Existence of such waves was first recognized by Hannes Alfvèn in 1942. Careful experiments conducted in plasma confirmed subsequently the existence of such waves which are now known as **Alfven waves**.

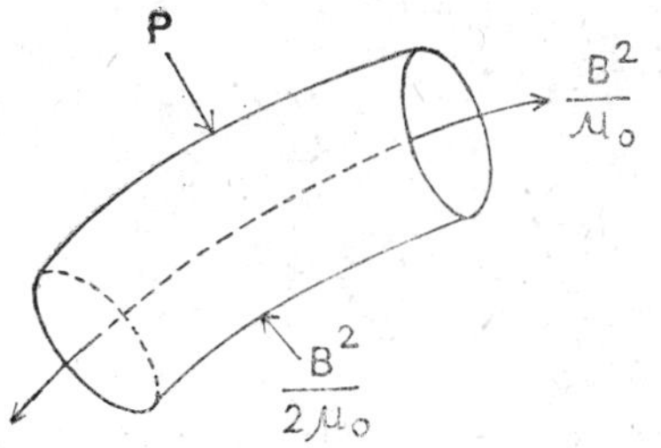

Fig. 12.13

In the case of strings, the wave velocity is given in terms of the tension T and the linear mass density ρ of the string as

$$v = \left(\frac{T}{\rho}\right)^{1/2} \tag{12.41}$$

Making the analogy to hydromagnetic plasma waves, we replace T by the magnetic tension $\frac{B^2}{\mu_0}$ and ρ becomes the plasma mass density. Thus for the hydromagnetic waves we have for the phase velocity

$$v_A = \frac{B}{(\mu_0 \rho)^{1/2}} \tag{12.42}$$

The velocity v_A is often called **Alfven velocity**.

PROBLEMS

12.1 A current flows axially in a cylindrical tube of plasma. The plasma pressure vanishes at $r = r_0$ and $P = N(r)kT$. Show that

$$2\,NkT = \frac{\pi}{2\mu_0}\left(rB\right)^2_{r=r_0}$$

where $$N = \int_0^{r_0} n(r)\ 2\pi r dr$$

12.2 Using the result of the preceding example, prove the following Bennett's relation for the linear pinch

$$I^2 = \frac{16\pi}{\mu_0} NkT.$$

12.3 The solar corona consists of ionized hydrogen with approximately 10^{12} particles/m³. Assuming the field in the corona to be $\mathbf{B} = 1000$ gauss, calculate the phase velocity of Alfven waves.

Appendix A

Vectors

Vector analysis is a powerful mathematical machinery devised to handle the actual physical processes easily and rapidly. It has several other advantages. The equations of electrodynamics become more concise if written in vector notation and the physical content becomes more clear.

A 1 The concept of a vector

In physics, we have to handle various kinds of quantities, two of which are more important. One of these is characterized by the assignment of a single number. Typical examples of this kind are: mass, temperature, density etc. These can be specified by giving their magnitude alone. Such quantities are called **scalars.** There are other quantities in physics which cannot be specified completely by a single number. Consider, for example, a particle moving from a point 0 to a point P distant 'r' cm from 0. This statement does not give the exact location of P, since P can be anywhere over the surface of a sphere of radius r. To specify the position of P uniquely, we have to keep track of which way the particle is going and state the direction of the displacement. Such quantities which require, besides the magnitude, the direction for their characterization are called **vectors.** It is customary to represent vectors by letters in **bold-face** type and scalars in *italics*. The displacement OP is represented by the symbol $\mathbf{r}$ which gives both magnitude and direction of the displacement.

Suppose, now, the point 0 at which the particle is initially located is the origin of a coordinate system. The location of the point P in this system is given by three numbers—the coordinates x, y, z. The symbol $\mathbf{r}$, therefore, is **not** a single number; it represents a set of three numbers considered as a whole. The three numbers x, y, z, are called the **components** of the vector $\mathbf{r}$ in the coordinate system considered. If a different coordinate system with the same origin 0 is used, the coordinates of P and, hence, the components of $\mathbf{r}$ will be different, say x', y', z'; but $\mathbf{r}$ still represents the displacement. In other words, the vector $\mathbf{r}$ is independent of the coordinate system used. If we change the coordinate system, we do not have to change the symbols of our equations if written in vector notation. This point of view results in an economy of symbolism which is yet another advantage of using vector technique.

A vector may be represented graphically by an arrow (Fig. A.1). The length and the direction of the arrow give the magnitude and the direction of the vector.

Fig. A.1

A vector having unit magnitude is called a **unit vector**. A unit vector in the direction r is represented by the symbol $\hat{\mathbf{e}}_r$.

A.2 Vector algebra

Mathematical operations with scalars obey the rules of ordinary algebra. Vectors being directed quantities, cannot be added or subtracted by ordinary methods and, hence, we have to formulate appropriate rules for combining vectors in various ways

(1) Addition and subtraction of vectors.

Two displacements of a point may be combined into a single equivalent displacement by following a simple procedure. Suppose there is a displacement OP represented by a vector **A** and this is followed by another displacement PP' represented by the vector **B** (Fig. A.2a). The final position of the point could have been obtained if the point had been directly displaced along OP'. We represent this displacement by the vector **C** which we call the sum of **A** and **B**, i.e. $\mathbf{C} = \mathbf{A} + \mathbf{B}$.

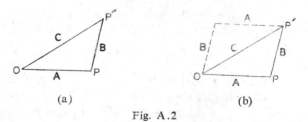

Fig. A.2

The resultant vector can be represented in an equally effective manner by using a parallelogram with sides **A** and **B** (Fig. A.2b). The diagonal of the parallelogram gives the vector **C**. The addition of vectors is commutative, i.e.

$$\mathbf{A} + \mathbf{B} = \mathbf{B} + \mathbf{A}. \tag{A1}$$

The law of vector addition, therefore, is the parallelogram law of addition which states that the sum of two vectors **A** and **B** is given in magnitude and direction by the diagonal of the parallelogram formed by the sides representing the vectors **A** and **B**.

We now define a vector more explicitly: A vector is a physical quantity which has both magnitude and direction and which follows the parallelogram law of addition.

The sum of two vectors having the same direction and sense, is obtained by adding their magnitudes, the direction remaining the same. The vector $m\mathbf{A}$, where 'm' is a pure number, is a vector having the same direction as that of $\mathbf{A}$, but magnitude m times larger. This leads to an important representation for a vector. If A is the magnitude of the vector $\mathbf{A}$ and $\hat{\mathbf{e}}_A$ is the unit vector in the direction of $\mathbf{A}$, then we can write

$$\mathbf{A} = A\hat{\mathbf{e}}_A \qquad \text{(A2)}$$

Further, if A_x, A_y, A_z are the components of the vector $\mathbf{A}$ along the coordinates and $\hat{\mathbf{e}}_x$, $\hat{\mathbf{e}}_y$, $\hat{\mathbf{e}}_z$ are the unit vectors in the direction of the axes, then the vector $\mathbf{A}$ can be represented as

$$\mathbf{A} = A_x\hat{\mathbf{e}}_x + A_y\hat{\mathbf{e}}_y + A_z\hat{\mathbf{e}}_z. \qquad \text{(A3)}$$

The difference of two vectors $\mathbf{A}$ and $\mathbf{B}$ can be obtained by adding the vector $-\mathbf{B}$ to $\mathbf{A}$, i.e.

$$\mathbf{A} - \mathbf{B} = \mathbf{A} + (-\mathbf{B}) \qquad \text{(A4)}$$

where $-\mathbf{B}$ is the vector having the same magnitude as $\mathbf{B}$ but direction opposite to that of $\mathbf{B}$. Hence, to obtain the difference graphically, reverse the direction of $\mathbf{B}$ and complete the parallelogram as before. In Fig. A.3 OP gives the sum of the vectors $\mathbf{C} = \mathbf{A} + \mathbf{B}$ and OQ the difference $\mathbf{D} = \mathbf{A} - \mathbf{B}$.

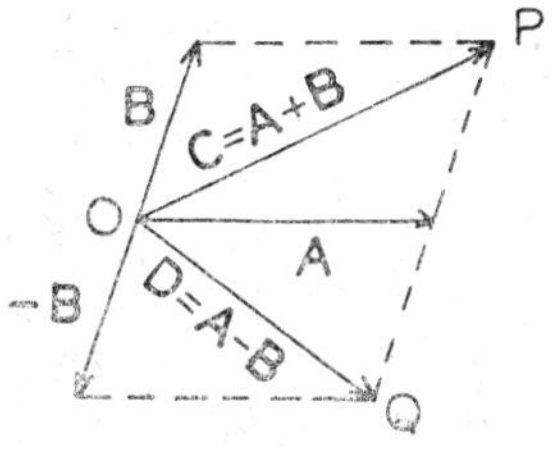

Fig A.3

(2) Product of vectors

We often come across, in physics, certain combination of vectors which have the properties of products. The product of two vectors may be a scalar or a vector quantity depending upon how the product is defined.

(a) Scalar product or dot product of two vectors: $\mathbf{A}\cdot\mathbf{B}$

We define the **scalar product** or **dot product** of two vectors as the number equal to the product of the magnitudes of the two vectors multiplied by the cosine of the angle between them. Thus,

$$\mathbf{A}\cdot\mathbf{B} = AB\cos\theta. \qquad \text{(A5)}$$

Clearly, the scalar product is the product of the magnitude of one vector and the projection of the other on it.

One can easily verify that the scalar product has the following properties of ordinary numbers:

(i) $\mathbf{A}\cdot\mathbf{B} = \mathbf{B}\cdot\mathbf{A}$ (A6)

(ii) $\mathbf{A}\cdot(\mathbf{B} + \mathbf{C} + \mathbf{D} + \ldots) = \mathbf{A}\cdot\mathbf{B} + \mathbf{A}\cdot\mathbf{C} +$ (A7)

It may be noted that the scalar product vanishes even when none of the factors is zero, provided the angle between them is 90°.

Thus,

$$\hat{\mathbf{e}}_x \cdot \hat{\mathbf{e}}_y = \hat{\mathbf{e}}_y \cdot \hat{\mathbf{e}}_z = \hat{\mathbf{e}}_z \cdot \hat{\mathbf{e}}_x = 0 \tag{A8}$$

$$\hat{\mathbf{e}}_x \cdot \hat{\mathbf{e}}_x = \hat{\mathbf{e}}_y \cdot \hat{\mathbf{e}}_y = \hat{\mathbf{e}}_z \cdot \hat{\mathbf{e}}_z = 1. \tag{A9}$$

You can, further, verify that if $\mathbf{A} = A_x\hat{\mathbf{e}}_x + A_y\hat{\mathbf{e}}_y + A_z\hat{\mathbf{e}}_z$ and $\mathbf{B} = B_x\hat{\mathbf{e}}_x + B_y\hat{\mathbf{e}}_y + B_z\hat{\mathbf{e}}_z$, then,

$$\mathbf{A} \cdot \mathbf{A} = A^2 \tag{A10}$$

$$\mathbf{A} \cdot \mathbf{B} = A_xB_x + A_yB_y + A_zB_z. \tag{A11}$$

(b) Vector product or cross-product: $\mathbf{A} \times \mathbf{B}$

The **vector product** or **cross product** **C** of two vectors **A** and **B** is defined as a vector whose magnitude is equal to the product of the magnitudes of **A** and **B** multiplied by the sine of the angle between them and whose direction is normal to the plane containing **A** and **B** and in a sense such that the vectors **A**, **B**, **C** form a right-handed system (Fig. A.4)

$$\mathbf{C} = \mathbf{A} \times \mathbf{B} = AB \sin\theta\, \hat{\mathbf{e}}_c. \tag{A12}$$

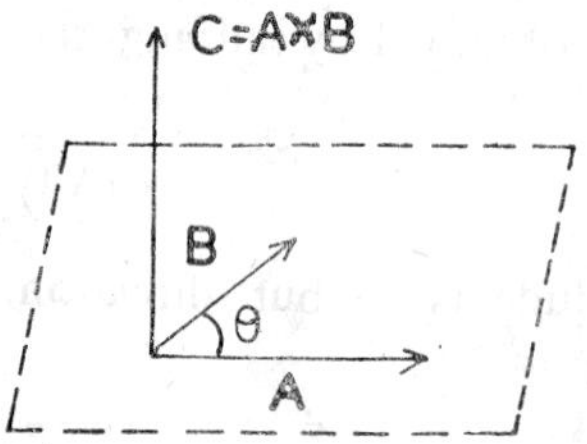

Fig. A.4

From geometrical considerations it can be shown that the magnitude of the cross-product of two vectors is equal to the area of the parallelogram having the two vectors as its sides. This suggests that it may be useful to represent areas by vectors. The area is represented by a vector perpendicular to the area having magnitude equal to the area. The sense of the vector is related to the sense in which the contour of the area is described.

As an example of vector product consider a torque produced by force **F** acting at a point P about a point 0. Let the position vector $OP = \mathbf{r}$ make an angle θ with the direction of **F**. (Fig. A.5). Then the magnitude of the torque is given by $N = rF \sin\theta$.

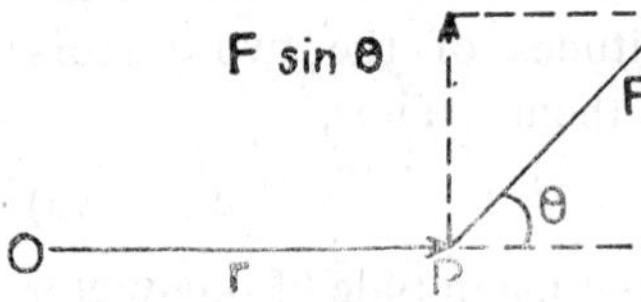

Fig. A.5

The torque produces rotation and, hence, is regarded as a vector quantity, its direction being determined by the right-hand rule,

$$\mathbf{N} = \mathbf{r} \times \mathbf{F} = rF \sin\theta\, \hat{\mathbf{e}}_n \tag{A13}$$

The following properties of vector products are easily verified:

(a) $\mathbf{A} \times \mathbf{B} = -\mathbf{B} \times \mathbf{A}$ (A14)

(b) $\mathbf{A} \times (\mathbf{B} + \mathbf{C} + \mathbf{D} +) = \mathbf{A} \times \mathbf{B} + \mathbf{A} \times \mathbf{C} +$ (A15)

(c) $\mathbf{A} \times \mathbf{A} = 0$ (A16)

(d) $\hat{\mathbf{e}}_x \times \hat{\mathbf{e}}_x = \hat{\mathbf{e}}_y \times \hat{\mathbf{e}}_y = \hat{\mathbf{e}}_z \times \hat{\mathbf{e}}_z = 0$ (A17)

(e) $\hat{\mathbf{e}}_x \times \hat{\mathbf{e}}_y = \hat{\mathbf{e}}_z;\ \hat{\mathbf{e}}_y \times \hat{\mathbf{e}}_z = \hat{\mathbf{e}}_x;\ \hat{\mathbf{e}}_z \times \hat{\mathbf{e}}_x = \hat{\mathbf{e}}_y$ (A18)

(f)
$$\mathbf{A} \times \mathbf{B} = (\mathbf{A}_y\mathbf{B}_z - \mathbf{B}_y\mathbf{A}_z)\hat{\mathbf{e}}_x + (\mathbf{A}_z\mathbf{B}_x - \mathbf{A}_x\mathbf{B}_z)\hat{\mathbf{e}}_y + (\mathbf{A}_x\mathbf{B}_y - \mathbf{A}_y\mathbf{B}_x)\hat{\mathbf{e}}_z$$

$$= \begin{vmatrix} \hat{\mathbf{e}}_x & \hat{\mathbf{e}}_y & \hat{\mathbf{e}}_z \\ A_x & A_y & A_z \\ B_x & B_y & B_z \end{vmatrix} \tag{A19}$$

(c) Multiple products.

With the aid of dot and cross products of two vectors, it is possible to build multiple products involving several vectors. We shall discuss here two kinds of triple products which are specially important.

(1) Scalar triple product: $\mathbf{A}\cdot(\mathbf{B} \times \mathbf{C})$

In terms of cartesian components

$$\mathbf{A}\cdot(\mathbf{B} \times \mathbf{C}) = (A_x\hat{\mathbf{e}}_x + A_y\hat{\mathbf{e}}_y + A_z\hat{\mathbf{e}}_z)\cdot \begin{vmatrix} \hat{\mathbf{e}}_x & \hat{\mathbf{e}}_y & \hat{\mathbf{e}}_z \\ B_x & B_y & B_z \\ C_x & C_y & C_z \end{vmatrix}$$

$$= A_x(B_yC_z - B_zC_y) + A_y(B_zC_x - B_xC_z) + A_z(B_xC_y - B_yC_x)$$

$$= \begin{vmatrix} A_x & A_y & A_z \\ B_x & B_y & B_z \\ C_x & C_y & C_z \end{vmatrix} \tag{A20}$$

Consider a parallelogram drawn with the three vectors **A**, **B**, **C**, as its edges (Fig. A.6). We have seen that the product $\mathbf{B} \times \mathbf{C}$ is perpendicular

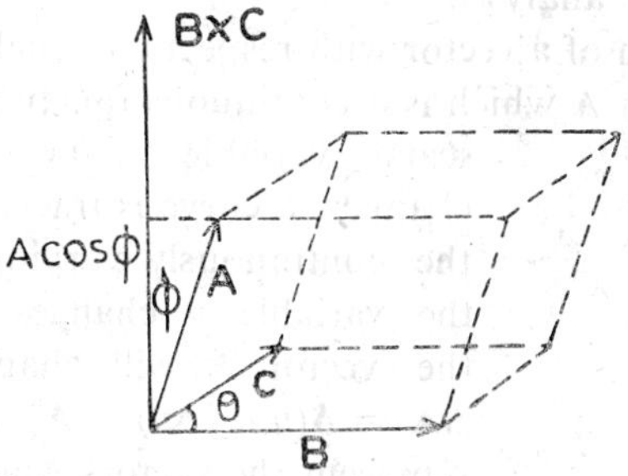

Fig. A.6

to the base and its magnitude is equal to the area of the base. Therefore,

$$\mathbf{A}\cdot(\mathbf{B}\times\mathbf{C}) = [\text{Projection of } \mathbf{A} \text{ on } \mathbf{B}\times\mathbf{C}] \times [\text{Area of the parallelogram}]$$
$$= \text{Height of the parallelopiped} \times \text{Area of the base of the parallelopiped}$$
$$= \text{Volume of the parallelopiped.} \tag{A21}$$

By taking various faces in turn we find

$$\mathbf{A}\cdot(\mathbf{B}\times\mathbf{C}) = \mathbf{B}\cdot(\mathbf{C}\times\mathbf{A}) = \mathbf{C}\cdot(\mathbf{A}\times\mathbf{B})$$
$$= \text{Volume of the parallelopiped.} \tag{A22}$$

This shows that if a cyclic change is made in the sequence of **A, B, C**, the scalar triple product remains the same. The parentheses in this product is superfluous because the product $\mathbf{A}\cdot\mathbf{B}\times\mathbf{C}$ always means the scalar product of **A** with $\mathbf{B}\times\mathbf{C}$ since $(\mathbf{A}\cdot\mathbf{B})\times\mathbf{C}$ is meaningless.

The relation

$$\mathbf{A}\cdot\mathbf{B}\times\mathbf{C} = \mathbf{A}\times\mathbf{B}\cdot\mathbf{C} \tag{A23}$$

shows that the cross and dot may be exchanged at will, provided the cyclic order of the vectors is retained. Hence, for convenience we can drop the dot and cross and represent the product by the symbol (ABC). If $(ABC) = 0$, and none of the vectors is zero, we conclude that the vectors are coplanar.

(2) Vector triple product: $\mathbf{A}\times(\mathbf{B}\times\mathbf{C})$

In this product, the vector $\mathbf{B}\times\mathbf{C}$ is perpendicular to the plane containing **B** and **C**. By the same reasoning, the vector $\mathbf{A}\times(\mathbf{B}\times\mathbf{C})$ is perpendicular to the plane containing **A** and $\mathbf{B}\times\mathbf{C}$, i.e. the vector $\mathbf{A}\times(\mathbf{B}\times\mathbf{C})$ will lie in the plane containing **B** and **C** and further it will be perpendicular to **A**.

The following most frequently used relation can be proved geometrically or analytically

$$\mathbf{A}\times(\mathbf{B}\times\mathbf{C}) = \mathbf{B}(\mathbf{A}\cdot\mathbf{C}) - \mathbf{C}(\mathbf{A}\cdot\mathbf{B}) \tag{A24}$$

A.3 Vector calculus

We shall now discuss the operations of differentiation and integration of vectors, as these concepts are necessary for defining the various operators useful in vector analysis.

(i) Differentiation of a vector with respect to a scalar.

Consider a vector **A** which is a continuous function of a continuous scalar variable u, i.e. $\mathbf{A} = \mathbf{A}(u)$. As u changes, a curve is traced by the terminus of the continuously varying **A** (Fig. A.7). As the variable u changes from u to $u + \Delta u$, the vector **A** will change by an amount $\Delta\mathbf{A} = \mathbf{A}(u + \Delta u) - \mathbf{A}(u)$. Let OP, OQ represent the vectors **A** and $\mathbf{A} + \Delta\mathbf{A}$ respectively. By the law of vector addition, PQ represents the increment $\Delta\mathbf{A}$ in vector **A**.

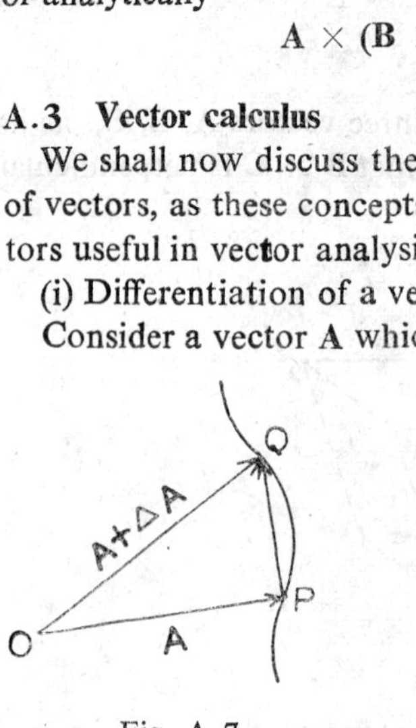

Fig. A.7

In analogy with scalar functions, we define the derivative $\frac{d\mathbf{A}}{du}$ as

$$\frac{d\mathbf{A}}{du} = \lim_{\Delta u \to 0} \frac{\Delta \mathbf{A}}{\Delta u} = \frac{\mathbf{A}(u + \Delta u) - \mathbf{A}(u)}{\Delta u} \tag{A25}$$

The derivative $\frac{d\mathbf{A}}{du}$ is a vector whose direction is the limiting direction of $\Delta\mathbf{A}$ as $\Delta u \to 0$. That is, the direction of the derivative lies along the tangent to the curve at the point A as $\Delta u \to 0$.

The rules for the differentiation of products also correspond to those for ordinary scalar functions except that the order of the vectors must not be changed in cases involving vector products. Thus,

$$\frac{d(\mathbf{A}\cdot\mathbf{B})}{du} = \frac{d\mathbf{A}}{du}\cdot\mathbf{B} + \mathbf{A}\cdot\frac{d\mathbf{B}}{du} \tag{A26}$$

$$\frac{d(\mathbf{A}\times\mathbf{B})}{du} = \frac{d\mathbf{A}}{du}\times\mathbf{B} + \mathbf{A}\times\frac{d\mathbf{B}}{du} \tag{A27}$$

If $\mathbf{A} = \mathbf{B}$ equation (A26) gives

$$\frac{d(\mathbf{A}\cdot\mathbf{A})}{du} = \frac{d}{du}(A^2), \text{ i.e. } 2\mathbf{A}\cdot\frac{d\mathbf{A}}{du} = 2A\frac{dA}{du}$$

or

$$\mathbf{A}\cdot\frac{d\mathbf{A}}{du} = A\frac{dA}{du} \tag{A28}$$

If $\mathbf{A}$ is constant in magnitude, we have $\frac{dA}{du} = 0$. Hence, $\mathbf{A}\cdot\frac{d\mathbf{A}}{du} = 0$. Since neither $\mathbf{A}$ nor $\frac{d\mathbf{A}}{du}$ vanishes, this means that the derivative of a vector of constant length is perpendicular to the vector.

(ii) Integration of vectors

The usual procedure of integration of scalar quantities can be used for integration of vectors. Generally, a vector integral is converted into a scalar integral and then evaluated by usual methods.

(a) Line integrals

The line integrals we often come across in physics are of the type

$$\int_C \phi d\mathbf{r}, \int_C \mathbf{V}\cdot d\mathbf{r}, \int_C \mathbf{V}\times d\mathbf{r}$$

where $\phi(x,y,z)$ and $\mathbf{V}(x, y, z)$ are scalar and vector point functions respectively and c a curve along which the integration is to be carried out.

Since $d\mathbf{r} = dx\hat{\mathbf{e}}_x + dy\hat{\mathbf{e}}_y + dz\hat{\mathbf{e}}_z$

$$\int_C \phi d\mathbf{r} = \hat{\mathbf{e}}_x \int \phi dx + \hat{\mathbf{e}}_y \int \phi dy + \hat{\mathbf{e}}_z \int \phi dz, \tag{A29}$$

$$\int_C \mathbf{V}\cdot d\mathbf{r} = \int_C V_x dx + \int_C V_y dy + \int_C V_z dz \tag{A30}$$

and

$$\int_C \mathbf{V} \times d\mathbf{r} = \hat{\mathbf{e}}_x \int_C (V_y dz - V_z dy) + \hat{\mathbf{e}}_y \int_C (V_z dx - V_x dz)$$
$$+ \hat{\mathbf{e}}_z \int_C (V_x dy - V_y dx). \quad \text{(A31)}$$

The integrals on the right-hand side can be evaluated by the usual methods.

(b) Surface integrals

The surface integrals of interest in physics are of the type

$$\int_S \phi d\sigma, \int_S \mathbf{V} \cdot d\boldsymbol{\sigma}, \int_S \mathbf{V} \times d\boldsymbol{\sigma}$$

where $d\sigma$ is an element of area which, as we have shown before can be represented as a vector. As in the case of line integrals, these integrals are first written in scalar form and then integrated.

(c) Volume integral

This is simpler to evaluate as the volume element $d\tau$ is a scalar quantity

$$\int_\tau \mathbf{V} d\tau = \hat{\mathbf{e}}_x \int_\tau V_x d\tau + \hat{\mathbf{e}}_y \int_\tau V_y d\tau + \hat{\mathbf{e}}_z \int_\tau V_z d\tau \quad \text{(A32)}$$

which can be readily integrated.

(iii) Scalar and Vector fields

If a certain quantity Q has a definite value at every point within a certain region of space, then this region is called the field of Q. If Q is a scalar quantity, the field is a **scalar field**; if Q is a vector quantity, the field is a **vector field**. Consider, for example, the region of space around a heated body. To every point in the region, there corresponds a definite value of temperature. This region of space is a scalar field-temperature field. As an example of vector field, you may consider the region surrounding an electric charge. To every point in the region there is associated an electric field with definite magnitude and direction.

(iv) Gradient of a scalar

In the study of a scalar field, it is sometimes necessary to know the rate of change of the scalar point function as we move from one point of the field to a neighbouring point. Let $\phi(\mathbf{r})$ be a scalar point function at a point defined by the position vector $\mathbf{r}$

$$\phi(\mathbf{r}) = \phi(x, y, z).$$

The change in the value of this function corresponding to a displacement of amount dr depends on the direction of the displacement. The change in $\phi(x, y, z)$ corresponding to a movement from the point (x, y, z) to the point $(x + dx, y + dy, z + dz)$ is given by the total differential $d\phi$

$$d\phi = \left(\frac{\partial \phi}{\partial x}\right) dx + \left(\frac{\partial \phi}{\partial y}\right) dy + \left(\frac{\partial \phi}{\partial z}\right) dz \quad \text{(A33)}$$

Since $\qquad d\mathbf{r} = dx\hat{\mathbf{e}}_x + dy\hat{\mathbf{e}}_y + dz\hat{\mathbf{e}}_z$

we can express the right-hand side of (A33) as the scalar proauct of two vectors

$$d\phi = \left(\hat{\mathbf{e}}_x \frac{\partial \phi}{\partial x} + \hat{\mathbf{e}}_y \frac{\partial \phi}{\partial y} + \hat{\mathbf{e}}_z \frac{\partial \phi}{\partial z}\right)\cdot(\hat{\mathbf{e}}_x dx + \hat{\mathbf{e}}_y dy + \hat{\mathbf{e}}_z dz).$$

The vector in the first bracket is called the **gradient** of scalar ϕ and is denoted by grad ϕ

$$\text{grad } \phi = \hat{\mathbf{e}}_x \frac{\partial \phi}{\partial x} + \hat{\mathbf{e}}_y \frac{\partial \phi}{\partial y} + \hat{\mathbf{e}}_z \frac{\partial \phi}{\partial z} \qquad \text{(A34)}$$

$$\therefore \quad d\phi = (\text{grad } \phi)\cdot d\mathbf{r}. \qquad \text{(A35)}$$

If we define a vector differential operator by the formula

$$\boldsymbol{\nabla} = \hat{\mathbf{e}}_x \frac{\partial}{\partial x} + \hat{\mathbf{e}}_y \frac{\partial}{\partial y} + \hat{\mathbf{e}}_z \frac{\partial}{\partial z} \qquad \text{(A36)}$$

we can write

$$\text{grad } \phi = \left(\hat{\mathbf{e}}_x \frac{\partial}{\partial x} + \hat{\mathbf{e}}_y \frac{\partial}{\partial y} + \hat{\mathbf{e}}_z \frac{\partial}{\partial z}\right)\phi = \boldsymbol{\nabla}\phi. \qquad \text{(A37)}$$

It may be noted that the vector differential operator $\boldsymbol{\nabla}$ (read as 'del') is not a vector in geometrial sense as it does not have any scalar magnitude; rather it is a vector operator which can operate both on scalar point function and vector point function. Thus the gradient of a scalar field generates a vector field.

The physical significance of the gradient may be seen in the following way: In a scalar field of ϕ we can connect all points in the region having the same value of ϕ. These points will lie on a surface. In Fig. A.8 we have shown two surfaces close to each other corresponding to the values ϕ and $\phi + d\phi$ of the scalar function. The displacement wholly within the surface does not alter the value of ϕ. Thus, if the displacement $d\mathbf{r}_0$ (AB) lies on one of these surfaces, we have

$$d\phi = (\text{grad } \phi)\cdot d\mathbf{r}_0 = 0$$

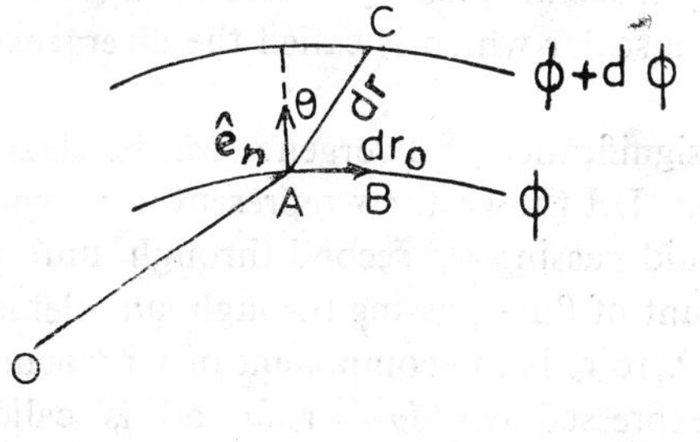

Fig. A.8

This means the vector grad ϕ is perpendicular to the surface.

$$\therefore \quad \text{grad}\,\phi = |\,\text{grad}\,\phi\,|\,\hat{\mathbf{e}}_n.$$

Consider a displacement $d\mathbf{r}$ along AC. We have

$$d\phi = (\text{grad}\,\phi)\cdot d\mathbf{r} = |\,\text{grad}\,\phi\,|\,\hat{\mathbf{e}}_n\cdot d\mathbf{r} = |\,\text{grad}\,\phi\,|\,dr\cos\theta \tag{A38}$$

where θ is the angle between $d\mathbf{r}$ and $\hat{\mathbf{e}}_n$.

$$\therefore \quad \frac{d\phi}{d\mathbf{r}} = |\,\text{grad}\,\phi\,|\cos\theta. \tag{A39}$$

The left-hand side gives the rate of change of ϕ with distance. If $\theta = 0$. $d\phi$ will have its greatest value and, hence, the maximum rate of change of ϕ is given by

$$\frac{d\phi}{dn} = |\,\text{grad}\,\phi\,|. \tag{A40}$$

Thus, the magnitude of the gradient of ϕ is equal to the maximum space rate of change of ϕ and its direction is the same as that along which the maximum space rate of change occurs i.e. perpendicular to the surface

$$\nabla\phi = \hat{\mathbf{e}}_n\frac{\partial\phi}{\partial n}. \tag{A41}$$

The space rate of change of ϕ along AC as given by equation (A39) is clearly the component of the gradient in that direction and is known as the **directional derivative** in that direction.

(v) Divergence of a vector

We have seen above that ∇ can be treated formally as a vector. Let us now find the scalar product of this operator and a vector **A**.

$$\nabla\cdot\mathbf{A} = \left(\hat{\mathbf{e}}_x\frac{\partial}{\partial x} + \hat{\mathbf{e}}_y\frac{\partial}{\partial y} + \hat{\mathbf{e}}_z\frac{\partial}{\partial z}\right)\cdot(\hat{\mathbf{e}}_xA_x + \hat{\mathbf{e}}_yA_y + {}_zA_z)$$

$$= \frac{\partial A_x}{\partial x} + \frac{\partial A_y}{\partial y} + \frac{\partial A_z}{\partial z}. \tag{A42}$$

This is obviously a scalar quantity. Thus, the scalar product of ∇ and a vector **A** gives a scalar which is called the **divergence** of **A** and is denoted by div **A**.

The physical significance of divergence will be clear if we consider a specific example. Let the vector **v** represent in magnitude and direction the volume of fluid passing per second through unit area perpendicular to **v**. The amount of fluid passing through an element of area $d\sigma$ per second is $v_x d\sigma$ where v_x is the component of **v** perpendicular to $d\sigma$. This amount can be expressed as $\mathbf{v}\cdot d\sigma = v_x d\sigma$ and is called the **flux** of the vector **v** through the area $d\sigma$. Consider as a volume element a rectangular parallelopiped with edges dx, dy, dz (Fig. A.9). At x, an amount $v_x dydz$ enters the volume element through the face $dydz$. At $x + dx$, the amount

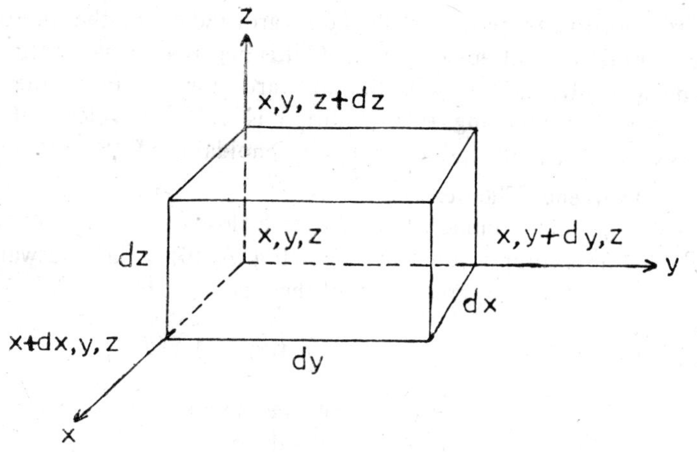

Fig. A.9

$\left(v_x + \frac{\partial v_x}{\partial x}\right) dydz$ streams out through the opposite face. The net flow outwards in the x-direction is

$$\left(v_x + \frac{\partial v_x}{\partial x}\, dx - v_x\right) dydz = \frac{\partial v_x}{\partial x}\, dxdydz$$

Similarly, the net contributions to the flow by the other two pairs of opposite faces are

$$\frac{\partial v_y}{\partial y}\, dxdydz \text{ and } \frac{\partial v_z}{\partial z}\, dxdydz.$$

Therefore, the total flux through the volume element is

$$\left(\frac{\partial v_x}{\partial x} + \frac{\partial v_y}{\partial y} + \frac{\partial v_z}{\partial z}\right) dxdydz.$$

Hence, the divergence of a vector $\mathbf{v}$ at any point in the field is the net outward flux of $\mathbf{v}$ per unit time per unit volume at the point

$$\text{div } \mathbf{v} = \nabla \cdot \mathbf{v} = \frac{\partial v_x}{\partial x} + \frac{\partial v_y}{\partial y} + \frac{\partial v_z}{\partial z}. \tag{A43}$$

We can express the divergence of a vector in a different form. Consider a small volume element $d\tau$ enclosed by an arbitrary closed surface. Let $d\sigma$ be an element on the surface so oriented that its normal is directed outwards. The amount flowing through this element is $\mathbf{v} \cdot d\sigma$ or $\mathbf{v} \cdot \hat{\mathbf{e}}_n d\sigma$. The flux through the entire surface is $\int \mathbf{v} \cdot \hat{\mathbf{e}}_n d\sigma$. This must be equal to $\nabla \cdot \mathbf{v} d\tau$

$$\therefore \quad \nabla \cdot \mathbf{v} d\tau = \int \mathbf{v} \cdot \hat{\mathbf{e}}_n d\sigma$$

and

$$\nabla \cdot \mathbf{v} = \lim_{d\tau \to 0} \frac{\int \mathbf{v} \cdot \hat{\mathbf{e}}_n d\sigma}{d\tau} \tag{A44}$$

If $\nabla \cdot \mathbf{v}$ is positive there is a net flow outwards indicating the presence of **sources** within the volume element. If it is negative, **sinks** exist in the volume element. If $\nabla \cdot \mathbf{v} = 0$, the outward flow balances the inward flo. A vector satisfying this condition is called a **solenoidal** vector. For example, magnetic induction **B** is solenoidal for $\nabla \cdot \mathbf{B} = 0$.

(vi) Divergence Theorem

Consider a finite volume τ bounded by a closed surface. Let this be subdivided into volume elements $d\tau_i$. (Fig. A.10). The outward flux from each of these is given by

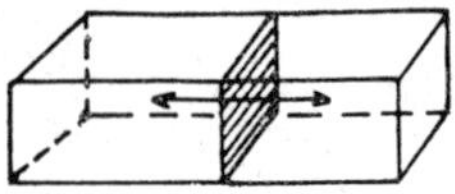

Fig. A.10

$$\nabla . \mathbf{v} d\tau_i = \int \mathbf{v} \cdot d\boldsymbol{\sigma}. \tag{A45}$$

Let both sides be summed over all elements. The left-hand side becomes $\int \nabla \cdot \mathbf{v} d\tau_i$. In taking the sum of the right-hand side, we see that the outflow for one element across the common surface is the inflow for the neighbouring element. Hence, these cancel each other and the total outflow from all elements is equal to the outflow across the closed surface surrounding the volume τ

$$\therefore \quad \int_\tau \nabla \cdot \mathbf{v} d\tau_i = \int_\sigma \mathbf{v} \cdot d\boldsymbol{\sigma} \tag{A46}$$

That is, the surface integral of a vector **v** taken over a closed surface is equal to the volume integral of the divergence of the vector taken over the enclosed volume τ. This is known as **Divergence theorem.**

(vii) Curl of a vector

We have seen that the divergence of a vector generates a scalar field. We shall now consider a differential operation which leads from one vector field to another vector field. Consider the vector product of ∇ and a vector **A**

$$\nabla \times \mathbf{A} = \left(\hat{\mathbf{e}}_x \frac{\partial}{\partial x} + \hat{\mathbf{e}}_y \frac{\partial}{\partial y} + \hat{\mathbf{e}}_z \frac{\partial}{\partial z}\right) \times \left(\hat{\mathbf{e}}_x A_x + \hat{\mathbf{e}}_y A_y + \hat{\mathbf{e}}_z A_z\right)$$

$$= \begin{vmatrix} \hat{\mathbf{e}}_x & \hat{\mathbf{e}}_y & \hat{\mathbf{e}}_z \\ \frac{\partial}{\partial x} & \frac{\partial}{\partial y} & \frac{\partial}{\partial z} \\ A_x & A_y & A_z \end{vmatrix}$$

$$= \hat{\mathbf{e}}_x \left(\frac{\partial A_z}{\partial y} - \frac{\partial A_y}{\partial z}\right) + \hat{\mathbf{e}}_y \left(\frac{\partial A_x}{\partial z} - \frac{\partial A_z}{\partial x}\right) + \hat{\mathbf{e}}_z \left(\frac{\partial A_y}{\partial x} - \frac{\partial A_x}{\partial y}\right). \tag{A47}$$

In order to understand the meaning of the curl, consider a small region of a vector field. Assume that the given vector **v** has the same direction at all points but different magnitudes at different points (Fig. A.11). Imagine a rectangular path $ABCD$ in this region. If the plane of the rectangle is perpendicular to the direction of **v**, the line integral $\oint \mathbf{v} . d\mathbf{l}$

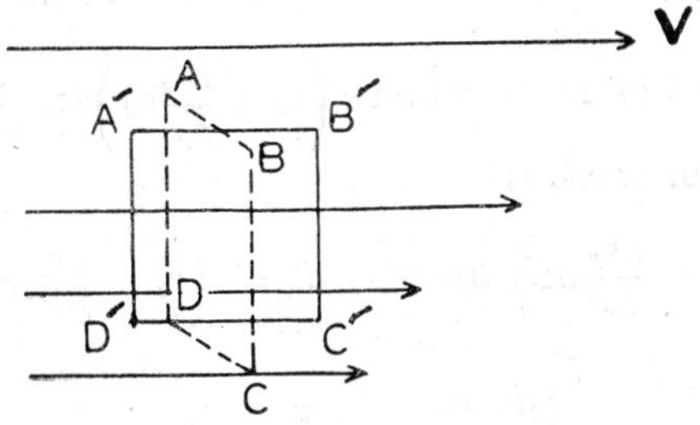

Fig. A.11

along the path is zero, because $\mathbf{v}$ is everywhere perpendicular to $d\mathbf{l}$. If on the other hand, the plane of the rectangle is parallel to the field $\mathbf{v}$ ($A'B'C'D'$), the line integral will not be zero, because the contribution from side $A'B'$ is different from that of the side $C'D'$, since the magnitude of $\mathbf{v}$ is different along these sides. The contribution from the other two sides is zero. For intermediate orientations of the rectangle, the line integral will have non-zero value but its magnitude will depend upon its orientation with respect to the field. The line integral will be maximum for a particular orientation of this area. This maximum line integral divided by the area is called the **curl** of the vector $\mathbf{v}$. The direction of the curl is given by the positive normal to the area when it is in the position of the maximum line integral

$$\text{curl } \mathbf{v} = \nabla \times \mathbf{v} = \lim_{d\sigma \to 0} \frac{\oint \mathbf{v} \cdot d\mathbf{l}}{d\sigma} \tag{A48}$$

We will now obtain the expression for curl $\mathbf{v}$ in terms of Cartesian components and show that it is the same as given by equation (A.47). This can be done by calculating the line integral round an infinitesimal area in each of the three coordinate planes. Each line integral will then represent the component of the curl along the corresponding axis.

Consider a small rectangular area $abcd$ (Fig. A.12) in the y-z plane plane with sides dy, dz. At a the components of $\mathbf{v}$ are v_x, v_y, v_z. The line integral along ab, therefore, will be $v_y dy$. Along cd, it will be

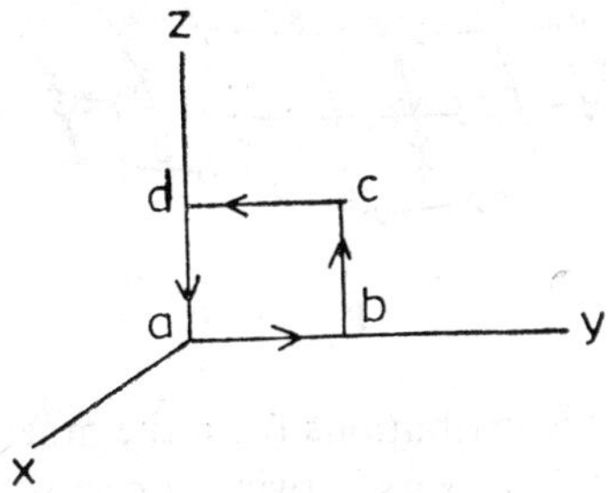

Fig. A.12

$-\left(v_y + \frac{\partial v_y}{\partial z} dz\right) dy$. The contribution of the line integral along *da* will be $- v_z dz$ and that along *bc* will be $\left(v_z + \frac{\partial v_z}{\partial y} dy\right) dz$. Hence, the integral around the contour *abcda* is

$$v_y\, dy - \left(v_y + \frac{\partial v_y}{\partial z} dz\right) dy - v_z\, dz + \left(v_z + \frac{\partial v_z}{\partial y} dy\right) dz$$

$$= \left(\frac{\partial v_z}{\partial y} - \frac{\partial v_y}{\partial z}\right) dy\, dz.$$

This integral is maximum since the paths *ab, bc, cd, da,* are taken either parallel or anti-parallel to the vector. The line integral per unit area is a vector along the positive normal to the area

$$\therefore \quad \text{curl}_x\, \mathbf{v} = \frac{\partial v_z}{\partial y} - \frac{\partial v_y}{dz}$$

In a similar manner we can find

$$\text{curl}_y\, \mathbf{v} = \frac{\partial v_x}{\partial z} - \frac{\partial v_z}{\partial x}$$

$$\text{curl}_z\, \mathbf{v} = \frac{\partial v_y}{\partial x} - \frac{\partial v_x}{\partial y}$$

$$\therefore \quad \text{curl}\, \mathbf{v} = \hat{\mathbf{e}}_x \left(\frac{\partial v_z}{\partial y} - \frac{\partial v_y}{\partial z}\right) + \hat{\mathbf{e}}_y \left(\frac{\partial v_x}{\partial z} - \frac{\partial v_z}{\partial x}\right) + \hat{\mathbf{e}}_z \left(\frac{\partial v_y}{\partial x} - \frac{\partial v_x}{\partial y}\right)$$

$$= \nabla \times \mathbf{v}$$

(viii) Stokes' Theorem

Consider a portion of a surface bounded by a curve (Fig. A.13). Imagine the surface to be subdivided into infinitely small elements. The boundary of each element is traversed in the same sense as the outer curve. For each element we have

$$\oint \mathbf{v} \cdot d\mathbf{l} = (\nabla \times \mathbf{v}) \cdot d\boldsymbol{\sigma}$$

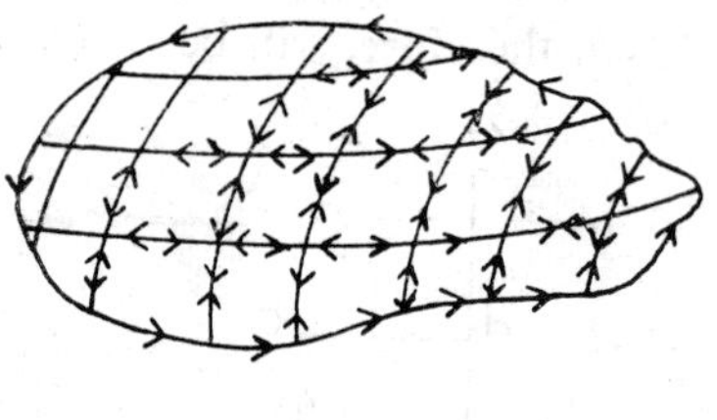

Fig. A.13

In taking the sum, the contributions from the inner dividing lines cancel out since each segment is traversed twice in opposite directions. Hence, the left-hand side becomes merely the integral along the bounding curve. We, thus, have

$$\oint_c \mathbf{v}\cdot d\mathbf{l} = \int_\sigma (\nabla \times \mathbf{v})\cdot d\boldsymbol{\sigma} \tag{A49}$$

This is known as **Stokes' theorem.** It states that the line integral of a vector $\mathbf{v}$, taken over a closed curve c is equal to the surface integral of the curl of $\mathbf{v}$ taken over any surface σ having c as the boundary.

(ix) Line to surface integral of scalar function:

By Stokes' theorem

$$\int_S (\nabla \times \mathbf{A})\cdot \hat{\mathbf{e}}_n \, dS = \oint_C \mathbf{A}\cdot d\mathbf{l} \tag{A50}$$

Let $\mathbf{A} = \phi\mathbf{B}$ where ϕ is a scalar point function and $\mathbf{B}$ a vector constant in space, i.e. $\nabla \times \mathbf{B} = 0$.

$$\therefore \quad \int_S (\nabla \times \phi\mathbf{B})\cdot \hat{\mathbf{e}}_n \, dS = \oint_C \phi\mathbf{B}.d\mathbf{l} \tag{A51}$$

Now,

$$\nabla \times (\phi\mathbf{B}) = \phi\,(\nabla \times \mathbf{B}) - (\mathbf{B} \times \nabla\phi)$$

$$\therefore \quad \int_S [\phi\,(\nabla \times \mathbf{B}) - (\mathbf{B} \times \nabla\phi)]\cdot \hat{\mathbf{e}}_n \, dS = \oint_C \phi\mathbf{B}\cdot d\mathbf{l}$$

i.e.
$$\oint_C \phi\mathbf{B}\cdot d\mathbf{l} + \int_S (\mathbf{B} \times \nabla\phi)\cdot \hat{\mathbf{e}}_n \, dS = 0$$

or
$$\oint_C \mathbf{B}\cdot \phi d\mathbf{l} + \int_S \mathbf{B}\cdot \nabla\phi \times \hat{\mathbf{e}}_n \, dS = 0$$

Hence,
$$\mathbf{B}\cdot\left[\oint_C \phi dl + \int_S \nabla\phi \times \hat{\mathbf{e}}_n \, dS\right] = 0 \tag{A52}$$

Since $\mathbf{B}$ is any arbitrary vector

$$\oint_C \phi d\mathbf{l} = -\int_S \nabla\phi \times \hat{\mathbf{e}}_n \, dS = \int \hat{\mathbf{e}}_n \times \nabla\phi \, dS \tag{A53}$$

(x) Fields in different coordinate systems

It is usually necessary to write equations in terms of coordinates suitable for a particular problem. Formulae for the fields resulting from operating with differential operators are given below in Cartesian, spherical polar and cylindrical polar coordinates.

(a) Divergence of $\mathbf{V}$

Cartesian:
$$\text{div}\,\mathbf{V} = \frac{\partial V_x}{\partial x} + \frac{\partial V_y}{\partial y} + \frac{\partial V_z}{\partial z} \tag{A54}$$

Spherical polar:
$$\text{div}\,\mathbf{V} = \frac{\partial V_r}{\partial r} + \frac{2}{r} V_r + \frac{1}{r}\frac{\partial V_\theta}{\partial \theta} + \frac{\cot\theta}{r} V_\theta + \frac{1}{r\sin\theta}\frac{\partial V_\phi}{\partial \phi} \tag{A55}$$

Cylindrical polar:

$$\text{div } \mathbf{V} = \frac{\partial V_r}{\partial r} + \frac{V_r}{r} + \frac{1}{r}\frac{\partial V_\phi}{\partial \phi} + \frac{\partial V_z}{\partial z} \tag{A56}$$

(b) The components of grad ψ

$$\text{Cartesian: grad}_x\ \psi = \frac{\partial \psi}{\partial x};\ \text{grad}_y\ \psi = \frac{\partial \psi}{\partial y};\ \text{grad}_z\ \psi = \frac{\partial \psi}{\partial z} \tag{A57}$$

$$\text{Spherical polar: grad}_r\ \psi = \frac{\partial \psi}{\partial r};\ \text{grad}_\theta\ \psi = \frac{1}{r}\frac{\partial \psi}{\partial \theta};$$

$$\text{grad}_\phi\ \psi = \frac{1}{r \sin\theta}\frac{\partial \psi}{\partial \phi} \tag{A58}$$

Cylindrical polar:

$$\text{grad}_r\ \psi = \frac{\partial \psi}{\partial r};\ \text{grad}_\phi\ \psi = \frac{1}{r}\frac{\partial \psi}{\partial \phi};\ \text{grad}_z\ \psi = \frac{\partial \psi}{\partial z} \tag{A59}$$

(c) The components of curl V

$$\text{Cartesian: curl}_x\ \mathbf{V} = \frac{\partial V_z}{\partial y} - \frac{\partial V_y}{\partial z};\ \text{curl}_y\ \mathbf{V} = \frac{\partial V_x}{\partial z} - \frac{\partial V_z}{\partial x};$$

$$\text{curl}_z\ \mathbf{V} = \frac{\partial V_y}{\partial x} - \frac{\partial V_x}{\partial y} \tag{A60}$$

Spherical polar:

$$\text{curl}_r\ \mathbf{V} = \frac{1}{r}\frac{\partial V_\phi}{\partial \theta} + \frac{\cot\theta}{r} V_\phi - \frac{1}{r\sin\theta}\frac{\partial V_\theta}{\partial \phi};$$

$$\text{curl}_\theta\ \mathbf{V} = \frac{1}{r\sin\theta}\frac{\partial V_r}{\partial \phi} - \frac{\partial V_\phi}{\partial r} - \frac{1}{r} V_\phi$$

$$\text{curl}_\phi\ \mathbf{V} = \frac{V_\theta}{r} + \frac{\partial V_\theta}{\partial r} - \frac{1}{r}\frac{\partial V_r}{\partial \theta} \tag{A61}$$

Cylindrical polar:

$$\text{curl}_r\ \mathbf{V} = \frac{1}{r}\frac{\partial V_z}{\partial \phi} - \frac{\partial V_\phi}{\partial z}$$

$$\text{curl}_\phi\ \mathbf{V} = \frac{\partial V_r}{\partial z} - \frac{\partial V_z}{\partial r}$$

$$\text{curl}_z\ \mathbf{V} = \frac{\partial V_\phi}{\partial r} + \frac{V_\phi}{r} - \frac{1}{r}\frac{\partial V_r}{\partial \phi} \tag{A62}$$

(d) Laplacian operator

Some of the equations in electrodynamics involve a second-order differential operator called **Laplacian.** This is a scalar operator defined by $\nabla \cdot \nabla = \nabla^2$

$$\text{Cartesian: } \nabla^2 = \frac{\partial^2}{\partial x^2} + \frac{\partial^2}{\partial y^2} + \frac{\partial^2}{\partial z^2} \tag{A63}$$

Spherical polar:

$$\nabla^2 = \frac{\partial^2}{\partial r^2} + \frac{2}{r}\frac{\partial}{\partial r} + \frac{1}{r^2}\frac{\partial^2}{\partial \theta^2} + \frac{\cot\theta}{r^2}\frac{\partial}{\partial \theta} + \frac{1}{r^2 \sin^2\theta}\frac{\partial^2}{\partial \phi^2} \quad \text{(A64)}$$

Cylindrical polar:

$$\nabla^2 = \frac{\partial^2}{\partial r^2} + \frac{1}{r}\frac{\partial}{\partial r} + \frac{1}{r^2}\frac{\partial^2}{\partial \phi^2} + \frac{\partial^2}{\partial z^2} \quad \text{(A65)}$$

(xi) Green's Theorem

Using the fundamental relations of line integral, Divergence theorem and Stokes' theorem, a number of formulae for the transformation of integrals can be obtained, two of which are specially important and are known as Green's theorems.

Let a vector **v** be defined by $\mathbf{v} = \phi\nabla\psi$ where ϕ and ψ are scalar functions. Then by Divergence theorem

$$\int(\phi\nabla\psi)\cdot d\sigma = \int\nabla\cdot(\phi\nabla\psi)\, d\tau = \int(\nabla\phi\cdot\nabla\psi + \phi\nabla^2\psi)\, d\tau \quad \text{(A66)}$$

This the first form of Green's theorem.

Interchanging ϕ and ψ we have

$$\int(\psi\nabla\phi)\cdot d\sigma = \int(\nabla\psi\cdot\nabla\phi + \psi\nabla^2\phi)\, d\tau \quad \text{(A67)}$$

Subtracting (A67) from (A66), we get

$$\int(\phi\nabla\psi - \psi\nabla\phi)\cdot d\sigma = \int(\phi\nabla^2\psi - \psi\nabla^2\phi)\, d\tau \quad \text{(A68)}$$

This the second form of Green's theorem.

A.4 Some useful vector relations

(1) $\nabla\cdot(\phi\mathbf{A}) = \mathbf{A}\cdot\nabla\phi + \phi\nabla\cdot\mathbf{A}$ (A69)

(2) $\nabla\cdot(\mathbf{A}\times\mathbf{B}) = \mathbf{B}.(\nabla\times\mathbf{A}) - \mathbf{A}\cdot(\nabla\times\mathbf{B})$ (A70)

(3) $\nabla\times(\phi\mathbf{A}) = \phi\nabla\times\mathbf{A} - \mathbf{A}\times\nabla\phi$ (A71)

(4) $\nabla\times(\nabla\times\mathbf{A}) = \nabla(\nabla\cdot\mathbf{A}) - \nabla^2\mathbf{A}$ (472)

(5) $\nabla\times(\mathbf{A}\times\mathbf{B}) = \mathbf{A}(\nabla\cdot\mathbf{B}) - \mathbf{B}(\nabla\cdot\mathbf{A}) + (\mathbf{B}\cdot\nabla)\mathbf{A} - (\mathbf{A}\cdot\nabla)\mathbf{B}$ (A73)

(6) $\nabla(\mathbf{A}\cdot\mathbf{B}) = (\mathbf{A}\cdot\nabla)\mathbf{B} + (\mathbf{B}\cdot\nabla)\mathbf{A} + \mathbf{A}\times(\nabla\times\mathbf{B})$
$+ \mathbf{B}\times(\nabla\times\mathbf{A})$ (A74)

Appendix B

Transformations of Quantities from Gaussian System to SI System*

Quantity	*Gaussian*	*SI*
Charge	q	$\frac{1}{(4\pi\epsilon_0)^{1/2}}\, q(\rho, I, j, P)$
(charge density, current, current density, polarization)	(ρ, I, j, P)	
Potential	Φ	$(4\pi\epsilon_0)^{1/2}\ \Phi\ (\mathbf{E})$
(electric field)	$(\mathbf{E})$	
Vector potential	$\mathbf{A}$	$\left(\frac{4\pi}{\mu_0}\right)^{1/2} \mathbf{A}\ (\mathbf{B})$
(Magnetic flux density)	$(\mathbf{B})$	
Magnetization	$\mathbf{M}$	$\left(\frac{\mu_0}{4\pi}\right)^{1/2} \mathbf{M}$
Electric displacement	$\mathbf{D}$	$\left(\frac{4\pi}{\epsilon_0}\right)^{1/2} \mathbf{D}$
Magnetic field	$\mathbf{H}$	$(4\pi\mu_0)^{1/2}\ \mathbf{H}$
Dielectric constant	ϵ	$\frac{\epsilon}{\epsilon_0}$
Permeability	μ	$\frac{\mu}{\mu_0}$
Electrical conductivity	σ	$\frac{\sigma}{4\pi\epsilon_4}$
Capacitance	C	$\frac{1}{4\pi\epsilon_0}\, C$

*These transformations are reversible.

Appendix C

Conversion of Electric and Magnetic Units

Symbol	*Quantity*	*SI*	*Gaussian*
$q(e)$	Charge	1 coulomb	3×10^9 statcoul
ρ	Charge density	1 coul/m^3	3×10^3 statcoul/cm^3
I	Current	1 amp	3×10^9 statamp
J	Current density	1 amp/m^2	3×10^5 statamp/cm^2
$\Phi(v)$	Potential	1 volt	(1/300) statvolt
E	Electric field	1 volt/m	$(1/3)\times 10^{-4}$ statvolt/cm
D	Electric displacement	1 coul/m^2	$12\pi\times 10^5$ statvolt/cm
A	Vector potential	1 weber/m	$(1/3)\times 10^{-10}$ gauss cm
B	Magnetic flux density	1 weber/m^2 (tesla)	10^4 gauss
H	Magnetic field	1 amp/m	$4\pi\times 10^{-3}$oersted
M	Magnetisation	1 amp/m	$(1/4\pi)\times 10^4$ gauss
P	Polarization	1 coul/m^2	3×10^5 statvolt/cm
ϵ	Dielectric permittivity	1 farad/m	$36\pi 10^9$ statfarad/cm
μ	Permeability	1 henry/m	$(1/4\pi)\times 10^7$ gauss/oersted
σ	Electrical conductivity	1 mho/m	9×10^9/sec
C	Capacity	1 farad	9×10^{11} statfarad

Appendix D

Physical Constants

Electron charge	1.602×10^{-19} coul
Electron mass	9.109×10^{-31} kg
Proton mass	1.673×10^{-27} kg
Vacuum permittivity	8.854×10^{-12} farad/m
Vacuum permeability	1.257×10^{-6} henry/m
Velocity of light	2.998×10^{8} m/sec
Classical electron radius	2.818×10^{-15} m
Electron volt	1.602×10^{-19} joule
Boltzmann constant	1.381×10^{-23} j/°K
Planck's constant	6.626×10^{-34} J sec
Bohr's radius	5.292×10^{-11} m
Avogadro's number	6.023×10^{23}/mole

Answers to Problems

Chapter 1

1.2 $mg \cdot \frac{a}{\sqrt{2}l} = \frac{q^2}{4\pi\epsilon_0 a^2}(\sqrt{2} + \frac{1}{2}) \frac{\sqrt{l^2 - a^2/2}}{l}$.

This gives 'a'; $m = m'$.

1.3 3.45×10^{-14} metres.

1.4 $Q_m = 603 \times 10^{10}\, C$; $Q_e = 517 \times 10^{12}\, C$.

1.6 (a) $2\pi\rho_0 a^3$; (b) $\frac{8\pi}{15}\rho_0 a^3$

1.7 $1.1 \times 10^{-12}\, C$; Positive.

1.10 $3.23 \times 10^{15}\, V$.

1.11 $W = 750$ MeV; Energy released $= 280$ MeV.

1.14 $T = \frac{2\pi}{Q}\sqrt{2\pi\epsilon_0 m a^3}$.

Chapter 2

2.1 $\frac{1}{2}\left(V + \frac{qd}{\epsilon_0 A}\right)$

2.2 $\frac{C_1 C_2}{2(C_1 + C_2)}(V_2 - V_1)^2$

2.4 Capacitance per unit area $= \frac{\epsilon_1 \epsilon_2}{d(\epsilon_1 + \epsilon_2)}$

2.5 $\mathbf{E}_{\text{(inside)}} = -\frac{\mathbf{P}}{3\epsilon_0}$

$\mathbf{E}_{\text{(outside)}} = \frac{R^3 (\mathbf{P}\cdot\mathbf{r})\mathbf{r}}{\epsilon_0 r^5} - \frac{R^3 \mathbf{P}}{3\epsilon_0 r^5}$

2.8 $V = \frac{q}{2\pi(\epsilon_1 + \epsilon_2) r}$; $\sigma_1 = \frac{q\epsilon_1}{2\pi(\epsilon_1 + \epsilon_2) R^2}$; $\sigma_2 = \frac{q\epsilon_2}{2\pi(\epsilon_1 + \epsilon_2) R^2}$

2.9 $q^2 = \frac{8\pi\epsilon_0 a^2 Mg(1 + \epsilon_r)^2}{\epsilon_r - 1}$

2.10 $p = 4.84 \times 10^{-30}$ cm; $r = 0.18$ nm

2.12 $\frac{\pi\epsilon_0 V^2}{\ln b/a}\chi = \pi(b^2 - a^2)h\rho g$

Chapter 3

3.4 $$V_i = \sum_l \frac{\sigma}{2\epsilon_0 R^{l-1}}(2l+1)^{-1}[P_{l+1}(\cos\alpha) - P_{l-1}(\cos\alpha)]\, r^l P_l(\cos\theta)$$

$$V_0 = \sum_l \frac{\sigma R^{l+2}}{2\epsilon_0}(2l+1)^{-1}[P_{l+1}(\cos\alpha) - P_{l-1}(\cos\alpha)]\, r^{-l-1} P_l(\cos\theta)$$

3.5 $$V = \frac{1}{2\pi\epsilon_0}\left(r^2 - \frac{a^5}{r^3}\right)\sin\theta\cos\theta\, e^{i\phi}$$

3.6 $$F = \frac{9}{4}\epsilon_0\,\pi\, a^2 E_0^{\,2}$$

3.7 $$V = -E_0\left(r - \frac{a^2}{r}\right)\cos\theta$$

3.9 $$V = \frac{q}{2\pi\epsilon_0}\ln\frac{r_2}{r_1}$$

3.10 $$V = \frac{1}{4\pi\epsilon_0}\left(\frac{q}{r_1} - \frac{q'}{r_2} + \frac{q'}{r_3} - \frac{q}{r_4}\right)$$

where $q' = \frac{qa}{b}$

3.15 $$V_{(\text{inside})} = \frac{qr^n \sin n\theta}{2\epsilon_0\, n\, a^{n-1}},\quad V_{(\text{outside})} = \frac{qa^{n+1}\sin n\theta}{2\epsilon_0\, n\, r^n}$$

Chapter 4

4.1 $$Q = \frac{I_0}{\alpha}(1 - e^{-\alpha t_0})$$

4.2 9×10^{-5} ms^{-1}.

4.4 $$F = -\frac{3}{2}\frac{\pi\mu_0\, a^4\, I^2\, z}{(a^2 + z^2)^{5/2}}$$

4.6 $B = 10^{-4}$ volts sec/m^2.

4.7 (i) $B = \dfrac{\mu_0 I r}{2\pi a^2}$; (ii) $B = \dfrac{\mu_0 I}{2\pi r}$ (iii) $B = 0$

4.8 $$I = \frac{3}{2}\pi J_0\, a^2; \qquad B_{(\text{inside})} = \frac{\mu_0 J_0 r}{2}\left[1 + \frac{r^2}{2a^2}\right]$$

$$B_{(\text{outside})} = \frac{3\mu_0\, J_0\, a^2}{4r}$$

4.12 $$H = \frac{Ic}{2\pi\,(a^2 - b^2)}\,\hat{\mathbf{e}}_y$$

Chapter 5

5.1 $\pi r^2\omega\, B_0\cos\omega t\cos\theta$

5.3 3.532 Nm.

5.6 $$A = \frac{\mu_0 I}{2\pi}\ln\frac{r_2}{r_1}$$

Chapter 6

6.2 The head of the resultant electric field vector describes an ellipse, the semi-axes of which are:

$$a = \sqrt{E_{01}^2 \cos^2 \alpha + E_{02}^2 \cos^2 (\alpha - \phi)}$$

$$b = \sqrt{E_{01}^2 \sin^2 \alpha + E_{02}^2 \sin^2 (\alpha - \phi)}$$

where

$$\tan 2\alpha = \frac{E_{02}^2 \sin 2\phi}{E_{01}^2 + E_{02}^2 \cos 2\phi}$$

6.4 10^{-4}.

6.5 0.796 m.

6.6 $\epsilon = 1 - \dfrac{n_0 e^2}{m\omega^2 \epsilon_0}$

Chapter 7

7.4 6.6 V m^{-1}

7.6 2.3

Chapter 8

8.1 $k^2 = \dfrac{\omega^2 \epsilon_1 |\epsilon_2|}{c^2 (|\epsilon_2| - \epsilon_1)}$

8.2 For an ideal conductor $\mathbf{E}_T = \sqrt{\dfrac{\mu\mu_0 \omega}{\sigma_t}} \left[\mathbf{H}_T \times \hat{\mathbf{e}}_n\right]$

8.3 The minimum value of $a = \lambda/2$

The maximum value of $a = \lambda/\sqrt{2}$

8.5 Three

Chapter 9

9.2 $\mathbf{E}(\mathbf{r}, t) = \dfrac{e}{4\pi\epsilon_0} \left[\dfrac{(1 - \beta^2)(\hat{\mathbf{e}}_n - \boldsymbol{\beta})}{R^2 (1 - \hat{\mathbf{e}}_n \cdot \boldsymbol{\beta})^3}\right]$

9.6 $\dfrac{128\pi \epsilon_0 |\mathbf{E}|^4}{3m^2c^3e^2}$ where E is the particle energy

Chapter 10

10.7 3.2×10^9 V/m

10.9 The path of the particle is given by

$$x = \frac{1}{e|\mathbf{E}|} \left[E_{01}^2 + (ce|\mathbf{E}|t)^2\right]^{1/2} + x_0$$

$$y = \frac{p_0\, c}{e\,|\,\mathbf{E}\,|} \ln\left[t + \sqrt{t^2 + \left(\frac{E_{01}}{ce\,|\,\mathbf{E}\,|}\right)^2}\right] + y_0$$

where

$$E_{01} = c(m^2c^2 + p_0^2)^{1/2}$$

For a slow particle $x = \dfrac{e\,|\,\mathbf{E}\,|}{2m} t^2 + x_0$; $y = \dfrac{p_0}{m} t$

10.10 $$\omega = \sqrt{\omega_0^2 + \frac{1}{4}\left(\frac{eH}{mc}\right)^2} \pm \frac{eH}{2mc}$$

For weak field

$$\omega = \omega_0 \pm \frac{eH}{2mc}$$

Chapter 11

11.2 $$d\sigma = \left(\frac{e^2}{4\pi\,\epsilon_0\, mc^2}\right)^2 \frac{[\mathbf{A} \times \hat{\mathbf{e}}_n]^2 + [\mathbf{B} \times \hat{\mathbf{e}}_n]^2}{[(\omega_0^2 - \omega^2)^2 + \omega^2\gamma^2]\,(A^2 + B^2)}\, \omega^4 d\Omega$$

where $\hat{\mathbf{e}}_n$ is the unit vector in the direction of scatter and $\mathbf{E} = \mathbf{A}\cos\omega t + \mathbf{B}\sin\omega t$; $\mathbf{A} \perp \mathbf{B}$

11.4 The radiation intensity J is given by:

$$J = \frac{p^2\omega^4}{6\pi\,\epsilon_0\, c^3}\left(1 - \frac{\beta}{4\,\pi\, r^3}\right)^2$$

Bibliography

Andrews, C.L., *Optics of the Electromagnetic Spectrum.* Prentice-Hall, Englewood Cliffs, (1960)

Argence, E. and Kahan, T., *Theory of Waveguides and Cavity Resonators,* Blackie, London (1967)

Band, W., *Introduction to Mathematical Physics,* East West Press (1964)

Bates, L.F., *Modern Mannetism,* 4th Ed. Cambridge University Press, (1961)

Becker, R. and Sauter, F., *Electromagnetic Fields and Interactions,* Vol. 1, Blaisdell, New York, (1964)

Bergmann, P.G., *Introduction to the Theory of Relativity,* Asia Publishing House, (1960)

Bleaney, B.I., and Bleaney, B. *Electricity and Magnetism* 3rd. *Ed.*, Oxford University Press, (1978)

Born, M. and Wolf, E., *Principles of Optics,* 4th Ed Pergamon, New York, (1970)

Boyd, T.J.M. and Sanderson, J.J., *Plasma Dynamics,* Nelson, London and Barnes and Noble, New York, (1969)

Brailsford, F., *Physical Principles of Magnetism,* Van Nostrand, London (1966)

Carter, G.W., *The Electromagnetic Fields in its Engineering Aspects,* 2nd Ed., Longmans, Green, London (1967)

Chambers Ll.G., *An Introduction to Mathematics of Electricity and Magnetism,* Chapman and Hall, London (1973)

Chirgwin B.H., Plumpton, C. and Kilmister, C.W. *Elementary Electromagnetic Theory* in 3 volumes. Pergamon Press, Oxford, (1972)

Clemmow, P.C., *The Plane Wave Spectrum Representation of Electromagnetic Fields,* Pergamon, Oxford (1966)

Clemmow, P.C. and Dougherty, J.P. *Electrodynamics of Particles and Plasmas,* Addison-Wesley, Reading, Mass., (1969)

Collin, R.E., *Field Theory of Guided Waves,* McGraw-Hill, New York, (1966)

Corson, D.R., and Lorrain, P., *Introduction to Electromagnetic Fields and Waves,* Freeman, San Francisco, (1962)

Coulson, C.A. and Boyd, T.J.M., *Electricity,* 2nd Ed., Longman Group Limited, London, (1979)

Cullwick, E.G, *The Fundamentals of Electromagnetism,* 3rd Ed. Cambridge University Press, (1966) de Groot, S.R., *The Maxwell Equations,* North-Holland, Amsterdam (1969)

Ecker, G., *Theory of Fully Ionized Plasmas,* Academic, New York, (1972)

Fano, R.M., Chu, L.J. and Adler, R.B., *Electromagnetic Fields, Energy and Forces.* Wiley, New York, (1960)

Ferrari, R.L., *An Introduction to Electromagnetic Fields,* Van Nostrand, Reinhold Company (1975)

Feynman, R.P. *The Electromagnetic Field,* Lectures on Physics, Vol.II, Addison-Wesley, (1964)

French, A.P., *Special Relativity,* Norton, New York, (1968)

Fröhlich, H. *Theory of Dielectrics,* Clarendon Press (1958)

Grant I.S. and Phillips, W.R., *Electromagnetism,* John Wiley & Sons Ltd. (1978)

Hallen, E., *Electromagnetic Theory,* Chapman-Hall, London (1962)

Harrington, R.F., *Time Harmonic Electromagnetic Fields,* McGraw-Hill, New York, (1961)

Heitler, W., *Quantum Theory of Radiation,* 3rd Ed. Oxford University Press, (1954)

Hammond, P., *Applied Electromagnetism,* Pergamon Press (1971)

Holt, E.H. and Haskell, R.E., *Foundations of Plasma Dynamics,* Macmillan, New York, (1965)

Hsu, F.H., Am. J. Phys. *40,* 492, (1972)

Hughes, W.F. and Young, F.J., *The Electrodynamics of Fluids,* Wiley, New York (1966)

Jackson, J.D., *Classical Electrodynamics,* 2nd Ed., Wiley Eastern Limited, (1978)

Jeans, J.H., *Mathematical Theory of Electricity and Magnetism,* 5th Ed., Cambridge Univ. Press (1948)

Jeffreys, H. and Jeffreys, B S., *Methods of Mathematical Physics,* 3rd. Ed., Cambridge University Press (1956)

Johnson, C.C., *Field and Wave Electrodynamics,* McGraw-Hill, New York, (1965)

Jordan E.C., and Balmain K.G., *Electromagnetic Waves and Radiating Systems,* 2nd Ed., Prentice-Hall, Englewood Cliffs (1968)

Kilmister, C.W., *Special Theory of Relativity,* Pergamon, Oxford, (1970)

Kittel, C., *Introduction to Solid State Physics,* 4th Ed. Wiley, New York (1971)

Kraus, J.D. and Carver, K.R., *Electromagnetics,* 2nd Ed., McGraw-Hill, (1973)

Landau, L.D. and Lifshitz, E.M., *The Classical Theory of Fields,* 4th Ed. Pergamon Press, (1975)

Laud, B.B., *Introduction to Statistical Mechanics,* Macmillan, India, (1981)

Linhart, J.G., *Plasma Physics,* North Holland Amsterdam (1960)

Marion, J.B. *Classical Electromagnetic Radiation,* Academic Press, New York, (1968)

Moller, C., *The Theory of Relativity*, 2nd Ed. Clarendon Press, Oxford. (1972)

Panofsky, W.K.H. and Phillips, M., *Classical Electricity and Magnetism*, 2nd Ed. Addison-Wesley, Reading, Mass (1962)

Penfield, P. and Haus, H.A., *Electrodynamics of Moving Media*, M.I.T. Press, (1967)

Popovie, B.D., *Introductory Engineering Electrodynamics*, Addison-Wesley, Reading, Mass. (1971)

Pugh, E.M. and Pugh E.W., *Principles of Electricity and Magnetism*, Addison-Wesley, Reading, Mass. (1960)

Ramo, S., Whinnery, J.R., and Van Duzer, T., *Fields and Waves in Communication Electronics*, Wiley, New York (1965)

Reitz, J.R. and Milford, F.J , *Foundations of Electromagnetic Theory*, 2nd Ed. Addison-Wesley (1967)

Robinson, F.N.H., *Macroscopic Electromagnetism*, Pergamon, Oxford, (1973)

Rosenfeld, L. *Theory of Electrons*, Dover (1966)

Sastry, G.P., Am. J. Phys., *38*, 267, (1970)

Schwartz, W.M., *Intermediate Electromagnetic Theory*, Wiley, New York, (1964)

Smythe, W.R., *Static and Dynamic Electricity*, McGraw-Hill, New York (1969)

Stone, J.M., *Radiation and Optics*, McGraw-Hill New York (1963)

Whittaker, E.T., *A History of the Theories of Aether and Electricity* 2 Volumes Nelson, London, Reprinted by Harper, New York (1960)

Index